OPTICAL FIBER NEW DEVELOPMENTS

OPTICAL FIBER NEW DEVELOPMENTS

Editor

Brian Ismay

scitus academics

Optical Fiber New Developments
Edited by **Brian Ismay**

ISBN: 978-1-68117-598-0
Library of Congress Control Number: 2016946453

SCITUS Academics LLC,
616, Corporate Way, Suite 2, 4766,
Valley Cottage, NY 10989

www.scitusacademics.com

Notice

Printed in United States of America on Acid Free Paper

PREFACE

Optical fiber technology benefits applications for lighting, sensing and communication systems extensively. In the past decades, many improvements were made to reduce fiber attenuation and increase fiber performance. New applications have now been created throughout the medical community. New optical fiber types will be developed after many decades of business use to improve some aspects of communications. Fiber improvements are available in ways that provide higher bandwidth, support longer distances or reduce the entire cost of transport in conjunction with optoelectronic devices at both ends. The development of a broadband multimode fiber (WBMMF) can be a particular development in the LAN fiber world. This new fiber type increases the number of usable wavelengths from one to many and also helps to multiply shortwave wavelengths between 850 and 950 nanometers.

This book shows recent advances in optical fiber technology allied to optical communication, network and system generation, and characterization. Innovative developments in optical fiber technology transform world communications. New techniques for data processing can be found via optical fibers. In next decades, the effect of fiber materials, devices and systems on communications will develop much of the need and primary literature for up-to-date review. This guide must be a valuable tool for researchers and practitioners working not only in academic institutions but also in industry.

TABLE OF CONTENTS

1

ADVANCES ON OPTICAL FIBER SENSORS

Luciano Mescia and Francesco Prudenzano

Dipartimento di Ingegneria Elettrica e dell'Informazione,
Politecnico di Bari Via E. Orabona 4, 70125 Bari, Italy

ABSTRACT

In this review paper some recent advances on optical fiber sensors are reported. In particular, fiber Bragg grating (FBG), long period gratings (LPGs), evanescent field and hollow core optical fiber sensors are mentioned. Examples of recent optical fiber sensors for the measurement of strain, temperature, displacement, air flow, pressure, liquid-level, magnetic field, and the determination of methadone, hydrocarbons, ethanol, and sucrose are briefly described.

KEYWORDS

laser; optical fiber; microstructured optical fibers

1. INTRODUCTION

Almost four decades have passed since the research on optical fiber sensors began in earnest. Various approaches and technologies have been utilized for measuring a number of different physical parameters, but only some types of optical fiber sensors are commercially interesting. In fact, in many cases, optical fiber sensor systems are not available in a complete form, *i.e.*, including both detecting and signal-processing electronics. However, their future is very promising because they exhibit well known advantages such as compactness, immunity to electromagnetic interference and to ionizing radiation (γ-ray, X-ray *etc.*), high sensitivity, large bandwidth, and minimum weight. These properties, make optical fiber sensors key photonic devices in radiative environments, like nuclear power plants, where the detection and evaluation of radiation levels and temperature changes are very important, especially in case of accidental constraints [1]. Optical fiber sensors have been developed to measure strain, temperature, pressure, current, voltage, gas, chemical contaminant, rotation, vibration, acceleration, bending, torsion, displacement, and biomolecules [2].

Chemical optical fiber sensors for the detection of polluting substances in water, soil, and air, such as the biological sensors, optimized for medicine applications, have attracted noticeable interest because they are directly related to the quality of human life. They have been investigated as key elements of environmental monitoring systems, with the aim of preventing environmental catastrophes, and also as novel techniques for medical diagnosis. In particular, optical fiber sensor systems can avoid the collection of samples to be investigated in appositely equipped laboratories. They provide means of overcoming the drawbacks exhibited by *ex-situ* techniques which are invasive, time consuming and expensive. In fact, optical fiber chemical/biological sensors, enable pollution detection and/or medical diagnostic monitoring via an *in-situ*and slightly invasive technique.

Optical spectroscopy of water, atmosphere, soil or biological samples, is commonly performed in chemical laboratories and is based on Laser Induced Fluorescence (LIF) or Raman effects [3]. LIF or Raman equipment is generally expensive due to both the utilized light sources (and related optical components) and to the sophisticated processing electronics. In order to develop low-cost pollution and bio-medical monitoring, a promising approach concerns the optimization of microstructured optical fibers (MOFs) for sensing. In fact, in recent years, a number of sensors made of suitable MOFs have been theoretically and experimentally investigated for the measurement of a large variety of physical and chemical parameters [4,5,6,7,8,9,10,11,12].

Both conventional fiber optic and MOF sensors can be based on fiber gratings, interferometers, scattering/reflecting, Faraday rotation, fiber-optic gyroscopes, fluorescence, luminescence, and interaction of evanescent electromagnetic field structures.

Most of the aforesaid fiber sensors employ optical gratings which, under phase matching conditions, can couple different propagation modes. Optical fiber gratings can be fabricated by employing different techniques, e.g., the exposure of the optical fiber core to an intense optical interference pattern

[13]. Permanent gratings in an optical fiber were first obtained by Hill *et al.* [14]. In [14] an Argon-ion laser radiation, at 488 nm, was launched into a germania-doped fiber; after some minutes most of the launched light was reflected. Successively, strain and temperature tuning of the fiber grating were investigated and suitable spectral measurements confirmed the fabrication of a fiber Bragg grating (FBG) via the photosensitivity effect. In fact, a coherent high power laser beam can interfere with the light reflected back from a discontinuity to produce a standing wave pattern within the optical fiber. If the optical fiber, after ultraviolet light exposure, is suitably annealed the grating becomes permanent. It was demonstrated that the gratings facilitated coupling to higher-order leaky modes. After this experiment, numerous scientific papers on FBGs and their applications were produced.

Different techniques have been utilized for grating writing. In the holographic technique, two overlapping ultraviolet light beams interfere producing a periodic interference pattern on the side of the fibers [13]. Therefore a periodic refractive index grating is written in the core because the fiber cladding is transparent to the ultraviolet light whereas the fiber core, generally germanium doped, highly absorbs the ultraviolet light. The phase mask technique is based on the fabrication of a flat slab of silica glass, transparent to ultraviolet light but having one of the flat surfaces with a one dimensional periodic corrugation. This corrugation, can be obtained via an etching technique and by using photolithographic techniques. The optical fiber is placed close to the corrugations of the phase mask and then irradiated by ultraviolet light which is incident normal to the phase mask [13].

FBGs are now commercially available and they find extensive application in many areas such as, filtering, routing, lasing and amplification of optical signals, and sensing of various measurands.FBG-based wavelength interrogation modules and broadband light sources, have been optimized for the safety monitoring of large structures, e.g., bridges, towers, dams, and railways [15,16,17,18,19,20,21].

Long period gratings (LPGs) have been exploited in the construction of filters, gain flatteners for erbium-doped fiber amplifiers, dispersion compensators, sensors, *etc.* [22,23,24,25,26,27,28,29,30,31, 32,33,34, 35,36, 37]. They have been fabricated in MOF, by employing different techniques, to obtain strain or refractive index sensors [28,29]. Moreover, LPGs were fabricated in all-solid photonic bandgap fibers (PBGFs) by UV illumination [36]. In particular, the exploitation of a tunable slit with a high-precision motorized linear stage and a suitable PBGF laser pulse exposition allowed an optimized LPG fabrication process [36]. The first fabrication of femtosecond-laser-drilled structural LPGs in a large mode area PCF was reported in [37]. Different techniques allowing grating inscription into the

MOF core and cladding have been demonstrated in a number of experimental works [23,25,26,30,35,36,37].

In this paper, some recent advances on optical fiber sensors are reported. Specifically, only a few recent investigations are cited for FBGs, LPGs and evanescent field sensors.

2. OPTICAL FIBER GRATINGS AS KEY ELEMENTS OF SENSORS

The core refractive index perturbation $\Delta n(z)$ along the fiber longitudinal direction z, constituting a grating, can be expressed as reported in [31]:

$$\Delta n(z) = n_1(z) - n_1 = n_1\sigma(z)\left[1 + m\sin\left(\frac{2\pi z}{\Lambda}\right)\right] \quad (1)$$

where n_1 is the core refractive index, Λ is the grating period, m is the induced-index fringe modulation, and $\sigma(z)$ is the slowly varying envelope of the grating.

The well-known coupled mode theory CMT can be employed to find the equations describing the electromagnetic field propagation along the grating, by considering the slowly varying envelope approximation (SVEA) [31]. Coupled equations can be correctly written when grating inscription induces a weak perturbation on the propagation modes and the electromagnetic field amplitudes of the forward and backward modes can be calculated.

A strong mode interaction can be obtained in the grating at the wavelength where both mode phase matching (phase synchronism) and a sufficient mode overlap occur. Therefore, different modes can exchange power between them. The local reflectivity is the complex ratio of the forward and backward wave amplitudes [13]. FBGs, characterized by a short period, typically couple the forward propagating core mode to the backward propagating one. Chirped fiber gratings exhibit a large reflection spectrum. They are generally utilized when the phenomenon that each wavelength component is reflected at different positions in different delay times must be exploited for the fabrication of specific devices [2]. Tilted fiber gratings exhibit different characteristics and they can be used to couple the forward propagating core mode to the backward one and to a backward propagating cladding mode. LPGs, characterized by a long period, are generally designed to allow the power exchange of a forward propagating core mode with the power of selected forward propagating cladding modes.

When a core mode is coupled with a cladding mode the transferred electromagnetic energy is affected by the mode phase mismatch (displacement from the phase matching condition) which can be due to a variation of a number of different measurands such as wavelength, temperature, stress, fiber bending *etc.* Therefore optical sensors can be obtained.

2.1. Fiber Bragg Grating Sensors

An FBG is a distributed mirror at the Bragg wavelength, *i.e.*, at the phase-matching condition $\lambda_B = 2n_{eff}/\Lambda$, where λ_B is the Bragg wavelength, n_{eff} is the effective index of the interfering guided mode [16]. Fiber laser cavity can be typically obtained by employing two different mirror implementations: (i) distributed Bragg reflector (DBR) formed by two cascaded wavelength-matched FBGs; and (ii) distributed feedback Bragg (DFB) structure where a phase-shifted FBG is distributed in an active fiber [17,18,19]. DBR and DFB laser sensors attract considerable interest because of their excellent characteristics [20,21]. Moreover, fiber grating laser sensors can be based on wavelength encoding or on a polarimetric heterodyning sensor.

FBGs allow fine control of fiber laser cavity by finely tuning their distributed reflectivity. This potential is strategic in the optimal design of both conventional and MOF fiber lasers [38,39,40,41,42,43,44,45] and therefore in the design of DBR and DFB laser sensors.

A heavy limitation of FBG strain sensors lies in their high temperature cross-sensitivity, this drawback is worsened by the difficulty to measure strain and temperature simultaneously. In [46] a hybrid sensor overcoming the aforesaid limitation has been proposed. It operates in the intensity domain, by converting into intensity changes two different items: (i) the polarization information from a polarization-maintaining photonic crystal fiber PM-PCF fiber sensor; and (ii) the wavelength information from FBG. A 3 dB coupler has been used to split the signals between the FBG sensor and the polarimetric sensor. A polarizer and an analyzer at the input and output of the PM-PCF have been used to measure the polarization variation due to the strain and temperature. A macrobend fiber-based edge filter has been used to convert the FBG wavelength variations due to the strain and temperature into light intensity changes. The output powers were measured by employing two photodiodes [46]. Also in displacement measurements, the FBG sensor output can be obtained by using an edge filter to interrogate the resonant wavelength [47] or a cantilever to transfer/convert the displacement into the chirp profile of the FBG [48].

Optical fiber sensors based on cladding-mode backward recoupling constitute an interesting technique to obtain high sensitivities [49,50,51,52,53]. In fact, an optical fiber cladding mode shows much higher sensitivity to micro deformations, like structural bending, compared to a fundamental core mode. For example, in [54] a structure for cladding-mode backward recoupling has been fabricated by including the up-taper and the Bragg grating as shown in Figure 1. The up-taper microscope photo is reported in Figure 2. Cladding-mode backward recoupling is the key principle on which the sensor is based. The structure is illuminated by a broad-band source from the same interrogation fiber. The up taper is designed to allow mode coupling between copropagating modes. The Bragg grating allows wavelength-selective mode coupling between counter-propagating modes. At the up-taper, the core mode exhibits loss and the cladding modes are excited. The excited cladding modes travel to the fiber grating with low attenuation if the sensor is kept straight. As described in [54], the guided core mode is coupled to the backward-propagating cladding ones by the grating. Then the cladding modes are recoupled to the guided core mode at the up taper. The cladding modes excited at the up taper are recoupled to the guided core mode by the fiber grating. Following this, the guided core mode travels into the interrogation fiber. Mode coupling between counter-propagating cladding modes can be neglected. Figure 3 illustrates cladding mode power and core mode power *versus* displacement and cladding mode power *versus* temperature. An improvement of the sensor performance is obtained by employing suitable uniform chirped FBGs (CFBGs) as illustrated in Figure 4. With the same measurement set-up, by employing the CFBG the maximum reflection power is increased from 800 nW (for the FBG) to 176 μW (for the CFBG).

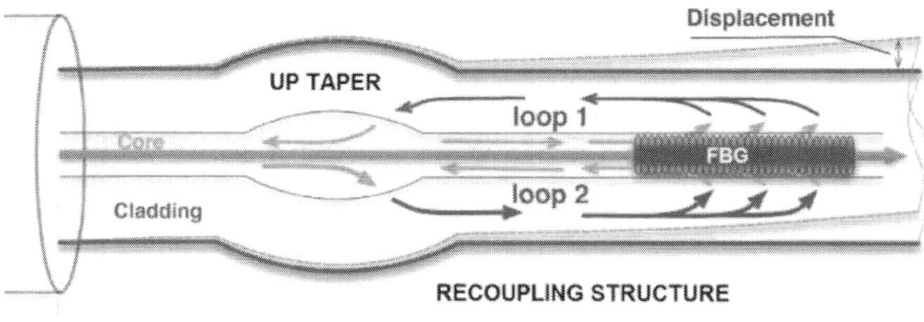

Figure 1. Schematic of cladding-mode backward recoupling structure including the up taper and the fiber Bragg grating proposed in [54].

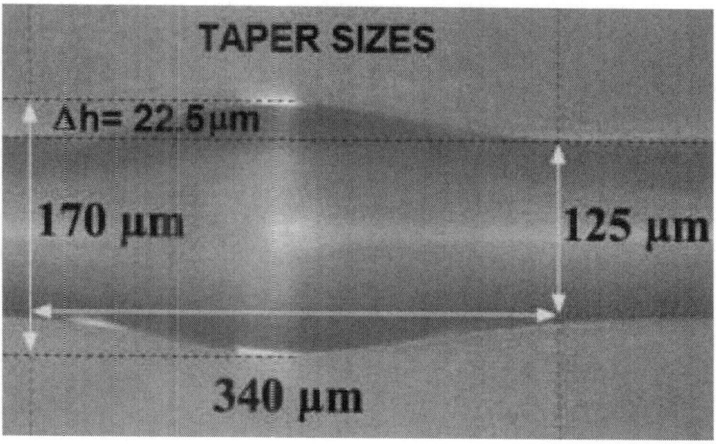

Figure 2. Microscope photo of the up taper constructed in [54].

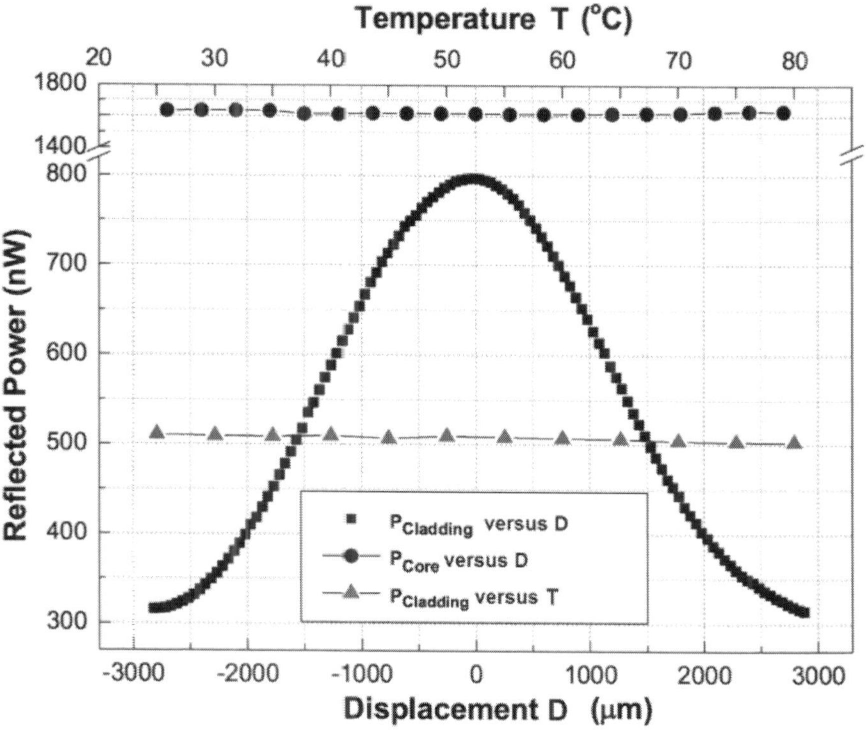

Figure 3. Cascade of up-taper and fiber Bragg grating (FBG) [54]. Behavior of cladding mode and core mode power *versus* displacement and cladding mode power *versus*temperature.

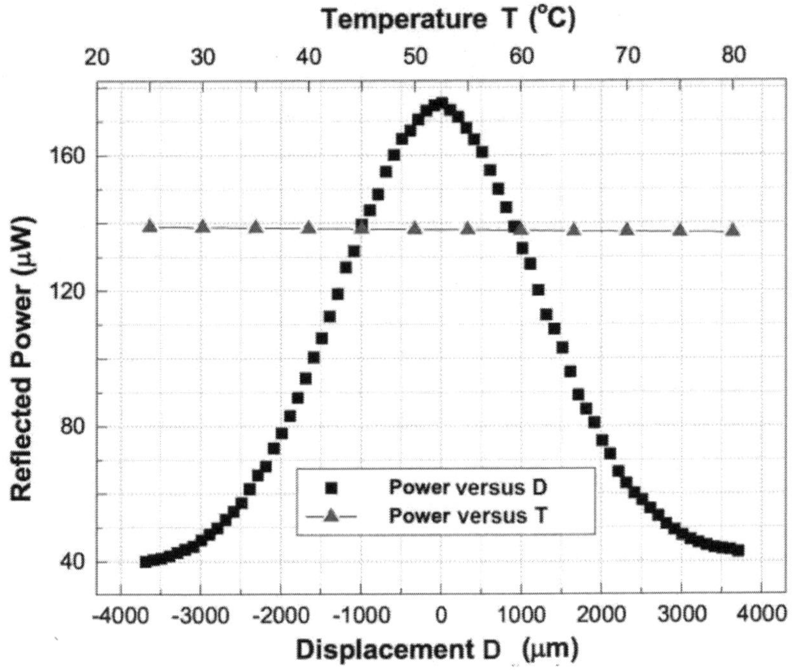

Figure 4. Cascade of up-taper and chirped FBG (CFBG) [54]. Behavior of cladding mode power and core mode power *versus* displacement and cladding mode power*versus* temperature.

Gas or liquid flow measurement is required in the automotive industry, in material processing, in marine applications and in a number of other areas. Fiber optic technology has been largely employed [55,56,57,58,59]. For example, an optical fiber thermal anemometer based on a short silver-film-coated FBG and a core-offset fusion splice is proposed in [60]. The set-up is schematically shown in Figure 5. The core-offset splice is performed in order to couple most of the laser power from the fiber core to the cladding, see photograph and sketch of the core-offset splicing point. The silver coating close to the FBG absorbs the pump laser and generates heat. A pump laser at 1450 nm with a maximum output power of 450 mW has been used to heat the FBG. Therefore, the generated heat increases the temperature and thus the Bragg wavelength of the FBG. The anemometer is cooled down by the air flow through the anemometer. As a consequence, the Bragg wavelength shifts toward the short wavelength. The velocity of the air flow can be measured from the amount of wavelength shift of the FBG with respect to the nominal Bragg wavelength. In this way, a highly sensitive anemometer is obtained. A linear response of 45.3 pm/m/s and a high resolution of 0.022 m/s were achieved for an airflow velocity below 6 m/s. The sensor characteristic is shown in Figure 6, where the Bragg wavelength *versus* the airflow velocity is illustrated.

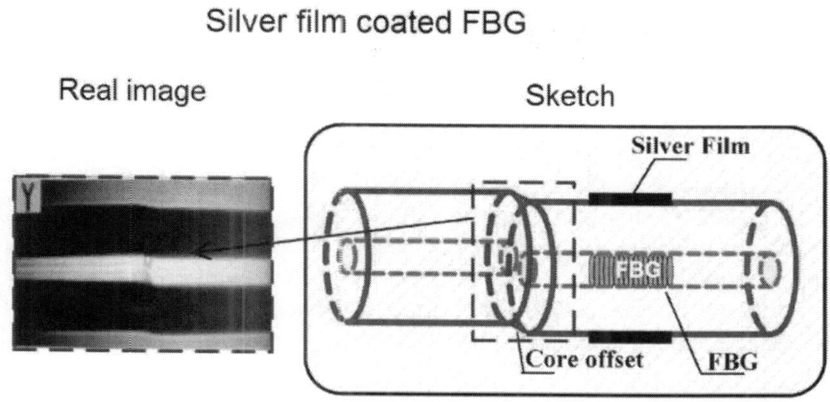

Figure 5. Schematic of the set-up proposed in [60]. FBG-based thermal anemometer.

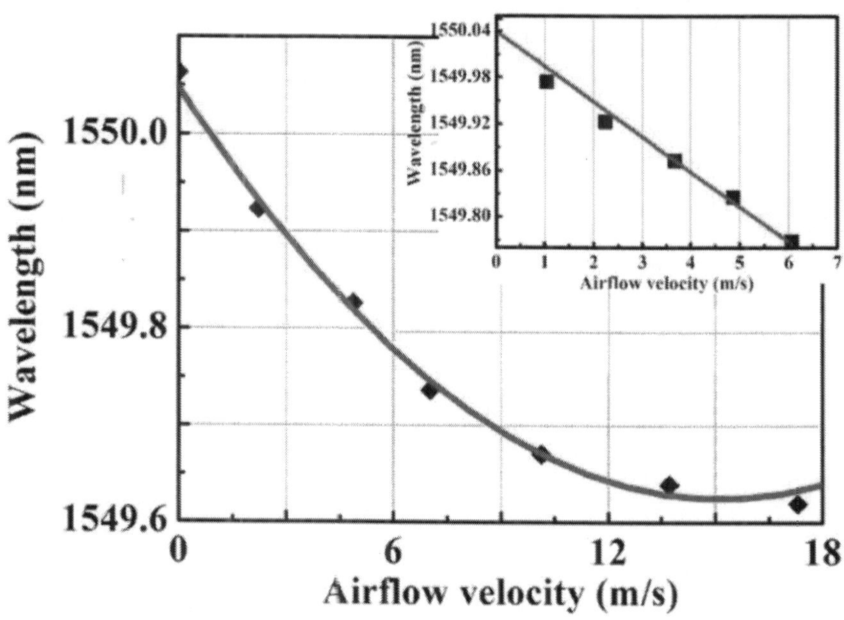

Figure 6. Resonant wavelength *versus* applied airflow velocity of the anemometer using silver-coated fiber Bragg grating [60].

Pressure sensors are desired in a number of areas e.g., industry, structural safety monitoring, oil wells, and power plants. Pressure sensors based on fiber Bragg gratings (FBGs) on side-hole fibers are very attractive [61,62,63,64]. In [65] a fiber-optic pressure sensor based on a π-phase-shifted fiber Bragg grating (πFBG) fabricated on a side-hole fiber is reported. A πFBG can be considered as aFabry-Perot (FP) resonator formed by two FBG mirrors. The side-hole fiber consists of two air channels running through the fiber cladding as shown in Figure 7. A hydrostatic pressure acting on the fiber surface induces an anisotropic stress distribution. Therefore a fiber birefringence via the photoelastic effect occurs. The fiber birefringence, proportional to the hydrostatic pressure, can be measured by evaluating the separation of the FBG peaks associated with the two polarization states. Therefore the FBG reflection spectrum gives a measurement of the pressure. The temperature can be simultaneously measured by monitoring the Bragg wavelength shift of the FBG peak. The performance can be enhanced by employing two πFBGs [65].

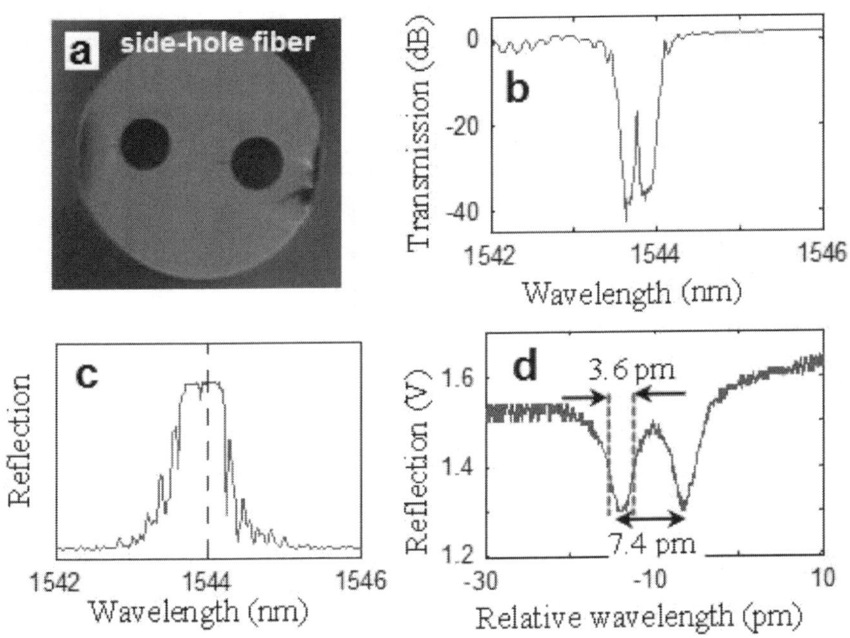

Figure 7. Fiber-optic pressure sensor based on a π-phase-shifted fiber Bragg grating (πFBG) fabricated on a side-hole fiber [65]. (**a**) Section of the side-hole fiber; (**b**) Transmission *versus* wavelength; (**c**) and (**d**) Reflection *versus* wavelength of a πFBG fabricated on the side-hole fiber.

The πFBG demonstrated in [65] is 8.3 mm long and the line width of each spectral notch is 3.6 pm, corresponding to a quality factor of 4.3×10^5. The spectral notch separation exhibited a sensitivity of 20 pm/kpsi to pressure. The Bragg wavelength shift exhibited a sensitivity of 11.4 pm/°C to temperature. A pressure detection limit of 1.4 psi and a temperature detection limit of 0.0025 °C were obtained.

2.2. Passive Long Period Grating Sensors

LPGs couple light very efficiently between two co-propagating modes in optical fibers. The resonance wavelength of an LPG, such as for the other gratings, is determined by the mode phase matching condition and by mode overlapping. It depends on the effective refractive index difference between these two coupled modes. Similarly to that occurring for FBGs, LPGs can be utilized to construct sensors for the detection of various physical parameters, including external refractive index, temperature, bending, loading, and liquid level [66,67]. LPG-based liquid-level sensor LLSs exhibit high sensitivity to external liquid (due to the light coupling to cladding modes) and high feasibility. In [68] an optical liquid-level sensor (LLS) based on long period fiber grating (LPG) interferometer was experimentally demonstrated. More precisely, an in-fiber Mach-Zehnder interferometer was obtained by employing two identical 3-dB LPGs. The sensing element, exposed to different liquids, is the fiber portion between two LPGs. The experimental sensor characteristic exhibited good linearity and a large measurement range. In Figure 8 the wavelength shift *versus* the liquid level is depicted for different refractive indices of the surrounding medium (liquids): 1.33 (rectangles), 1.398 (circles), and 1.413 (triangles) when the sensing element between two gratings is 250 mm long.

LPGs such as FBGs can be utilized to enhance particular physical effects, e.g., Surface Plasmon Resonance (SPR). SPR is employed in a high number of sensors [69,70,71,72,73,74,75], it is based on electron density oscillations, propagating as transverse-magnetic (TM) polarized waves at the interface between a metal and a dielectric medium. These waves are highly attenuated evanescent ones. LPG can be employed to achieve SPR of a single cladding mode [76,77,78,79]. In [80] the fabrication of a fiber-optic refractive index (RI) sensor has been reported. A LPG has been fabricated to achieve surface plasmon resonance of a single cladding mode at the gold-coated tip of a single-mode fiber. The set-up is shown in Figure 9.

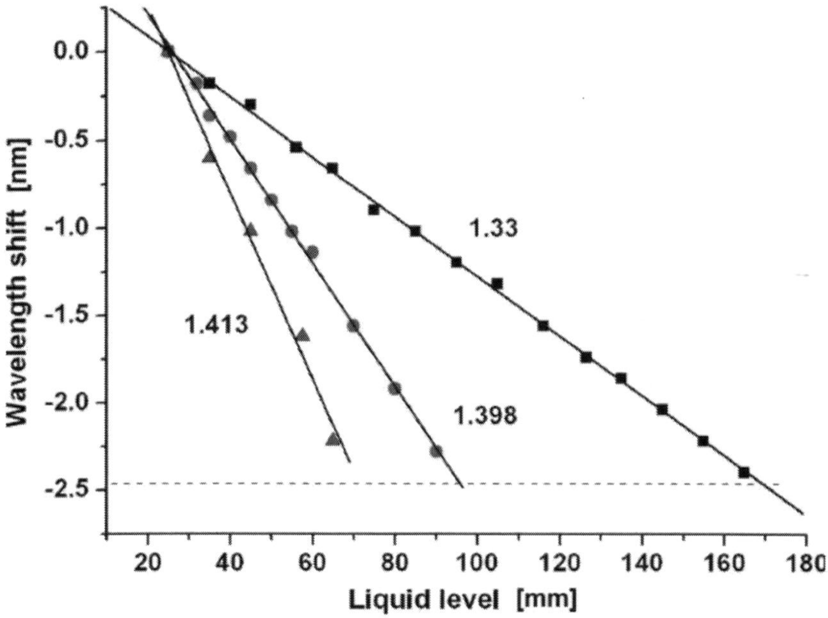

Figure 8. The relationship between peak wavelength and liquid-level with the surrounding liquid refractive indices [68].

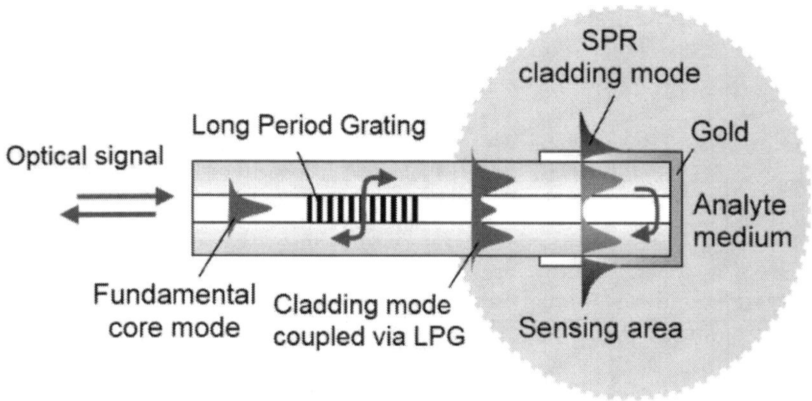

Figure 9. Set-up of the fiber-optic surface plasmon resonance (SPR) sensor proposed in [80].

The sensor is based on a single-mode fiber. The fiber tip is covered by a gold-plate, this part is the sensing area which is close to core LPG. A thin poly-directional coating supports the formation of SPR cladding modes. This structure has been proposed to obtain a biosensor. In fact, the power attenuation is strongly affected by the refractive index of the surrounding analyte medium. Small RI changes, e.g., caused by biomolecular binding processes, induce changes of the transmitted optical power. The LPG is designed for bidirectional coupling of the fundamental core mode with a cladding mode. Then the cladding modes are converted into an SPR cladding mode. The sensor has exhibited a high level of sensitivity in a miniaturized sensing area. LPGs with periods of 114 μm were fabricated to couple in to the designated cladding mode $HE_{1,20}$ at a wavelength of around 660 nm. LPG lengths of 30 mm enabled mode coupling of 15 dB.

Figure 10 shows the measured normalized transmission spectra for increasing refractive index of the surrounding medium (1–1.38). The LPG resonances are close to 660.6 nm. The vertical dashed line is the spectral location of the maximum SPR transmission loss.

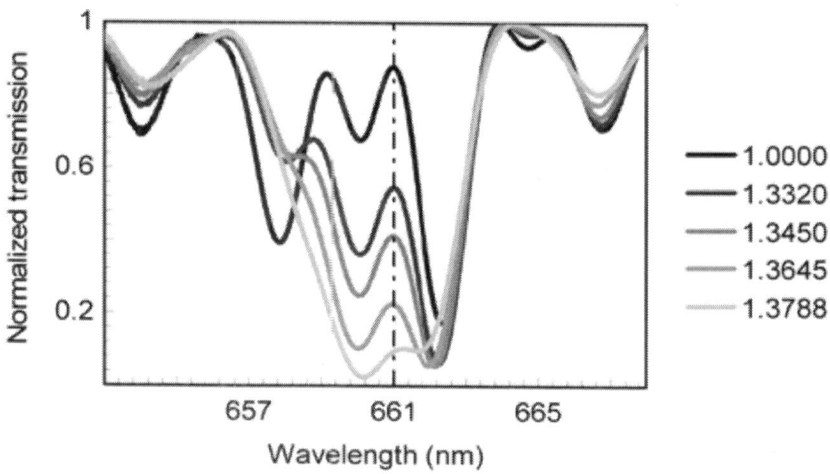

Figure 10. Normalized transmission spectra of the Miniaturized Long-Period Fiber Grating Assisted Surface Plasmon Resonance Sensor [80].

LPGs have been inscribed in MOFs filled with ferrofluid. Figure 11 shows the microscopic photograph of the Ferrofluid-Infiltrated Microstructured Optical Fiber Long-Period Grating reported in [81]. The cladding modes can be tuned more easily than the core mode. The mode dispersion curves are affected by the applied magnetic field. An enhancement of the applied magnetic field gradually reduces the refractive index of ferrofluid. This effect causes the decreasing of the effective refractive index difference between core and cladding modes, as shown in Figure 12. The MOF-based LPG has shown a sensitivity of 1.946 nm/Oe for a magnetic range of 0–300 Oe. The potential applications as a high-sensitivity magnetic-field sensor are apparent.

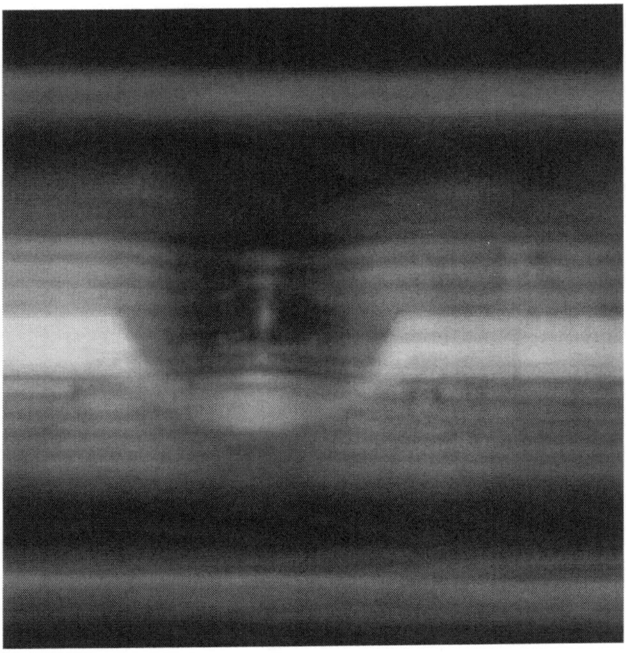

Figure 11. LPG inscribed in microstructured optical fiber (MOF) filled with ferrofluid [81].

2.3. Active Long Period Grating Sensors

Laser sensors, include active behavior (e.g., due to the rare earth doped materials) and grating transmission spectra (e.g., FBG or LPG), which can be highly sensitive to the variation of some measurands. In [82,83] the design criterion to choose the suitable spatial distribution of multiple long period gratings MLPG couplers with respect to a MOF laser cavity was identified. The reported model extensively illustrates the behavior of the LPG in active medium, via the coupled mode equation and the rate equation of the rare earth ion population. In fact, it describes an Yb^{3+}-doped MOF laser cavity by taking into account the presence of different MLPGs inscribed within the core. The MLPGs reported in [83] enable the interaction among the two-fold degenerate fundamental mode guided in the fiber core (HE_{11}) and the inner cladding modes, at the pump wavelength. The model includes the following kinds of couplers: (a) MLPGs-P, Multiple Long Period Gratings in Passive fibers, *i.e.*, grating cascades inscribed in the undoped region, just outside the laser cavity; (b) MLPGs-A, Multiple Long Period Gratings in Active fibers, *i.e.*, grating cascades inscribed within the laser cavity (see Figure 13); (c) DS-MLPGs, Double Stage Multiple Long Period Gratings, *i.e.*, a combination of passive grating cascades, inscribed outside the laser cavity, and active ones, inscribed within the laser cavity, are employed). Figure 14 shows the power in the cladding at pump wavelength $P_{clad}(\lambda_p)$, in the core at pump wavelength $P_{core}(\lambda_p)$, forward and backward in the core at signal wavelength$P^{+}_{core}(\lambda_s)$, $P^{-}_{core}(\lambda_s)$ *versus* the position along the fiber longitudinal axis *z* for the DS-MLPGs. The vertical lines represent the grating cascade along the fiber axis.

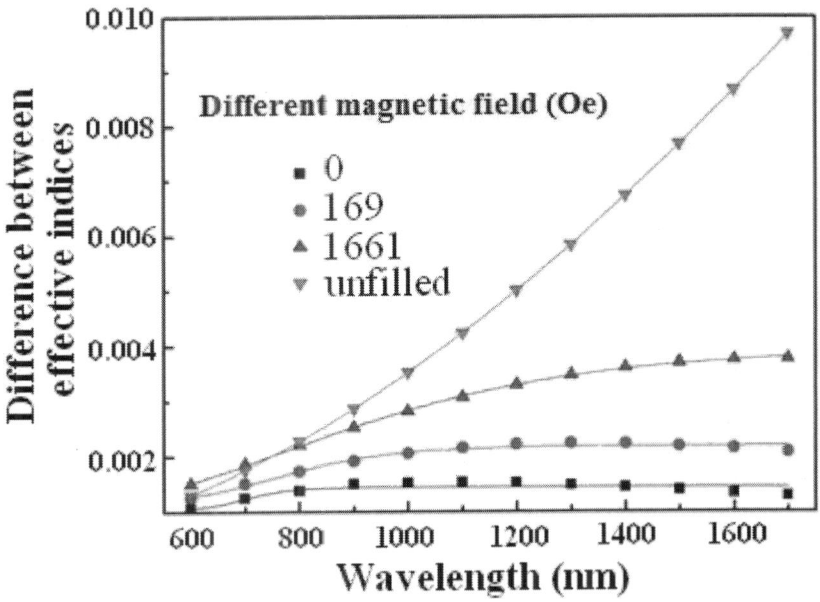

Figure 12. Difference between cladding-mode and core-mode effective indices for different applied magnetic fields [81].

Even if MLPGs-P, MLPGs-A and DS-MLPGs were designed in [83] to increase the pump power transfer from the inner cladding modes at the pump power towards the fundamental core mode of the MOF laser, the model can be employed for designing active MLPG sensors since both (i) the LPG transmission spectrum; and (ii) its efficiency as power coupler can be affected by different measurands (e.g., bending or strain).

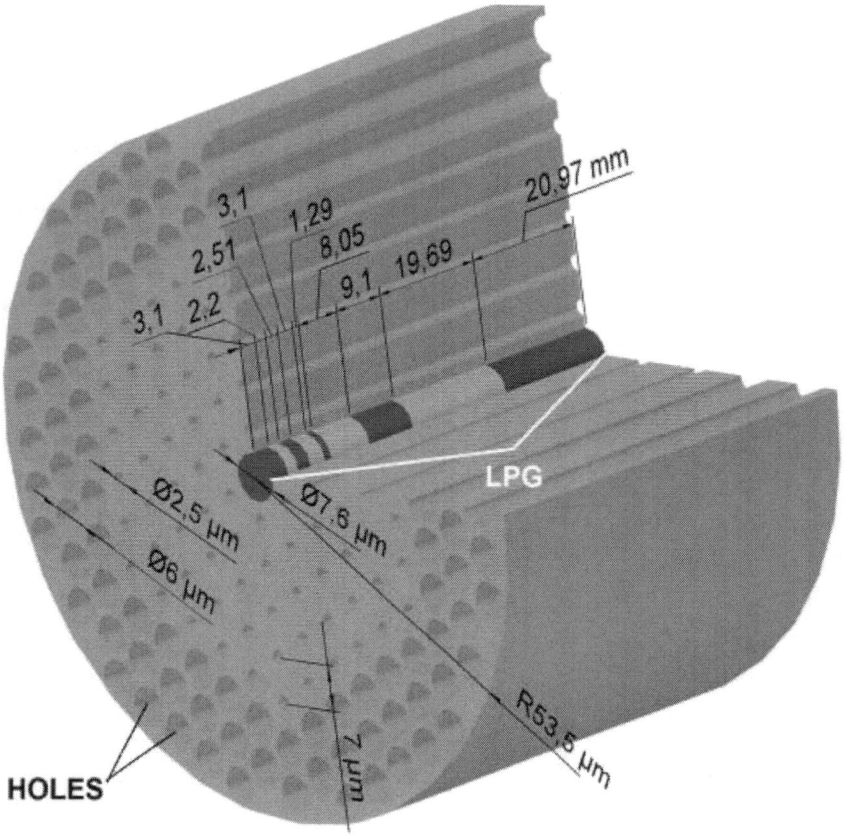

Figure 13. 3D sketch of the ytterbium-doped MOF laser with the MLPGs-A grating cascade [83].

Since the phase matching condition of the coupled modes in the LPG is strongly affected by small changes in the effective index difference, LPG refractive index sensors are feasible. It is well known that by filling the holes of MOFs with some functional material, their modal and dispersion properties may change strongly. This also affects the tuning properties of LPGs inscribed in such fibers [81]. Therefore, MOF based LPGs are highly sensitive/tunable devices [84,85,86].

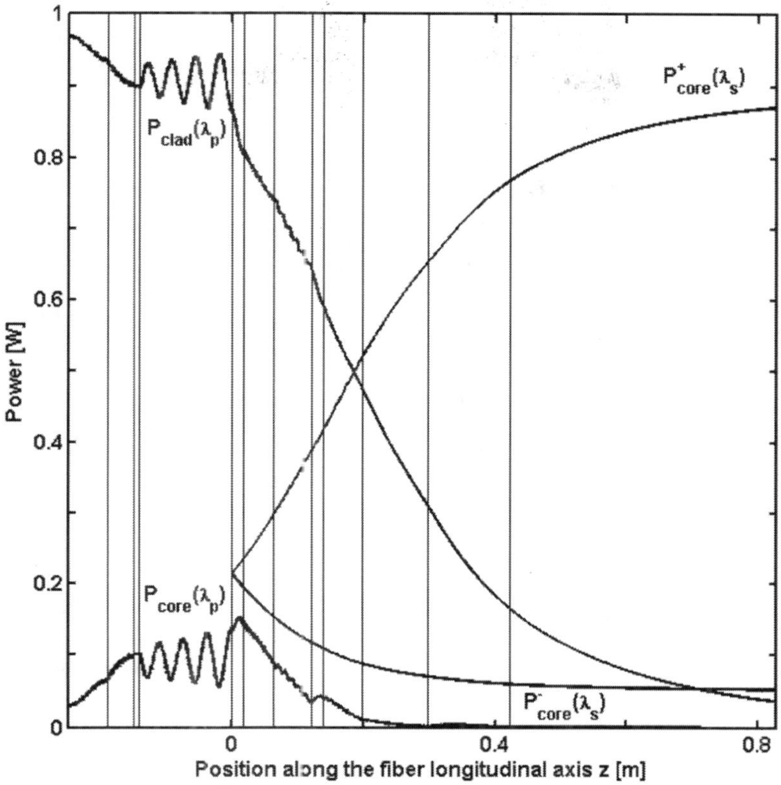

Figure 14. Power $P_{clad}(\lambda_p)$, $P_{core}(\lambda_p)$, $P^{+}_{core}(\lambda_s)$, $P^{-}_{core}(\lambda_s)$ *versus* the position along the fiber longitudinal axis *z*, DS-MLPGs [83].

Recently LPGs, similarly to FBG, were proposed for optical pulse shaping and coding techniques [87,88,89,90,91]. These topics are interesting within communication systems and, in addition, to develop a large interrogating optical network e.g., based on Optical Time Domain Reflectometer (OTDR) techniques and communication protocols.

3. EVANESCENT FIELD, HOLLOW CORE, LIQUID FILLED AND EXPOSED OPTICAL FIBER SENSORS

Optical fiber evanescent wave sensors are a valid alternative for chemical analysis [92]. Distributed optical fiber sensors based on evanescent field interactions enable the measurement of chemical distributions over very long distances. Their potential applications are in structural health monitoring, environmental monitoring, homeland security, and many other applications [93,94,95,96,97,98,99,100].

The optimization of evanescent field optical sensors plays an important role in the development of low-cost pollution monitoring. In this case, the detection is obtained via transmittance or absorption measurements at the wavelength where the contaminant exhibits the highest absorption peak. In [101,102,103,104,105,106,107] optical sensing of hydrocarbons in air or in water, using UV or NIR evanescent field optical sensors was reported. Figure 15 illustrates the example of the set-up reported in [106].

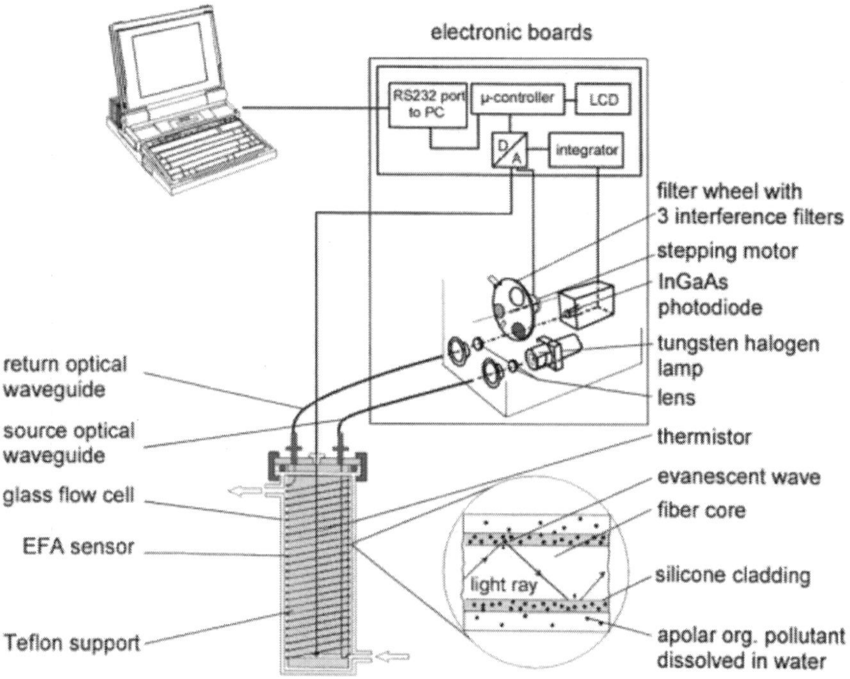

Figure 15. Evanescent field absorption sensing principle and instrumental set-up of coiled fiber-optic sensor installed in a flow cell and and NIR bandpass filter photometer unit [106].

In [108,109,110] fiber optic sensors for biosensing are reported. In particular in [108] the design of a photonic crystal fiber sensor for methadone detection in water was performed. The sensor feasibility was numerically investigated by employing a Finite Element Method (FEM) numerical code. The fiber cross section, including a polymeric cladding as sensitive layer, was optimized. In the design, the thickness of the polymeric layer, the hole diameter, the hole-to-hole spacing and the transversal distribution of the holes were parametrically varied. The design of the fiber geometry was performed to balance competing requirements such as the reduction of the number of the propagating modes and the improvement of the evanescent field optical power interacting with the sensitive layer. The polymeric sensitive layer interacts with the water polluted by the methadone. At equilibrium, the contaminant concentration (C_p) in the polymer and in the water (C_w) are linked by the distribution constant $K_{p/w} = C_p/C_w$, whose value is peculiar to that of the used polymer. The methadone induces the change of the polymer absorption loss which increases by increasing the contaminant concentration. Therefore, a change of the imaginary part of the complex effective refractive index of the guided modes occurs. In particular, the light guided in the fiber core suffers a power attenuation due to the methadone. The optimal UV operation wavelength is that at which the methadone absorption peak occurs. Figure 16 shows the simulated transverse profile of the electric field of the HE_{11} guided propagation mode (a) for square-lattice; and (b) for the hexagonal-lattice photonic crystal fibers.

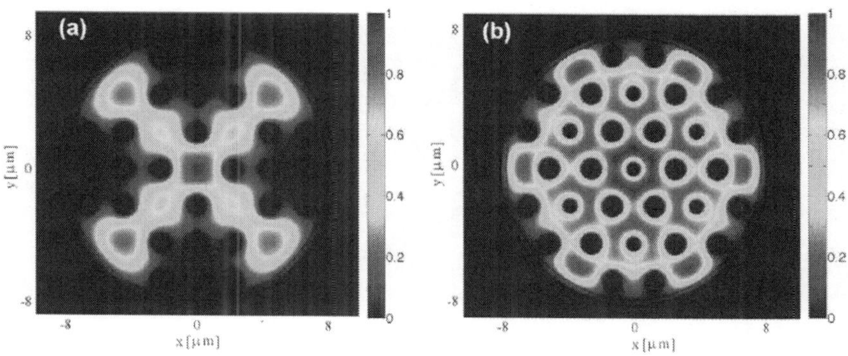

Figure 16. Simulated transverse profile of the electric field of the HE_{11} guided propagation mode (**a**) for square-lattice; and (**b**) for the hexagonal-lattice photonic crystal fibers [108].

Figure 17 depicts the absorbance A *versus* the methadone concentration for the square-lattice photonic crystal fibers (PCF) (full curve), hexagonal-lattice PCF (dash curve) and conventional step-index fiber (dot curve) having core radius $R = 8$ μm and PDMS thickness $t_c = 0.4$ μm, for the sensor length $L = 15$ cm.

In [109] the design of a silica exposed-core fiber sensor for methadone detection in water was reported; exposed-core fiber coated with polydimethylsiloxane PDMS was fabricated at the University of Adelaide in order to further enhance the sensitivity. Figure 18 reports the sketch of (a) exposed-core fiber; and (b) exposed-core fiber with PDMS coating. By simulation the absorbance sensitivity for a fiber length L = 10 cm for the exposed core fiber without PDMS is close to A = 0.3 × 10^{-6} [ppb−1] while it is close to A = 0.011 [ppb−1] for exposed core fiber with PDMS. Lead silicate glass (F2) was considered in the simulation as an alternative to silica glass and different contaminants (benzene, toluene *etc.*).

The air holes of *ad-hoc* designed MOFs can be filled with different materials [111,112,113]. Hollow-core MOFs having their core filled with gas or liquids were investigated for sensing and spectroscopic applications [114, 115,116]. In [117] a liquid-core air-clad microstructured fiber was developed for determination of ethanol and sucrose concentrations in aqueous solutions by Raman spectroscopy. The application of the sensor to white wine characterization has been proposed. Figure 19shows the SEM images of the fiber (left) and of the narrow silica bridges forming the air-clad (right) employed in [117].

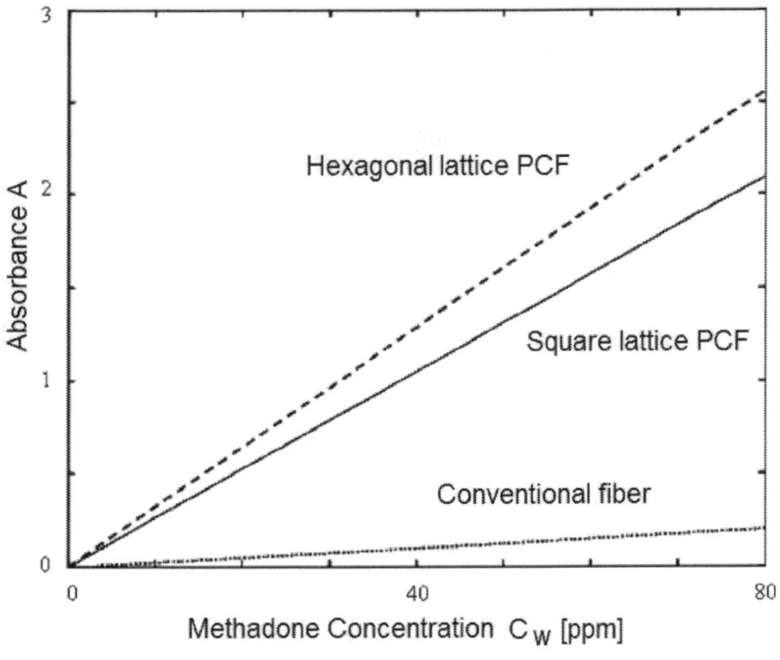

Figure 17. Absorbance A *versus* the methadone concentration for the square-lattice photonic crystal fibers (PCF) (full curve), hexagonal-lattice PCF (dash curve) and conventional step-index fiber (dot curve) [108].

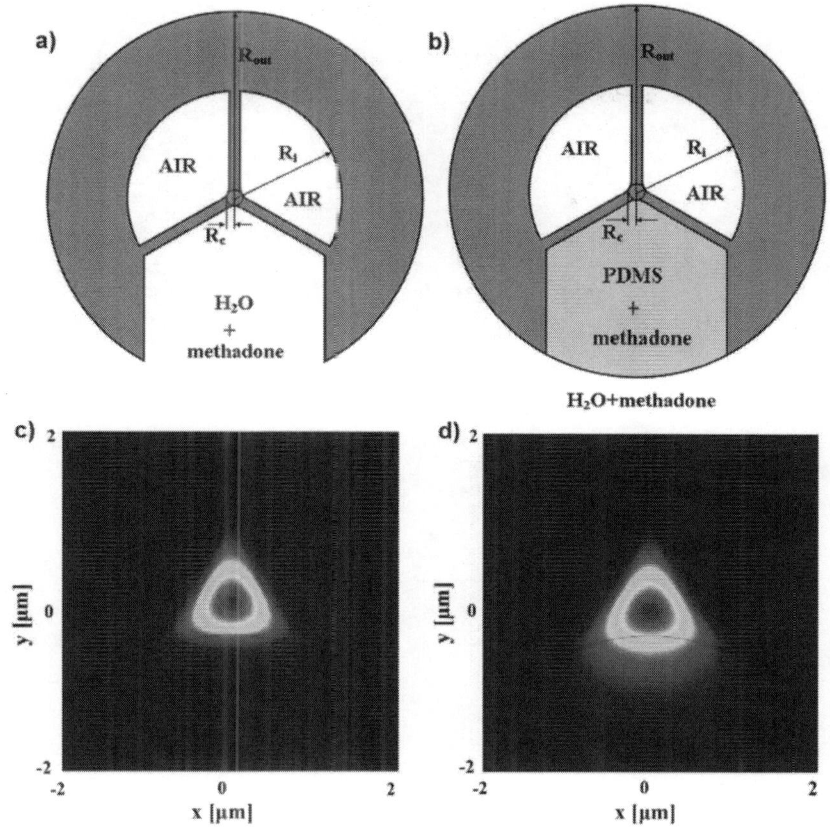

Figure 18. Sketch of the (**a**) exposed-core fiber; and (**b**) exposed-core fiber with polydimethylsiloxane PDMS coating. Simulated transverse electric field norm of the HE_{11} y-polarized guided mode at the wavelength λ = 292 nm for (**c**) exposed-core fiber; and (**d**) exposed-core fiber with PDMS coating [109].

Figure 20 shows the measured and reproduced Raman spectra of wine with the spectral contribution of water, ethanol, and sucrose.

By considering the optical fiber as a part of the sensing mechanism a number of other devices have been proposed. As further examples, passive and active microspheres coupled to taper fibers were proposed for biosensing applications [118,119,120,121,122,123,124, 125,126,127, 128, 129,130]. To conclude, the variety of e proposed optical fiber sensors as well as their potential is extremely impressive.

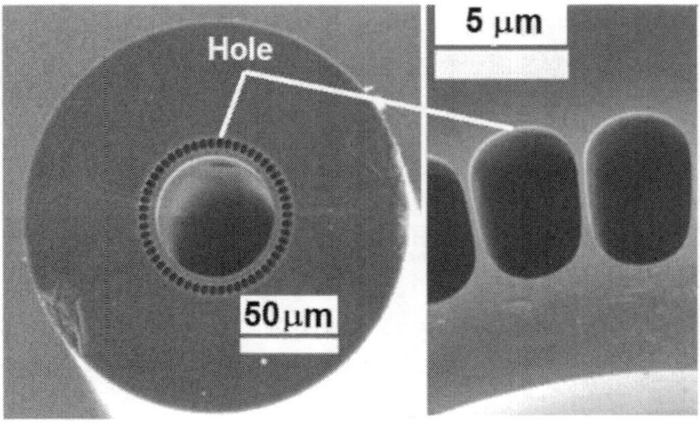

Figure 19. SEM images of the fiber (left) and of the narrow silica bridges forming the air-clad (right) [117].

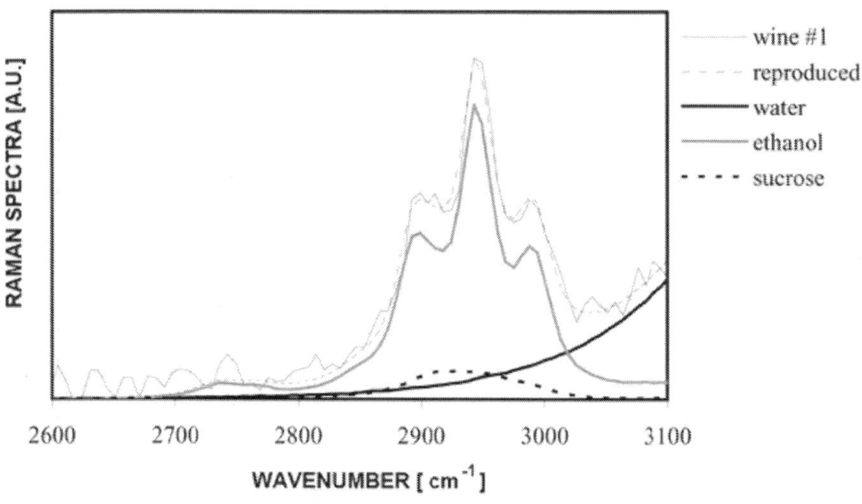

Figure 20. Raman spectra of wine [117].

4. CONCLUSIONS

FBG, LPG, evanescent field and hollow core optical fiber sensors are briefly illustrated here. More precisely, examples of recent optical fiber sensors for the measurement of strain, temperature, displacement, air flow, pressure, liquid-level, magnetic field, as well as the determination of methadone, hydrocarbons, ethanol, and sucrose are discussed.

CONFLICTS OF INTEREST

The authors declare no conflict of interest.

REFERENCES

1. Girard, S.; Mescia, L.; Vivona, M.; Laurent, A.; Ouerdane, Y.; Marcandella, C.; Prudenzano, F.; Boukenter, A; Robin, T.; Paillet, P; Goiffon, V.; Gaillardin, M.; Cadier, B.; Pinsard, E.; Cannas, M.; Boscaino, R. Design of radiation-hardened rare-earth doped amplifiers through a coupled experiment/simulation approach. J. Lightwave Technol. 2013, 31, 1247–1254.

2. Lee, B. Review of the present status of optical fiber sensors. Opt. Fiber Technol. 2003, 9, 57–79.

3. Pelli, S.; Chiasera, A.; Ferrari, M.; Righini, G.C. Spectroscopic Techniques for Sensors. In An Introduction to Optoelectronic Sensors; Series in Optics and Photonics; World Scientific Publishing Co. Pte. Ltd.: Singapore, 2009; Volume 7, ISBN: 978-981-283-412-6 981-283-412-5.

4. Cordeiro, C.M.B.; Franco, M.A.R.; Chesini, G.; Barretto, E.C.S.; Lwin, R.; Cruz, C.H.B.; Large, M.C.J. Microstructured-core optical fibre for evanescent sensing applications. Opt. Express 2006, 14, 13056–13066.

5. Jensen, J.; Hoiby, P.; Emiliyanov, G.; Bang, O.; Pedersen, L.H.; Bjarklev, A. Selective detection of antibodies in microstructured polymer optical fibers. Opt. Express 2005, 13, 5883–5889.

6. Jensen, J.B.; Pedersen, L.H.; Hoiby, P.E.; Nielsen, L.B.; Hansen, T.P.; Folkenberg, J.R.; Riishede, J.; Noordegraaf, D.; Nielsen, K.; Carlsen, A.; Bjarklev, A. Photonic crystal fiber based evanescent-wave sensor for detection of biomolecules in aqueous solutions. Opt. Lett. 2004, 29, 1974–1976.

7. Lehmann, H.; Brückner, S.; Kobelke, J.; Schwotzer, G.; Schuster, K.; Willsch, R. Toward photonic crystal fiber based distributed chemosensors. Proc. SPIE 2005, 5855, 419–422.

8. Rothwell, J.H.; Flavin, D.A.; MacPherson, W.N.; Jones, J.D.C.; Knight, J.C.; Russell, P.St.J. Photonic sensing based on variation of propagation properties of photonic crystal fibres. Opt. Express 2006, 14, 12445–12450.

9. Ma, J; Bock, W.J.; Wang, Z. Y.; Hao, W; MacKinnon, S. M. Fiber-optic membrane fluorescent sensor based on photonic crystal fiber with a glass rod in the fiber end. In IEEE Sensors 2005, Irvine, CA, USA, 30 October–November 2005; pp. 1096–1099.

10. Cox, F.M.; Argyros, A.; Large, M.C.J. Liquid-filled hollow core microstructured polymer optical fiber. Opt. Express2006, 14, 4135–4140.

11. Fini, J.M. Microstructure fibres for optical sensing in gases and liquids. Meas. Sci. Technol. 2004, 15, 1120–1128.

12. Monro, T.M.; Belardi, W.; Furusawa, K.; Baggett, J.C.; Broderick, N.G.R.; Richardson, D.J. Sensing with microstructured optical fibres. Meas. Sci. Technol. 2001, 12, 854–858.

13. Hill, K.O.; Meltz, G. Fiber Bragg grating technology fundamentals and overview. J. Lightwave Technol. 1997, 15, 1263–1276.

14. Hill, K.O.; Fujii, Y.; Johnson, D.C.; Kawasaki, B.S. Photosensitivity in optical fiber waveguides: Application to reflection filter fabrication. Appl. Phys. Lett. 1978, 32, 647–649.

15. Culshaw, B.; Kersey, A. Fiber-optic sensing: A historical perspective. J. Lightwave Technol. 2008, 26, 1064–1078.

16. Guan, B.; Jin, L.; Zhang, Y.; Tam, H. Polarimetric heterodyning fiber grating laser sensors. J. Lightwave Technol. 2012,30, 1097–1111.

17. Dong, L.; Loh, W.H.; Caplen, J.E.; Minelly, J.D.; Hsu, K.; Reekie, L. Efficient single-frequency fiber lasers with novel photosensitive Er/Yb optical fibers. Opt. Lett. 1997, 22, 694–696.

18. Hill, D.J.; Hodder, B.; DeFreitas, J.; Thomas, S.D.; Hickey, L. DFB Fibre-Laser Sensor Developments. In Proceedings of SPIE, the 17th International Conference on Optical Fibre Sensors, Bruges, Belgium, 23 May 2005; Volume 5855, pp. 904–907.

19. Li, J.; He, F.; Xu, T.; Wang, Y.; Liu, Y. High Performance Distributed Feedback Fiber Laser Sensor Array System. In Proceedings of Asia Communications and Photonics Conference Exhibition, Shanghai, China, 2–6 November 2009; Volume 7634, pp. 76340K:1–76340K:9.

20. Cranch, G.A.; Flockhart, G.; Kirkendall, C.K. Distributed feedback fiber laser strain sensors. IEEE Sens. J. 2008, 8, 1161–1172.

21. Beverini, N.; Maccioni, E.; Morganti, M.; Stefani, V.; Falciai, R.; Trono, C. Fiber laser strain sensor device. J. Opt. A: Pure Appl. Opt. 2007, 9, 958–962.

22. Saitoh, K.; Koshiba, M.; Hasegawa, T.; Sasaoka, F. Chromatic dispersion control in photonic crystal fibers: Application to ultra-flattened dispersion. Opt. Express 2003, 11, 843–852.

23. Westbrook, P.S.; Eggleton, B.J.; Windeler, R.S.; Hale, A.; Strasser, T.A.; Burdge, G.L. Cladding-mode resonances in hybrid polymer-silica microstructured optical fiber gratings. IEEE Photonics Technol. Lett. 2000, 12, 495–497.

24. Lee, B.H.; Liu, Y.; Lee, S.E.; Choi, S.S.; Jang, J.N. Displacements of the resonant peaks of a long-period fiber grating induced by a change of ambient refractive index. Opt. Lett. 1997, 22, 1769–1771.

25. Steivurzel, P.; Moore, E.D.; Mägi, E.C.; Kuhlmey, B.T.; Eggleton, B.J. Long period grating resonances in photonic bandgap fiber. Opt. Express 2006, 14, 3007–3014.

26. Lee, D.; Jung, Y.; Jeong, Y.S.; Oh, K.; Kobelke, J.; Schuster, K.; Kirchhof, J. Highly polarization-depend periodic coupling in mechanically induced long period grating over air-silica fibers. Opt. Lett. 2006, 31, 296–298.

27. Lim, J.H.; Lee, K.S.; Kim, J.C.; Lee, B.H. Tunable fiber gratings fabricated in photonic crystal fiber by use of mechanical pressure. Opt. Lett. 2004, 29, 331–333.

28. Petrovic, J.S.; Dobb, H.; Mezentsev, V.K.; Kalli, K.; Webb, D.J.; Bennion, I. Sensitivity of LPGs in PCFs fabricated by an electric Arc to temperature, strain, and external refractive index. J. Lightwave Technol. 2007, 25, 1306–1312.

29. Zhao, C.L.; Xiao, L.; Ju, J.; Demokan, M.S.; Jin, W. Strain and temperature characteristics of a long-period grating written in a photonic crystal fiber and its application as a temperature-insensitive strain sensor. J. Lightwave Technol.2008, 26, 220–227.

30. Eggleton, B.J.; Westbrook, P.S.; Windeler, R.S.; Spalter, S.; Strasser, T.A. Grating resonances in air-silica microstructured optical fibers. Opt. Lett. 1999, 24, 1460–1462.

31. Erdogan, T. Cladding-mode resonances in short-and long-period fiber grating filters. J. Opt. Soc. Am. A 1997, 14, 1760–1773.

32. Mescia, L. Design of long-period gratings in cladding-pumped microstructured optical fiber. J. Opt. Soc. Am. B 2008,25, 1883–1839.

33. Calò, G.; D'Orazio, A.; De Sario, M.; Mescia, L.; Petruzzelli, V.; Allegretti, L.; Palmisano, T.; Prudenzano, F. Improvement of the Pump Power Coupling in Double Cladding Photonic Crystal Fiber. In Proceedings of IEEE/LEOS Winter Topical Meeting Series, Sorrento, Italy, 14–16 January 2008; pp. 146–147.

34. Baek, S.; Roh, S.; Jeong, Y.; Lee, B. Experimental demonstration of enhancing pump absorption rate in cladding-pumped ytterbium-doped fiber laser using pump-coupling long-period gratings. IEEE Photonics Technol. Lett. 2006, 18, 700–702.

35. Allsop, T.; Kalli, K.; Zhou, K.; Smith, G.; Laia, V.; Smith, G.; Dubov, M.; Webb, D.; Bennion, V. Long period gratings written into a photonic crystal fibre by a femtosecond laser as directional bend sensors. Opt. Commun. 2008, 281, 5092–5096.

36. Long, J.; Zhi, W.; Yange, L.; Guiyun, K.; Xiaoyi, D. Ultraviolet-inscribed long period gratings in all-solid photonic bandgap fibers. Opt. Express 2008, 16, 21119–21131.

37. Shujing, L.; Long, J.; Wei, J.; Dongning, W.; Changrui, L.; Ying, W. Structural long period gratings made by drilling micro-holes in photonic crystal fibers with a femtosecond infrared laser. Opt. Express 2010, 18, 5496–5503.

38. D'Orazio, A.; De Sario, M.; Mescia, L.; Petruzzelli, V.; Prudenzano, F. Design of double-clad ytterbium doped microstructured fibre laser. Appl. Surf. Sci. 2005, 248, 499–502.

39. Prudenzano, F. Erbium-doped hole-assisted optical fiber amplifier: Design and optimization. J. Lightwave Technol.2005, 23, 330–340.

40. Sabaiean, M.; Nadgaran, H.; De Sario, M.; Mescia, L.; Prudenzano, F. Thermal effects on double clad octagonal Yb: glass fiber laser. Opt. Mater. 2009, 31, 1300–1305.

41. Prudenzano, F.; Mescia, V.; Allegretti, L.; De Sario, M.; Palmisano, T.; Smektala, F.; Moizan, V.; Nazabal, V.; Troles, J. Design of Er3+-doped chalcogenide glass laser for MID-IR application. J. Non-Cryst. Solids 2009, 355, 1145–1148.

42. Li, L.; Schülzgen, A.; Temyanko, V.L.; Morrell, M.M.; Sabet, S.; Li, H.; Moloney, J.V.; Peyghambarian, N. Ultracompact cladding-pumped 35-mm-short fiber laser with 4.7-W single-mode. Applied Physics Letters. 2006, 88, 161106:1–161106:2.

43. Prudenzano, F.; Mescia, L.; Allegretti, L.; Moizan, V.; Nazabal, V.; Smektala, F. Theoretical study of cascade laser in erbium-doped chalcogenide glass fibers. Opt. Mater. 2010, 33, 241–245.

44. Mescia, L.; Smektala, F.; Prudenzano, F. New trends in amplifiers and sources via chalcogenide photonic crystal fibers. Int. J. Opt. 2012, 2012, E212091.

45. Prudenzano, F.; Mescia, L.; Di Tommaso, A.; Surico, M.; De Sario, M. Design and refinement of rare earth doped multicore fiber lasers. Opt. Mater. 2013, 35, 1941–1946.

46. Rajan, G.; Ramakrishnan, M.; Semenova, Y.; Milenko, K.; Lesiak, P.; Domanski, A.W.; Wolinski T.R.; Farrell, G. A photonic crystal fiber and fiber Bragg grating-based hybrid fiber-optic sensor. Syst. IEEE Sens. J. 2012, 12, 39–43.

47. Zou, Y.; Dong, X.; Lin, G.; Adhami, R. Wide range FBG displacement sensor based on twin-core fiber filter. J. Lightwave Technol. 2012, 30, 337–343.

48. Zhu, Y.; Shum, P.; Lu, C; Lacquet, M.; Swart, P.; Chtcherbakov, A.; Spammer, S. Temperature insensitive measurements of static displacements using a fiber Bragg grating. Opt. Express 2003, 11, 1918–1924.

49. Guo, T.; Shao, L.; Tam, H.-Y.; Krug, A.; Albert, J. Tilted fiber grating accelerometer incorporating an abrupt biconical taper for cladding to core recoupling. Opt. Express 2009, 17, 20651–20660.

50. Shao, L.Y.; Albert, J. Compact fiber-optic vector inclinometer. Opt. Lett. 2010, 35, 1034–1036.

51. Zhou, W.; Zhou, Y.; Dong, X.; Shao, L.-Y.; Cheng, J.; Albert, J. Fiber-optic curvature sensor based on cladding-mode Bragg grating excited by fiber multimode interferometer. IEEE Photonics J. 2012, 4, 1051–1057.

52. Rong, Q.; Qiao, X.; Zhang, J.; Wang, R.; Hu, M.; Feng, Z. Simultaneous measurement for displacement and temperature using fiber Bragg grating cladding mode based on core diameter mismatch. J. Lightwave Technol. 2012, 30, 1645–1650.

53. Gouveia, C.; Jorge, P.; Baptista, J.M.; Frazao, O. Temperature-independent curvature sensor using FBG cladding modes based on a core misaligned splice. IEEE Photonics Technol. Lett. 2011, 23, 804–806.

54. Qi, T.; Xiao, S.; Shi, J.; Yi, L.; Zhou, Z.; Bi, M.; Hu, W. Cladding-mode backward-recoupling-based displacement sensor incorporating fiber up taper and Bragg grating. IEEE Photonics J. 2013, 5, 7100608.

55. Zhao, Y.; Chen, K.; Yang, J. Novel target type flowmeter based on a differential fiber Bragg grating sensor.Measurement 2005, 38, 230–235.

56. Jewart, C.; McMillen, B.; Cho, S.K.; Chen, K.P. X-probe flow using self-powered active fiber Bragg gratings. Sens. Actuators A Phys. 2006, 127, 63–68.

57. Byrne, G.D.; James, S.W.; Tatam, R.P. A Bragg grating based fibre optic reference beam laser Doppler anemometer.Meas. Sci. Technol. 2001, 12, 909–913.

58. Lien, V.; Vollmer, F. Microfluidic flow rate detection based on integrated optical fiber cantilever. Lab Chip 2007, 7, 1352–1356.

59. Lee, C.L.; Hong, W.Y.; Hsieh, H.J.; Weng, Z.Y. Air gap fiber Fabry-Pe´ rot interferometer for highly sensitive microairflow sensing. IEEE Photonics Technol. Lett. 2011, 23, 905–907.

60. Dong, X.; Zhou, Y.; Zhou, W.; Cheng, J.; Su, Z. Compact anemometer using silver-coated fiber Bragg grating. IEEE Photonics J. 2012, 5, 1381–1386.

61. Kreger, S.; Calvert, S.; Udd, E. High Pressure Sensing Using Fiber Bragg Gratings Written in Birefringent Side Hole Fiber. In Proceedings of the 15th Optical Fiber Sensors Conference Technical Digest, Portland, OR, USA, 10 May 2002; pp. 355–358.

62. Chen, T.; Chen, R.; Jewart, C.; Zhang, B.; Cook, K.; Canning, J.; Chen, K.P. Regenerated gratings in air-hole microstructured fibers for high-temperature pressure sensing. Opt. Lett. 2011, 36, 3542–3544.

63. Jewart, C.M.; Wang, Q.; Canning, J.; Grobnic, D.; Mihailov, S.J.; Chen, K.P. Ultrafast femtosecond-laser-induced fiber Bragg gratings in air-hole microstructured fibers for high-temperature pressure sensing. Opt. Lett. 2010, 35, 1443–1445.

64. Chmielewska, E.; Urbanczyk, W.; Bock, W.J. Measurement of pressure and temperature sensitivities of a Bragg grating imprinted in a highly birefringent side-hole fiber. Appl. Opt. 2003, 42, 6284–6291.

65. Zhang, Q.; Liu, N.; Fink, T.; Li, H.; Peng, W.; Han, M. Fiber-optic pressure sensor based on π-phase-shifted fiber Bragg grating on side-hole fiber. IEEE Photonics Technol. Lett. 2012, 24, 1519–1522.

66. Khaliq, S.; James, S.W.; Tatam, R.P. Fiber-optic liquid-level sensor using a long-period grating. Opt. Lett. 2001, 26, 1224–1226.

67. Yun, B.; Chen, N.; Cui, Y. Highly sensitive liquid-level sensor based on etched fiber Bragg grating. IEEE Photonics Technol. Lett. 2007, 19, 1747–1749.

68. Fu, H.; Shu, X.; Zhang, A.; Liu, W.; Zhang, L.; He, S.; Bennion, I. Implementation and characterization of liquid-level sensor based on a

long-period fiber grating mach-zehnder interferometer. IEEE Sens. J. 2011, 11, 2878–2883.

69. Homola, J. Present and future of surface plasmon resonance biosensors. Anal. Bioanal. Chem. 2003, 377, 528–539.

70. Räther, H. Surface Plasmons on Smooth and Rough Surfaces and on Gratings. In Springer Tracts in Modern Physics; Springer-Verlag: Berlin, Germany, 1988; Volume 11.

71. Kretschmann, E. Die Bestimmung optischer Konstanten vonMetallen durch anregung von oberflächenplasmaschwingungen. Zeitschrift für Physik A: Hadrons and Nuclei 1971, 241, 313–324.

72. Dostáleka, J.; Čtyrokýa, J.; Homolaa, J.; Bryndab, E.; Skalskýa, M.; Nekvindovác, P.; Špirkovác, J.; Škvord, J.; Schröfele, J. Surface plasmon resonance biosensor based on integrated optical waveguide. Sens. Actuators B Chem. 2001, 76, 8–12.

73. Gupta, B.D.; Verma, R.K. Surface plasmon resonance-based fiber optic sensors: Principle, probe designs, and some applications. J. Sens. 2009.

74. Obando, L.; Gentleman, D.; Holloway, J.; Booksh, K. Manufacture of robust surface plasmon resonance fiber optic based dip-probes. Sens. Actuators B Chem. 2004, 100, 439–449.

75. Esteban, Ó.; Díaz-Herrera, N.; Navarrete, M.; González-Cano, A. Surface plasmon resonance sensors based on uniform-waist tapered fibers in a reflective configuration. Appl. Opt. 2006, 45, 7294–7298.

76. He, Y.J.; Lo, Y.L.; Huang, J.F. Optical-fiber surface plasmon-resonance sensor employing long period fiber gratings in multiplexing. J. Opt. Soc. Am. B 2006, 23, 801–811.

77. Spacáková, B.; Homola, J. Theoretical analysis of a fiber optic surface plasmon resonance sensor utilizing a Bragg grating. Opt. Express 2009, 17, 23254–23264.

78. Lu, Y.C.; Huang, W.P.; Jian, S.S. Influence of mode loss on the feasibility of grating-assisted optical fiber surface plasmon resonance refractive index sensors. J. Lightwave Technol. 2009, 27, 4804–4808.

79. Caucheteur, C.; Shevchenko, Y.; Shao, L.-Y.; Wuilpart, M.; Albert, J. High resolution interrogation of tilted fiber grating SPR sensors from polarization properties measurement. Opt. Express 2011, 19, 1656–1664.

80. Schuster, T.; Herschel, R.; Neumann, N.; Schäffer, C.G. Miniaturized long-period fiber grating assisted surface Plasmon resonance sensor. J. Lightwave Technol. 2012, 30, 1003–1008.

81. Miao, Y.; Zhang, K.; Liu, B.; Lin, W.; Zhang, H.; Lu, Y.; Yao, J. Ferrofluid-infiltrated microstructured optical fiber long-period grating. IEEE Photonics Technol. Lett. 2013, 25, 306–309.

82. Mescia, L.; Palmisano, T.; Surico, M.; Prudenzano, F. Long-period gratings for the optimization of cladding-pumped microstructured optical fiber laser. Opt. Mater. 2010, 33, 236–240.

83. Prudenzano, F.; Mescia, L.; Palmisano, T.; Surico, M.; De Sario, M.; Righini, G.C. Optimization of pump absorption in MOF lasers via multi-long-period gratings: Design strategies. Appl. Opt. 2012, 51, 1410–1420.

84. Slavík, R.; Kulishov, M.; Park, Y.; Azaña, J. Long-period fiber-grating-based filter configuration enabling arbitrary linear filtering characteristics. Opt. Lett. 2009, 34, 1045–1047.

85. Kim, S.J.; Eom, T.J.; Lee, B.H.; Park, C.S. Optical temporal encoding/decoding of short pulses using cascaded long period fiber gratings. Opt. Express 2003, 11, 3034–3040.

86. Ashrafi, R.; Li, M.; Azaña, J. Tsymbol/s optical coding based on long-period gratings. IEEE Photonics Technol. Lett. 2013,25, 910–913.

87. Slavik, R.; Park, Y.; Kulishov, M.; Azaña, J. Terahertz-bandwidth high-order temporal differentiators based on phase-shifted long-period fiber gratings. Opt. Lett. 2009, 34, 3116–3118.

88. Ashrafi, R.; Li, M.; Azaña, J. Coupling-strength-independent long-period grating designs for THz-bandwidth optical differentiators. IEEE Photonics J. 2013, 5, 7100311.

89. Shu, X.; Allsop, T.; Gwandu, B.; Zhang, L.; Bennion, I. High-temperature sensitivity of long-period gratings in B-Ge codoped fiber. IEEE Photonics Technol. Lett. 2001, 13, 818–820.

90. Wang, Z.; Ramachandran, S. Ultrasensitive long-period fiber gratings for broadband modulators and sensors. Opt. Lett. 2003, 28, 2458–2460.

91. Steinvurzel, P.; Moore, E.D.; Mägi, E.C.; Eggleton, B.J. Tuning properties of long period gratings in photonic bandgap fibers. Opt. Lett. 2006, 31, 2103–2105.

92. Potyrailo, R.A.; Hobbs, S.E.; Hieftje, G.M. Optical waveguide sensors in analytical chemistry: Today's instrumentation, applications and trends for future development. Fresenius J. Anal.Chem. 1998, 362, 349–373.

93. McAdam, G.; Newman, P.J.; McKenzie, I.; Davis, C.; Hinton, B.R.W. Fiber optic sensors for detection of corrosion within Aircraft. Struct. Health Monit. 2005, 4, 47–56.

94. Mendoza, E.A.; Khalil, A.N.; Sun, Z.; Robinson, D.; Syracuse, S.J.; Egalon, C.O.; Gunther, M.F.; Lieberman, R.A. Embeddable distributed moisture and pH sensors for nondestructive inspection of aircraft lap joints. Proc. SPIE 1995,2455, 102–112

95. Mendoza, E.A.; Robinson, D.; Lieberman, R.A. Miniaturized integrated optic chemical sensors for environmental monitoring and remediation. Proc. SPIE 1996, 2836, 76–87.

96. Cordero, S.R.; Beshay, M.; Low, A.; Mukamal, H.; Ruiz, D.; Lieberman, R.A. A distributed fiber optic chemical sensor for hydrogen cyanide detection. Proc. SPIE 2005, 5993, 599302.

97. Cordero, S.R.; Mukamal, H.; Low, A.; Locke, E.P.; Lieberman, R.A. A fiber optic sensor for nerve agent. Proc. SPIE2006, 6378, 63780U–63783U.

98. Mukamal, H.; Cordero, S.R.; Ruiz, D.; Beshay, M.; Lieberman, R.A. Distributed fiber optic chemical sensor for hydrogen sulfide and chlorine detection. Proc. SPIE 2005, 6004, 600406.

99. Ghandehari, M.; Vimer, C.S. In situ monitoring of pH level with fiber optic evanescent field spectroscopy. NDT E Int.2004, 37, 611–616.

100. Sinchenko, E.; Gibbs, W.E.K.; Mazzolini, A.P.; Stoddart, P.R. The effect of the cladding refractive index on an optical fiber evanescent-wave sensor. J. Lightwave Technol. 2013, 31, 3251–3257.

101. Stewart, G.; Culshaw, B. Optical waveguide modelling and design for evanescent field chemical sensors. Opt. Quantum Electron. 1994, 26, S249–S259.

102. Schwotzer, G.; Latka, I.; Lehmann, H.; Willsch, R. Optical sensing of hydrocarbons in air or in water using UV absorption in the evanescnt field of fibers. Sens. Actuators B 1997, 38–39, 150–153.

103. Gupta, B.D.; Dodeja, H.; Tomar, A.K. Fibre-optic evanescent field absorption sensor based on a U-shaped probe. Opt. Quantum Electron. 1996, 28, 1629–1639.

104. Zimmerman, B.; Burck, J.; Ache, H.J. Studies on siloxane polymers for NIR-evanescent wave absorbance sensor. Sens. Actuators B 1997, 41, 45–54.

105. Qing, D.-K.; Yamaguchi, I. Analysis of the sensitivity of optical waveguide chemical sensor for TM modes by the group-index method. J. Opt. Soc. Am. B 1999, 16, 1359–1369.

106. Buerck, J.; Roth, S.; Kraemer, K.; Scholz, M.; Klaas, N. Application of a fiber-optic NIR-EFA sensor system for in situmonitoring of aromatic

hydrocarbons in contaminated groundwater. J. Hazard. Mater. 2001, 83, 11–28.

107. Prudenzano, F.; Mescia, L.; Allegretti, L.A.; Calò, G.; D'Orazio, A.; De Sario, M.; Palmisano, T.; Petruzzelli, V. Design of an optical sensor array for hydrocarbon monitoring. Opt. Quantum Electron. 2009, 41, 55–68.

108. Mescia, L.; Prudenzano, F.; Allegretti, L.; Calò, G.; De Sario, M.; D'Orazio, A.; Maiorano, L.; Palmisano, T.; Petruzzelli, V. Design of silica-based photonic crystal fiber for biosensing applications. J. Non-Cryst. Solids 2009, 355, 1163–1166.

109. Palmisano, T.; Prudenzano, F.; Warren-Smith, S.C.; Monro, T.M. Design of exposed-core fiber for methadone monitoring in biological fluids. J. Non-Cryst. Solids 2011, 357, 2000–2004.

110. Kostecki, R.; Ebendorff-Heidepriem, H.; Davis, C.; McAdam, G.; Warren-Smith, S.C.; Monro, T.M. Silica exposed-core microstructured optical fibers. Opt. Mater. Express 2012, 2, 1538–1547.

111. Eggleton, B.J.; Kerbage, C.; Westbrook, P.S.; Windeler, R.S.; Hale, A. Microstructured optical fiber devices. Opt. Express2001, 9, 698–713.

112. Bise, R.T.; Windeler, R.S.; Kranz, K.S.; Kerbage, C.; Eggleton, B.J.; Trevor., D.J. Tunable photonic band gap fiber. InProc. Opt. Fiber Commun. Conf. Exhib. OSA Trends Opt. Photon. (Opt. Soc.Am. Washington); 2002; Volume 70, pp. 466–468.

113. Wang, Y.; Tan, X.; Jin, W.; Ying, D.; Hoo, Y.L.; Liu, S. Temperature controlled transformation in fiber types of fluid-filled photonic crystal fibers and applications. Opt. Lett. 2010, 35, 88–90.

114. Yiou, S.; Delaye, P.; Rouvie, A.; Chinaud, J.; Frey, R.; Roosen, G.; Viale, P.; Février, S.; Roy, P.; Auguste, J.-L.; Blondy, J.-M. Stimulated Raman scattering in an ethanol core microstructured optical fiber. Opt. Express 2005, 13, 4786–4791.

115. Huang, Y.; Xu, Y.; Yariv, A. Fabrication of functional microstructured optical fibers through a selective-filling technique. Appl. Phys. Lett. 2004, 85, 5182–5184.

116. Cordeiro, C.M.B.; dos Santos, E.M.; Cruz, C.H.B.; de Matos, C.J.S.; Ferreira, D.S. Lateral access to the holes of photonic crystal fibers—Selective filling and sensing applications. Opt. Express 2006, 14, 8403–8412.

117. Meneghini, C.; Caron, S.; Poulin, A.C.J.; Proulx, A.; Émond, V.; Paradis, P.; Paré, C.; Fougères, A. Determination of ethanol concentration by Raman spectroscopy in liquid-core microstructured optical fiber. IEEE Sens. J. 2008, 8, 1250–1255.

118. Ren, H.C.; Vollmer, F.; Arnold, S.; Libchaber, A. High-Q microsphere biosensor-analysis for adsorption ofrodlike bacteria. Opt. Express 2007, 15, 17410–17423.

119. Vollomer, F.; Stephen, A. Whispering-gallery-mode biosensing: Label-free detection down to single molecules. Nat. Methods 2008, 5, 591–596.

120. Yang, L.; Vahala, K.J. Gain functionalization of silica microresonators. Opt. Lett. 2008, 28, 592–594.

121. Murugan, G.S.; Zervas, M.N.; Panitchob, Y.; Wilkinson, J.S. Integrated Nd-doped borosilicate glass microsphere laser.Opt. Lett. 2011, 36, 73–75.

122. Chen, S.Y.; Sun, T.; Grattan, K.T.V.; Annapurna, K.; Sen, R. Characteristics of Er and ErYbCr doped phosphate microsphere fibre lasers. Opt. Commun. 2009, 282, 3765–3769.

123. Feron, P. Whispering Gallery Mode Lasers in Erbium doped fluoride glasses. Annales de la Fondation Louis de Broglie2004, 29, 317–329.

124. Nunzi Conti, G.; Chiasera, A.; Ghisa, L.; Berneschi, S.; Brenci, M.; Dumeige, Y.; Pelli, S.; Sebastiani, S.; Feron, P.; Ferrari, M.; Righini, G.C. Spectroscopic and lasing properties of Er3+ doped glass microspheres. J. Non-Cryst. Solids2006, 352, 2360–2363.

125. Elliot, G.R.; Hewak, D.W.; Senthil Murugan, G.; Wilkinson, J.S. Chalcogenide glass microspheres; their production, characterization and potential. Opt. Express 2007, 15, 17542–17553.

126. Mescia, L.; Prudenzano, F.; De Sario, M.; Palmisano, T.; Ferrari, M.; Righini, G.C. Design of rare-earth-doped microspheres. IEEE Photonics Technol. Lett. 2010, 22, 422–424.

127. Kouki, T.; Makoto, T. Optical microsphere amplification system. Opt. Lett. 2007, 32, 3197–3199.

128. Boucher, Y.G.; Feron, P. Generalized transfer function: A simple model applied to active single-mode microring resonators. Opt. Commun. 2009, 282, 3940–3947.

129. Mescia, L.; Bia, P.; De Sario, M.; Di Tommaso, A.; Prudenzano, F. Design of mid-infrared amplifiers based on fiber taper coupling to erbium-doped microspherical resonator. Opt. Express 2012, 20, 7616–7629.

130. Mescia, L.; Bia, P.; Losito, O.; Prudenzano, F. Design of Mid-IR Er3+-doped microsphere laser. IEEE Photonics J. 2013, 5, 1501308.

2

STRESS SENSING BY AN OPTICAL FIBER SENSOR: METHOD AND PROCESS FOR THE CHARACTERIZATION OF THE SENSOR RESPONSE DEPENDING ON SEVERAL DESIGNS

Mustapha Remouche[1,2], Francis Georges[3], Patrick Meyrueis[3]

[1]Volumion Company SAS, Mulhouse, France

[2]Research Unit for Industrial Technologies URTI/CSC, Annaba University, Annaba, Algeria

[3]Icube Laboratory, Strasbourg University, Strasbourg, France

ABSTRACT

In this paper we propose an analyzing of the response of a stress optical fiber sensor of which we proposed several design. We show that an optical fiber sensor with these designs can covenanting allow the measuring the force/stress applied to a mechanical structure or which it is linked, by optimizing the uses of appropriate materials for constituting the sensor support. The experiment that we introduce to validate our approach based in principles includes design with a support bearing a multimode optical fiber organized in such a way that the transmitted light is attenuated when the fiber-bending angle coming from stitching in holes of the support is modified by the effects of the force/stress applied to the optical fiber sensor realized in this way. The tests realized concern the most relevant parameters that define the performances of the stress sensor that we propose. We present the problems that we to solved for the optimization of the sensor for selecting the more efficient material for the optical fiber sensor support related to a relevant choice of optical fibers.

KEYWORDS

Optical Waveguide; Optical Fiber; Sensor; Force; Stress; Strain; Microbending

1. INTRODUCTION

Macrobend and microbend losses in optical waveground, have been discussed for a large set of optical fiber sensors in related publications. These sensors were used to measure: temperature, pressure, chemical species, strain, moisture, force, displacement, acceleration, etc. [1-11]. Optical fiber, microbend sensors are based on the control and the analysis of the couplings and the leakages of modes that are propagating in a deformed microbend optical fiber [11-14] by fiber output light energy measurement. Some previous studies have discussed the losses due to the macrobending of the microbend optical fiber axis. These studies did not take into account the effects coming from the mechanical specificities of the material support that is bearing by a fiber stitching in a holes network the microbented optical fiber sensor, on the microbent optical fiber sensor response. These supports have high and low stress zones when they are bent.

Our purpose is to characterize the effects of some mechanical specificities of the support bearing on microbent optical fiber sensor. This support has a periodic array of holes inside which the optical fiber is threaded to constitute the optical fiber stress sensor. The period of the fiber microbending is resulting of the threading constant, The perforations are located only in the support high stress zones.

2. BACKGROUND: BASIC PRINCIPLES OF A MICROBENT OPTICAL FIBER SENSOR FOR CONSTRAINTS ANALYSES AND MEASURES

The refractive index profile distribution in the optical fiber is described by the following equation:

$$n^2(r) = n^2(0)\left[1 - 2\Delta(r/\rho)^\alpha\right] \tag{1}$$

where $\Delta = \left[n^2(0) - n^2(\rho)\right]/2n^2(0), n(0) = n_0, n(r)$ and $n(\rho)$ are respectively the refractive indices at distances 0, r and ρ from the fiber axis and ρ is the core radius it includes both parabolic $(\alpha = 2)$ and step $(\alpha = \infty)$ profiles. The modes propagating in an optical fiber with the core index described by Equation (1) may be associated into mode groups having a principal mode number m and a longitudinal propagation constant β_m given by the Equation (2) (WKBJ approximation method) [12-14]:

$$\beta_m = kn(0)\left[1 - 2\Delta(m/M)^{2\alpha/(\alpha+2)}\right]^{1/2} \tag{2}$$

where $k = 2\pi/\lambda$ is the wave vector module, λ is the optical wavelength, and M^2 is the total number of guided modes propagating inside the fiber,

$$M^2 = \left[\alpha/(\alpha+2)\right](n_0 k \rho)^2 \Delta \tag{3}$$

The spacing between two adjacent mode groups m and $m+1$ is given by the following equation:

$$\begin{aligned}\Delta\beta_m &= \beta_m - \beta_{m-1} \approx \mathrm{d}\beta/\mathrm{d}m \\ &= \left(2\sqrt{\Delta}/\rho\right)\left[\alpha/(\alpha+2)\right]^{1/2}(m/M)^{(\alpha-2)/(\alpha+2)}\end{aligned} \tag{4}$$

If there is a mechanical dynamic perturbation applied to the microbent fiber having a spatial wavelength of Λ such that:

$$\Delta\beta_m = \beta_m - \beta_{m-1} = \mp 2\pi/\Lambda \tag{5}$$

It may couple two modes groups by satisfying the selection rule that induce the coupling to the nearest neighbor.

With a parabolic index profile for the microbent optical fiber, adjacent mode groups are separated by a critical constant $\Delta\beta_c\left[=(2\Delta)^{1/2}/\rho\right]$; A modification of the spatial wavelength Λ_c can couple all the neighboring mode groups:

$$\Lambda_c = \sqrt{2}\pi\rho/(\Delta)^{1/2} \tag{6}$$

With a step-index profile of the optical fiber, the spacing $\Delta\beta_m$ is defined by Equation (4) with $\alpha = \infty$,

$$\Delta\beta_m = \left[(2\Delta)^{1/2}/\rho\right](m/M) \tag{7}$$

and

$$\Lambda = \left[2\pi\rho/(2\Delta)^{1/2}\right](M/m) \tag{8}$$

where $0 \le m \le M$, There are $(m+1)$ possible values of Λ to couple the modes. The (m/M) ratio corresponds to the normalized order of the group of modes. The upper limit corresponds to the highest pair of guided modes ($m = M, m/M = 1$ total coupling of modes). For an optical fiber having mode-independent intrinsic loss, the guided light power must be coupled across this highest $\Delta\beta_m$ to reach the radiation spectrum. A significant loss would appear for the following distortion of the periodicity of bent:

$$\Lambda \le \Lambda_c = \pi\rho/(\Delta)^{1/2} \tag{9}$$

3. EXPERIMENTAL SET-UP

The experimental setup is showed in **Figure 1**.

3.1. Optical Fiber System

The bent optical fiber by stitching that will constitute the sensor is integrated in a mechanical fiber holding structure that builds up the whole force/stress sensor as shown in Figures 1 and 2. The holding support of the fiber with microbends that constitute the sensor consists of a flexible support, typically a piece of rubber, on which a multimode optical fiber is woven with a period Λ (**Figure 2**).

Light propagates from one end of the optical fiber (light input) to the other end of the optical fiber (light output) trough series of microbending by stitching the optical fiber into the holes of support. At each bend (located in a support hole) along the optical fiber, some optical power "leaks" out of the optical fiber and so induces an optical energy loss of the light propagating inside the fiber. The magnitude of the optical loss depends mainly on the microbending mechanical wavelength (distance between support holes) and to some extend of the radius of curvature and the length of the microbent segment.

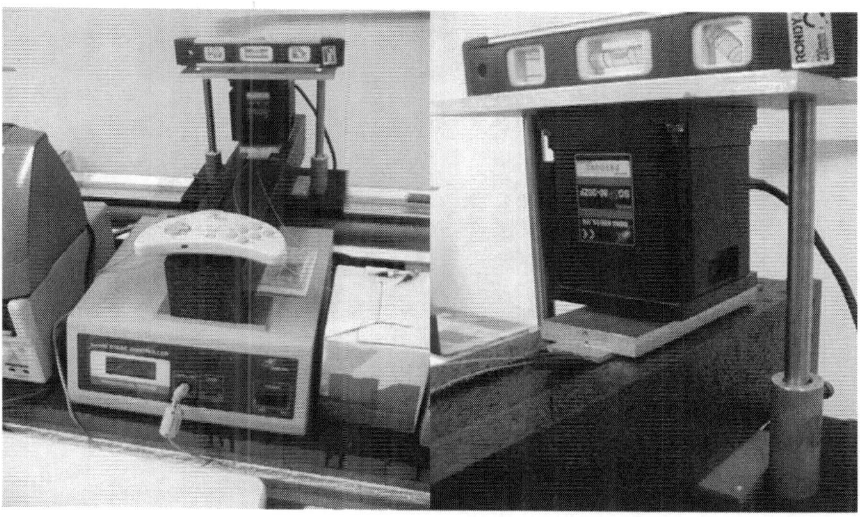

Figure 1. Experimental set up.

As it can be seen in **Figure 3**, the force/stress sensor is sandwiched between a front plate and a back support or between a deformable upper plate (e.g. in PVC) and a back piece of foam. If a force F is applied to the front plate, the thickness of the sensor decreases and thus the microbending period it also diminished. As the period decreases, the optical loss increases. As a result, the measurement of the attenuation of the light that propagates along the microbent segments at the output of the fiber yields to a direct measurement of the force/stress applied through an analysis of the sensor optical energy modulations response at its output.

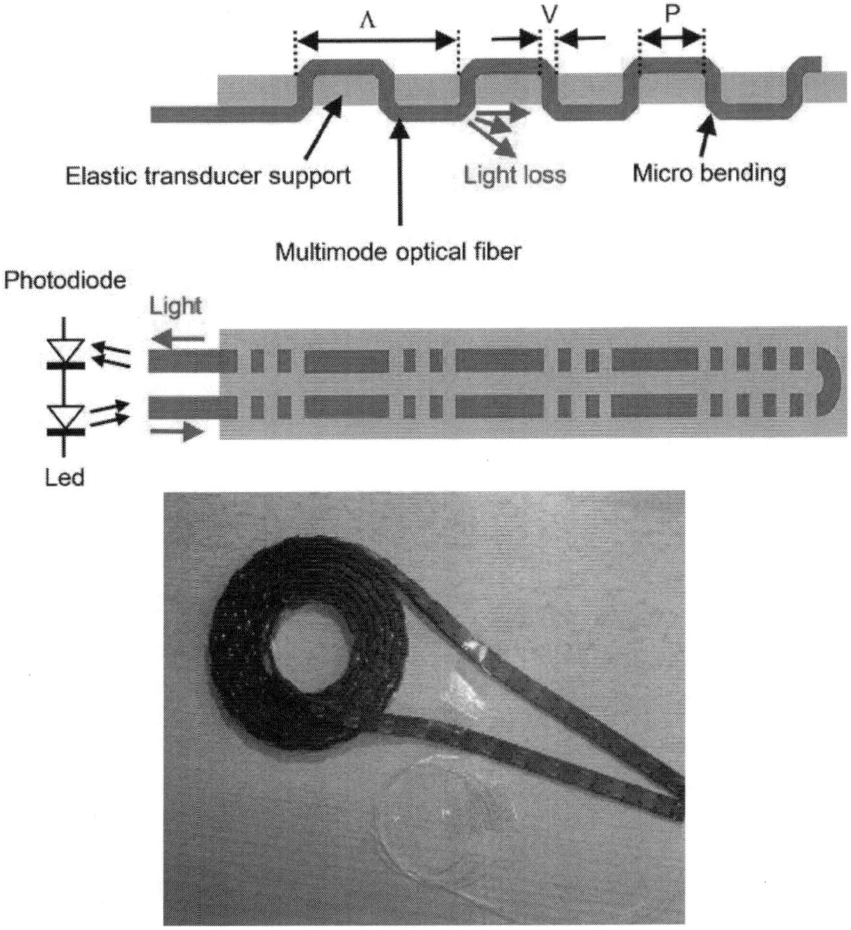

Figure 2. Schematic diagram of the proposed microbend optical fiber sensor design. Λ: mechanical wavelength of the periodic perturbation, V: perforated length zone, P: nonperforated zone.

3.2. The Temperature Chamber Used for the Experimental Tests of the Sensor

In order to run a temperature test of the response of the sensor to temperature modulations, a programmable temperature chamber that provides temperature modulation data to the computer through a RS-232 serial interface that can also allow to control the temperature modulation of the chamber (**Figure 4**). The temperature modulation was programmed to run temperature pre-defined cycles, the response to the stress of the optical fiber sensor located in the chamber was forwarded and recorded by the computer.

3.3. Step Motor Used for the Experimental Test

To apply the force, we use a step motor that has a 50 nm resolution, a 20 mm course and can move a load of 150 N maximum. It is driven via a GPIB interface by a PC computer. So repeatedly the step motor applies an effort (a calibrated force) to bent optical fiber the transducer. This set up can be also used to apply a torque.

A Labview software drives the motor according to the chosen test process. The user first defines the successive absolute positions that the motor has to reach. He defines this way a displacement cycle. The steps of this cycle are:

- A single displacement cycle;
- A finite number of displacement cycles;
- Continuous and repeated cycles until the user stop the execution.

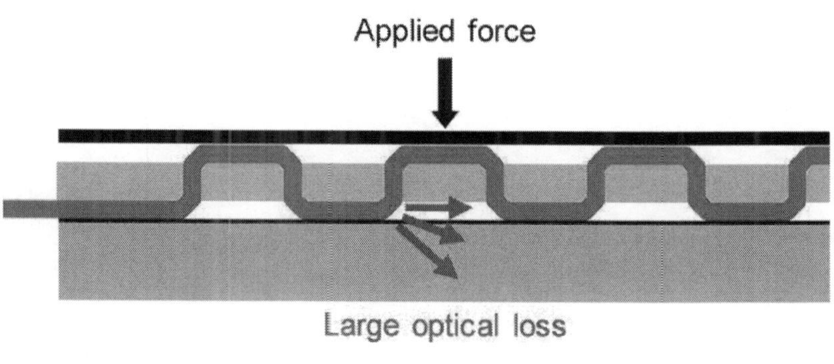

Figure 3. Optical fiber force sensor sandwiched between a deformable front plates.

3.4. Standard Force Sensor

A commercial force sensor with a dynamic range of 50 N is inserted in the chamber with the step motor to measure as a reference the actual effort applied to the sensor (Figure 5).

The analog signal from the reference is amplified and directly recorded by the acquisition system. To test correctly the proposed sensor, we need first to calibrate it. The calibration consists in setting several weights between 0 and 60 N applied to the sensor, and to measure the sensor light output as the voltage from a photoconverter (**Figure 6**).

Figure 4. Temperature chamber.

We observe that the response is almost linear with small fluctuations. These fluctuations are recorded during a 24 hours test (see **Figure 7**).

During 24 hours the reference sensor fluctuations can be neglected, (0.07% of FS) (**Figure 7**). A hysteresis phenomenon was also observed (**Figure 8**) that can be neglected. We conclude that we do not need to take in account these fluctuations for our evaluation tests.

4. INVESTIGATIONS WITH THE EXPERIMENTAL SET UP

Two different fibers have been implemented in sensor prototypes in order to determine among them the bestsuited fiber for the force/stress intrinsic optical fiber sensor.

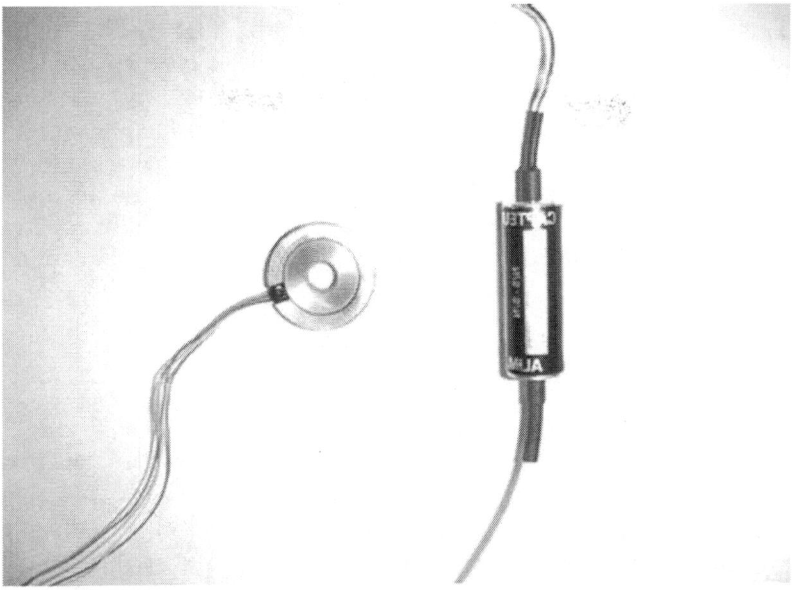

Figure 5. Commercial force sensor used.

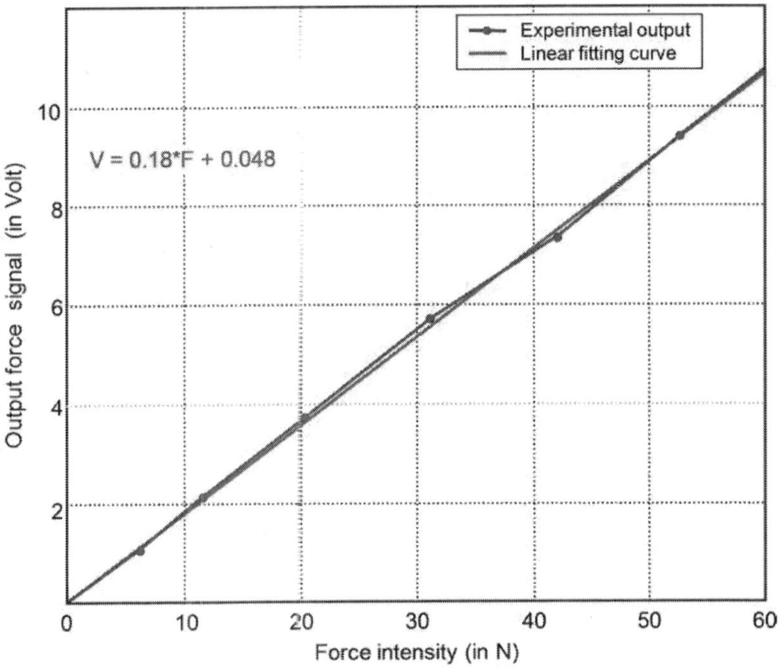

Figure 6. Standard reference force sensor response.

4.1. Step Index Optical Fiber: Silica Core/Hard Polymer Cladding Fiber (HCS200/230/500)

The core, cladding and coating diameters of the optical fibers we used are respectively 200 μm, 230 μm, 500 μm. These fibers were selected for their good mechanical behavior. When a fiber has to react to the actions of a network of microbendings of period Λ, the condition of coupling between the guided modes and the radiant modes is given by Equation (9) with $\Delta = N.A.^2/2n_0^2$:

$$\Lambda \le \Lambda_c = \sqrt{2}\pi\rho n_0 / N.A. \tag{10}$$

where $n_0 = 1.467, \rho = 100\ \mu\text{m}$ and $N.A. = 0.39$. What gives for the period: $\Lambda_c = 1.65\ \text{mm}$ Every optical fiber has a mechanical critical curvature radius Rc for which the fiber will get broken [14]. Hence, the space requirement of the stress optical fiber sensor that we propose is imposed by this minimum. The lifetime of the proposed optical fiber sensor will be shorter if the bend diameter of an optical fiber bend of this sensor is close to this critical fracture diameter.

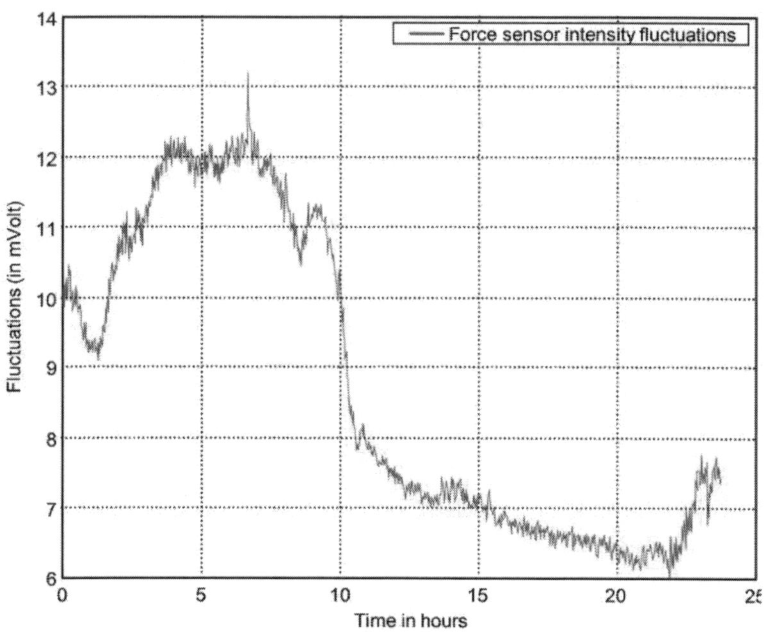

Figure 7. Reference force sensor fluctuations.

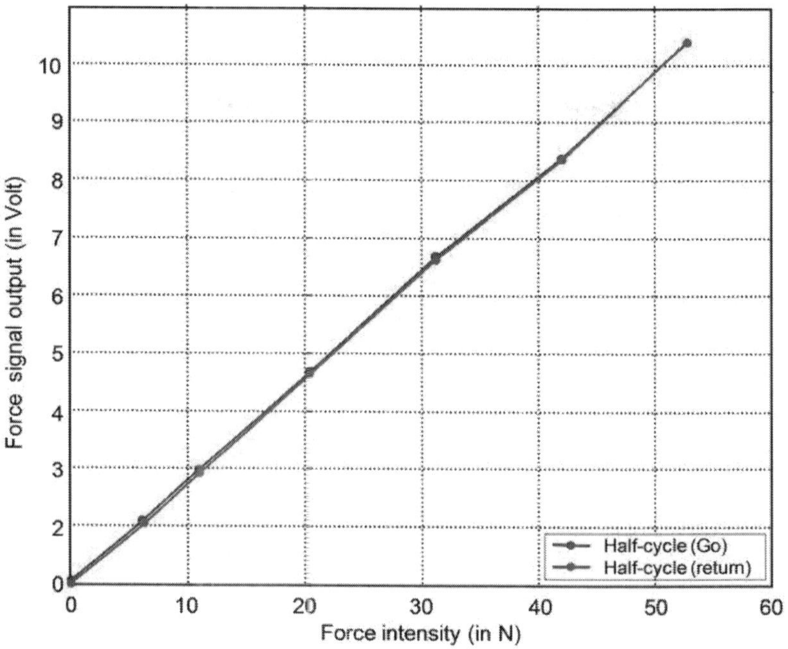

Figure 8. Calibration force sensor response (go/return cycle).

For reasons linked to the mechanical limits of the optical fiber constituting the sensor that we propose and the significant light energy losses of the signal, associated with extreme bending of the optical fiber it is obvious that we cannot select a period $(\Lambda = 1.65\ \text{mm})$. Close to the bending limit there is a critical bending range, which results from a phenomenon of energy exchange between modes, even if the inequality (Equation (10)) is not respected, that has to be avoided.

We chose periods that are multiple of Λ, with a duty cycle (ratio of dimensions between not perforated zone and perforated zone) that take into account the conditions (that are mentioned in**Table 1**).

The holes networks used to stitch the optical fiber were engraved for evaluation on two different materials: the first one in EPDM (rulber) with a thickness of 2 mm and the second one in PVC with a thickness of 1.5 mm.

We have selected four hardness for each material: "Shore A50", "Shore A60", "Shore A65" and "Shore A75".

The losses induced by a fiber bend curvature of radius R can be approached by the following expression [4]:

$$\alpha(R)_{dB} = 10\log_{10}\left[1-\left(2\rho n_0/RN.A._2^2\right)\right] \quad (11)$$

Table 1. Stitched networks mechanical parameters.

Silica core/hard polymer cladding fiber (HCS200/230/500)			
Period 1	5Λ = 7.22 mm		
Non-perforated zone in mm	P = 5.77	P = 1.45	P = 3.61
Perforated length in mm	V = 1.45	V = 5.77	V = 3.61
Duty cycle: P/V	P/V = 3.979	P/V = 0.251	P/V = 1
Period 2	5Λ = 10.1 mm		
Non-perforated zone in mm	P = 8.65	P = 1.45	P = 5.05
Perforated length in mm	V = 1.45	V = 8.65	V = 5.05
Duty cycle: P/V	P/V = 5.96	V = 8.65	P/V = 1
Parabolic index optical fiber 100/140/250			
Period 2	5Λ = 8.16 mm		
Non-perforated zone in mm	P = 6.56	P = 1.60	P = 4.08
Perforated length in mm	V = 1.60	V = 6.56	V = 4.08
Duty cycle: P/V	P/V = 4.1	P/V = 0.244	P/V = 1

This expression shows that the fiber holes related organization parameters should be adjusted to obtain the best sensitivity for the sensor. That we, in our case Rc is defined by the thickness of the support, the period and the duty cycle of the network of microbending.

4.2. Parabolic Index Optical Fiber 100/140/250 with Acrylate Coating Material

The core, cladding and coating diameters are 100 μm, 140 μm, 250 μm respectively. This fiber is chosen for its higher sensitivity, to stress by microbending. A network of bending induced by holes of, with a period Λ will determine the coupling conditions between guided modes and radiated modes it is given by Equation (6) with $\Delta = N.A.^2/2n_0^2$:

$$\Lambda_c = 2\pi\rho n_0 / N.A. \tag{12}$$

where $n_0 = 1.467, \rho = 50\ \mu\text{m}$ and $N.A. = 0.28$. What gives for the period:

$$\Lambda_c = 1.60\ \text{mm}$$

This relation remains valid when we choose the multiples of the period Λ it is the contrary with the step index fiber.

5. DISCUSSIONS ABOUT EXPERIMENTAL RESULTS

The experimental work consists in characterizing the sensor according to the force/stress and to the temperature and the support material used depending on the fiber characteristics.

5.1. Sensor Made with a Parabolic Index Optical Fiber 100/140/250

5.1.1. Effect of the Applied Force on the Response of the Sensor

The curves in **Figure 9** show a good sensitivity of the sensor to the applied force when the condition concerning the critical period is respected ($\Lambda = \Lambda_C$, where Λ_C is the critical period). This sensitivity decreases quickly when the period is going away from the critical period. The hard materials for the support which are the best suited to meet the range of forces that we used (30 N to 60 N) are the ones that correspond to the index "Shore A50". By choosing a high value for the period for the microbending of the optical fiber (for example Λ = 9.8 mm) or a lower hardness ("Shore A45" or "Shore

A40") or a lower thickness (1 to 1.5 mm), we obtain a larger range of response.

5.1.2. Temperature Effects on the Responses of the Sensor

The study of the behavior of the sensor during a temperature cycle between 50°C and +65°C shows that the temperature variation effect on the light energy at the output of the fiber represents about 30% of the total light energy output with the presence of a hysteresis phenomenon (**Figure 10**). With some optical fiber, when it is very strongly curved, the sensor is not very sensitive to the temperature. This is very probably due to some movements of the EPDM support resulting from a heating effect. The test cycle has lasted a short time (2 × 30 minutes.)

What can explain the hysteresis. It is possible to also explain it by the fact that the fiber is bended in way that can induce it to be sensitive to the small movements of the support providing a too important sensitivity to the sensor. So we have in this way some possibilities to improve the stability of the response of the sensor with the temperature evolutions. Another possibility remains to be choice of a more stable material as a support.

5.2. Sensor Realized with Step Index Optical Fiber Hcs-200/230/500

5.2.1. Effect of the Force on the Response of the Sensor

This fiber was used for the stress sensor that we propose for the interest it presents related to the sensor production costs. From a sensitivity point of view, it appears that it is the network having a $\Lambda = 7.22$ mm which gives the best results on a support in EPDM having a thickness of 2 mm and a Shore hardness: A75 (Figures 11 and 12). It should be noted that the dispersion of the sensor responses according to the parameters of the networks and according to the hardness of the support materials.

Another significant point has to be discussed when we compare the responses obtained by using two different materials the EPDM and PVC with identical networks design of microbending structure (**Figure 13**). The sensors using the PVC supports have very low sensitivities, we can suppose that the compression ratio of this material is much lower than the one of the EPDM. This difference makes us rejects the choice of such a material. It does not enable us to make a comparative study between two different thicknesses (1.5 mm and 2 mm) because it will not be relevant.

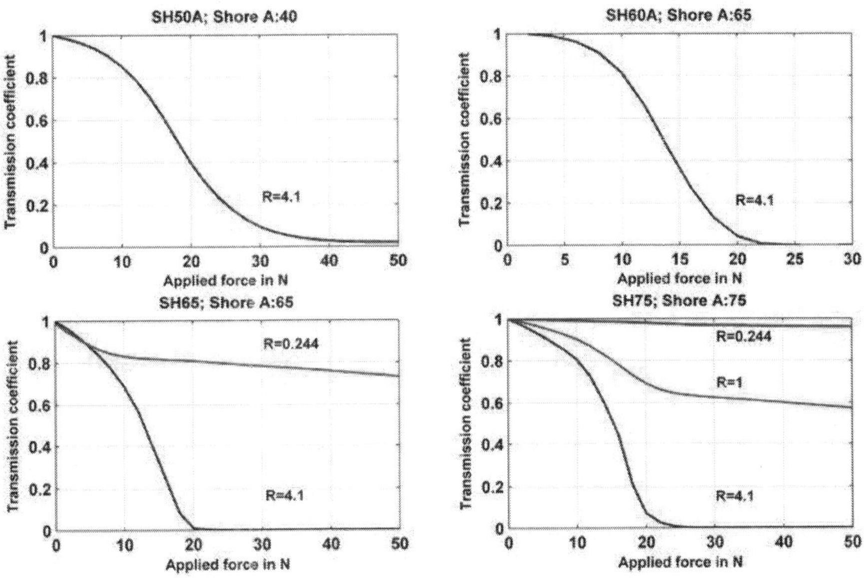

Figure 9. Response of optical fiber sensor with networks microbending of period Λ = 8.16 mm on a parabolic index optical fiber 100/140/250 (ON = 0.28) with duty cycles on supports in EPDM of thickness 2 mm, and with A50, A60, A65 and A75 of hardness. The force application length: 60 mm.

5.2.2. Effect of the Length or Which Forces Are Applied Surfaces

Another experiment consists to apply the same force to several lengths/surfaces what induce variations of the applied pressure. The results obtained are summarized in the **Figure 14**. They show, that, when the lengths/surface of the applied forces decreases, the sensitivity is better first, and then decreases quickly when the applied force increases. That is explained by the fact that the material is initially stiff and then becomes deformed and then reaches quickly a compression stationary phase, which limits the sensitivity.

For the important lengths/surfaces of forces application, with the same applied force, this stationary phase is not reached and the losses are added over several periods of microbending a better sensitivity results of that. So, we choose a material having a compression ratio proportional to the applied force in the range 0 - 100 N. The use of a fiber coating harder than the support, will also allow the deformation over a more important length. Moreover, it seems that the sensor responses will become uniform as soon as we exceed a limiting lengths/surfaces of contact threshold that will depend on the compression ratio of the support material.

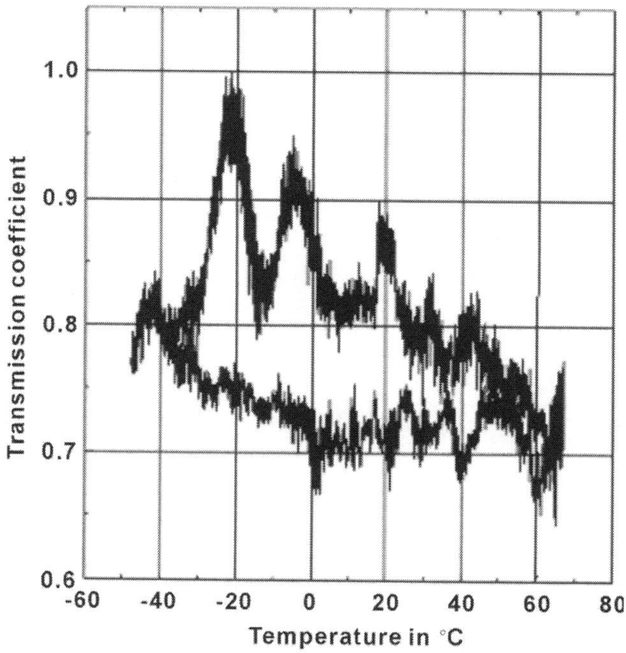

Figure 10. Behavior of the sensor as a function of the temperature. A parabolic index optical fiber 100/140/250 is woven with a microbending per hole on a network of holes having a period Λ = 8.16 mm on a support in EPDM thickness 2 mm and Shore hardness A50.

The use a hard coating, segmented over adapted lengths should, in theory improve, notably the uniformity of the sensor response. A threshold detector can be considered in this case the therefore, the threshold signal is reached for the lower stress.

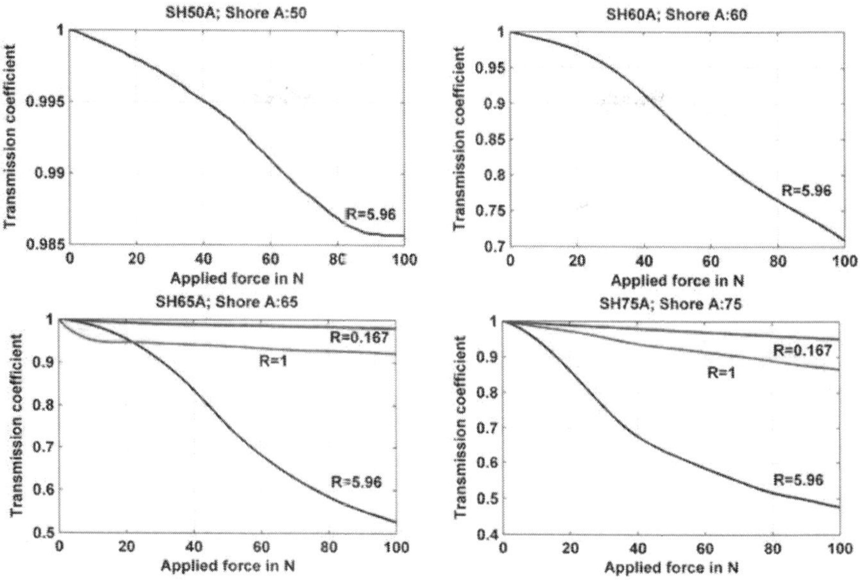

Figure 11. Responses of sensors realized with networks of hole microbending of period Λ = 1.01 mm on a step index optical with a hard cladding fiber 200/230/500 (ON = 0.39) with several duty cycles. The supports are made out of EPDM with a thickness of 2 mm, and a hardness value of A50, A60, A65 and A75. The force application length is: 60 mm.

5.2.3. Effect of the Temperature on the Sensor Response

The sensor, without a coating material on the optical fiber, was placed in a climatic room to a weight of 6 N was used to cause the stress. Jointly a cycle of temperature between −50°C and +100°C was applied to the sensor (**Figure 15**).

We observed a variation of about 30% of the response signal with a hysteresis phenomenon. This effect has two reasons, the first reason comes from the fact that an optical fiber when it is curved becomes sensitive to the temperature (**Figure 16**), the second one is related to support material (hysteresis), but with a smaller effect to coming from the size (0.5 mm) and of the rigidity of fiber. It should be noted that these variation present a good degree of linearity which could allow an easy compensation.

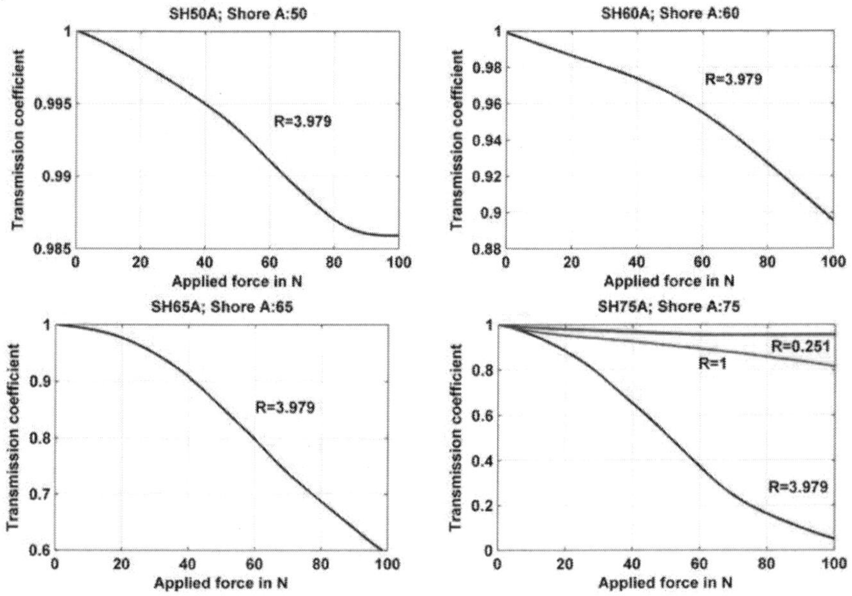

Figure 12. Response of sensors realized with network of hole with fiber microbending, the period is Λ = 7.22 mm a step index optical fiber used is: 200/230/500 (ON = 0.39) with a hard cladding with several duty cycles on supports made out of EPDM having a thickness of 2 mm, and a hardness of A50, A60, A65 and A75. The force application length is 60 mm.

6. CONCLUSIONS

We have identified the most important parameters in the optical fiber stress sensor realized with the technology that we proposed which plays a part on the performances of the sensor such as: the choice of fiber, the choice of the network of microbending induced by holes in a support and the choice of the material for the support. It was made possible to identify work of optimization, which remains to be done. First prototypes of validation were investigated with two fibers optics were associated with two different materials for constituting support:

1) HCS-200/230/500 optical fiber was associated with a support in EPDM having a hardness ranging of from 80 to 85 and a thickness of 1.5 mm.

2) The graded index fiber was associated with a support made out of EPDM with a hardness of 45 and a thickness of 1 mm.

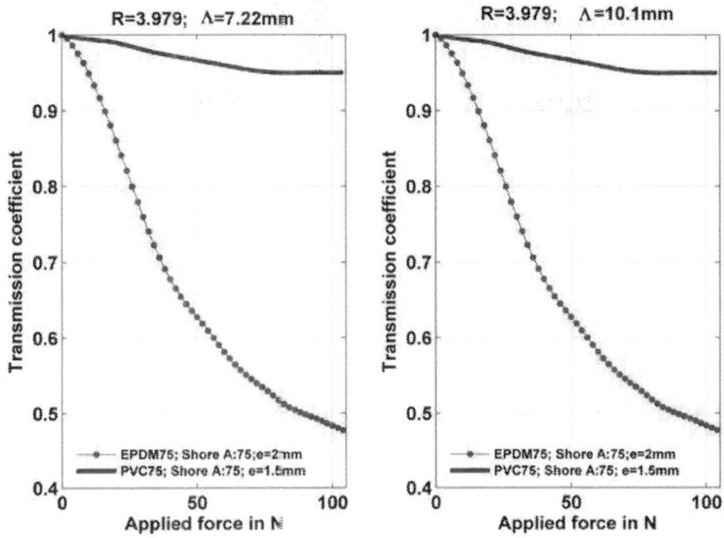

Figure 13. Response of sensor with networks of microbended optical fiber having periods Λ = 7.22 mm and Λ = 1.01 mm. The optical fiber has a hard cladding. They were step index optical fiber 200/230/500 (ON = 0.39) on a support made out of EPDM having a thickness of 2 mm. The PVC one has a thickness of 1.5 mm and a hardness Shore of a A75, the force was applicated on a length of 60 mm.

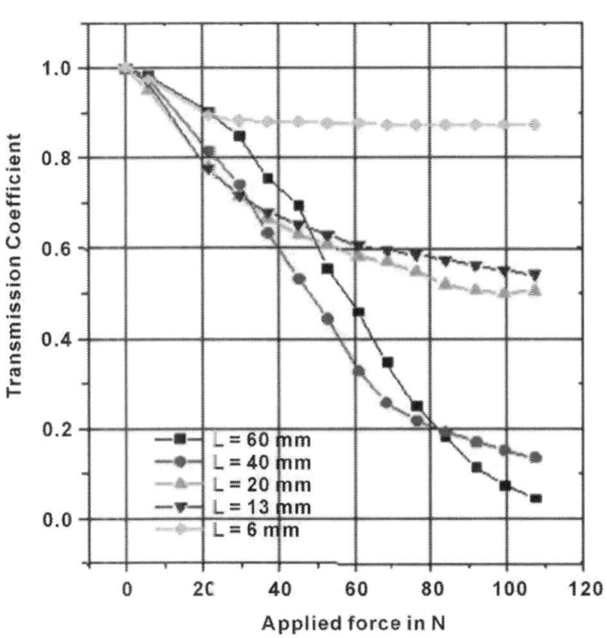

Figure 14. Influence of the length of application of the forces. The networks of microbending has periods of Λ = 7.22 mm a hard cladding optical fiber is used 200/230/500 (ON = 0.39) on supports made out of EPDM having a thickness of 2 mm and the PVC is thickness 1.5 mm. Shore hardness is A75, with several lengths of the force application: 60 mm; 40 mm; 19.5 mm; 13 mm and 6 mm.

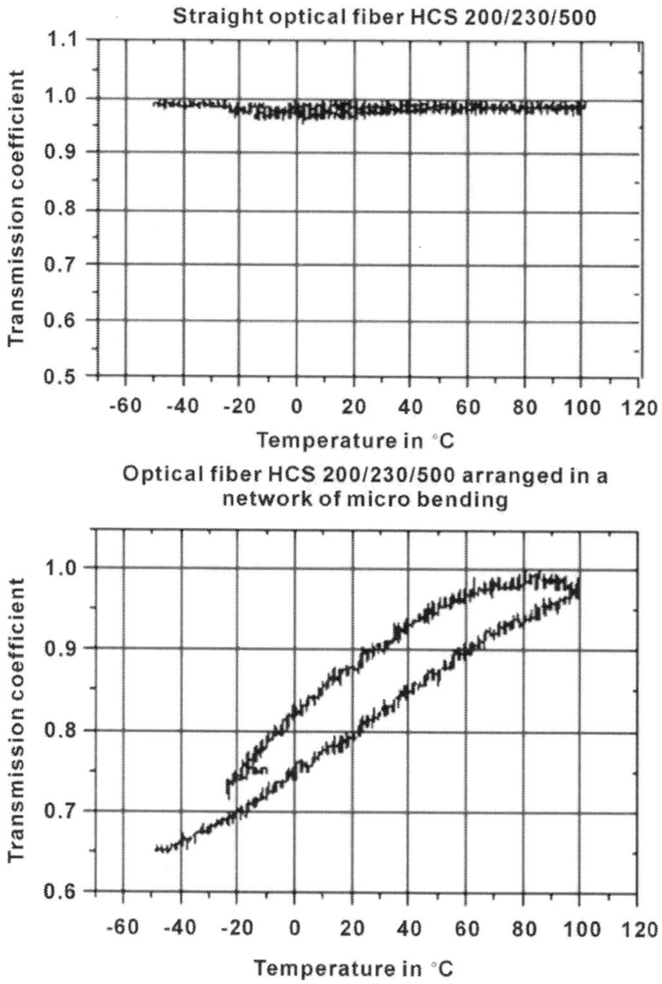

Figure 15. Response of the sensor as a function of the temperatures top: behavior of a sensor with a straight hardclad step index fiber profile 200/230/500 (ON = 0.39) bottom: the same fiber woven on a network of holes to form microbendings having a Λ = 7.22 mm period on a support in EPDM having a thickness 2 mm and shore hardness A7.5.

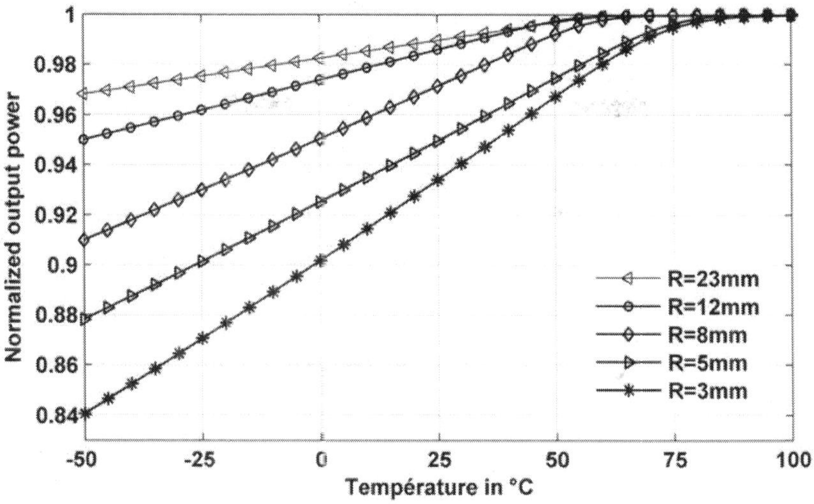

Figure 16. Behavior of the hard-clad optical fiber 200/230/ 500 (ON = 0.39) according to the temperature variations with different radii of curvatures.

There is still some optimization work to achieve for having a very good reproducibility of the sensor response for an applied force that stay local and then throughout the length of the sensor. Evaluation of the response times remains also to be made, it will depend on the choice of the design for support and coating.

REFERENCES

1. M. Remouche, R. Mokdad and A. Chakari, "Intrinsic Integrated Optical Sensor Based on Waveguide Bend Loss," Optics & Laser Technology, Vol. 39, No 7, 2007, pp. 1454-1460.

2. M. Remouche, R. Mokdad and M. Lahrashe, "Intrinsic Optical Fiber Temperature Sensor Operating by Modulation of the Local Numerical Aperture," Optical Engineering, Vol. 46, No. 2, 2007, Article ID: 024401.

3. M. Remouche, F. Georges and P. Meyrueis, "Flexible Optical Waveguide Bent Loss Attenuation Effects Analysis and Modeling Application to an Intrinsic Optical Fiber Temperature Sensor," Optics and Photonics Journal, Vol. 2, No. 1, 2012, pp. 1-7.

4. Tioman, A. Chakari and P. Meyrueis, "Capteur de Température à Fibre Optique par Effet Modal," Revue Pratique de Contrôle Industriel, Vol. 30, No. 168, 1991, pp. 64-68.

5. M. Vaziri and C. Chen, "Optical-Fiber Strain Sensors with Asymmetric Etched Structures," Applied Optics, Vol. 32, No. 31, 1992, pp. 6399-6406.

6. S. Miclos and T. Zisu, "Chalcogenide Fiber Displacement Sensor," Journal of Optoelectronics and Advanced Materials, Vol. 3, No. 2, 2001, pp. 373-376.

7. L. S. Thomas, N. A. Geoge, P. Sureshkumar, et al., "Chemical Sensing with Microbent Optical Fiber," Optics Letters, Vol. 26, No. 20, 2001, pp. 1541-1543.

8. J. B Freal, C. J Zarobila and C. M Davis, "A Microbend Horizontal Accelerometer for Borehole Deployment," Journal of Lightwave Technology, Vol. LT-5, No. 7, 1987, pp. 993-996.

9. S. G. Pierce, A. MacLean and B. Culshaw, "Optical Frequency-Domain Reflectometry for Microbend Sensor Demodulation," Applied Optics, Vol. 39, No. 25, 2000, pp. 4569-4581.

10. D. Yadav and A. K. Nadir, "Transformer Temperature Measurement Using Optical Fiber Microbend Sensor," Sensors & Transducers Journal, Vol. 84, No. 10, 2007, pp. 1651-1659.

11. N. Lagakos, J. H. Cole and J. A. Bucaro, "Microbend Fiber-Optic Sensor," Applied Optics, Vol. 26, No. 11, 1987, pp. 2171-2180.

12. L. Jeunhomme and J. P. Pocholle, "Mode Coupling in a Multimode Optical Fiber with Microbends," Applied Optics, Vol. 14, No. 4, 1975, pp. 2400-2405.

13. R. Olshansky and S. M. Oaks, "Differential Mode Attenuation Measurements in Graded-Index Fibers," Applied Optics, Vol. 17, No. 11, 1978, pp. 1830-1835.

14. R. Olshansky, "Mode Coupling Effects in Graded-Index Optical Fibers," Applied Optics, Vol. 14, No. 4, 1975, pp. 935-945.

3

NANOMECHANICAL OPTICAL FIBER WITH EMBEDDED ELECTRODES ACTUATED BY JOULE HEATING

Zhenggang Lian †, Martha Segura †,*, Nina Podoliak †, Xian Feng, Nicholas White and Peter Horak

Optoelectronics Research Centre, University of Southampton, Southampton SO17 1BJ, UK

ABSTRACT

Nanomechanical optical fibers with metal electrodes embedded in the jacket were fabricated by a multi-material co-draw technique. At the center of the fibers, two glass cores suspended by thin membranes and surrounded by air form a directional coupler that is highly temperature-dependent. We demonstrate optical switching between the two fiber cores by Joule heating of the electrodes with as little as 0.4 W electrical power, thereby demonstrating an electrically actuated all-fiber microelectromechanical system (MEMS). Simulations show that the main mechanism for optical switching is the transverse thermal expansion of the fiber structure.

KEYWORDS

micro-structured optical fibers; optical fiber sensors; all-optical fiber devices; optical switching

1. INTRODUCTION

Microstructured optical fibers (MOF) have been designed in a large variety of geometries. Most common structures are based on hexagonally stacked geometries with solid cores, but various specialty fibers, e.g., suspended-core [1], suspended-web [2] or hollow-core fibers [3,4], have also been demonstrated. Many of these efforts were focused on control of dispersion, enhancement of the nonlinearity, reduction of the fiber transmission loss, or increase of sensitivity to environmental conditions in single-material MOFs [5]. Multi-material MOFs have also been developed for added functionality, e.g., hollow-core fibers filled with liquids [6] or gases [7] for sensing, with gases for high-power pulse compression [8,9] and nonlinear effects [10,11], with semiconductors for electronic functionality [12], and with integrated electrodes [13,14] and liquid crystals [15] for optical switching.

Optical fibers with optical properties that can be externally controlled after fabrication could open up a new range of applications, in particular in telecommunications, sensing, and imaging, following the successful model of chip-based microelectromechanical systems (MEMS) [16]. An example of such a reconfigurable fiber is the dual suspended-core nanomechanical optical fiber [17], which consists of two independent cores suspended by thin membranes and surrounded by air. The cores interact with each other via their evanescent optical fields; thus, mechanically changing the core-to-core separation modifies the optical coupling length and allows for active control of the propagating light. Various mechanisms have already been proposed to achieve this actuation. In [17], optical switching was demonstrated by pressurizing the gas-filled holes of the fiber. With just an 8-nm change of the core separation optical switching of the light at the fiber output from one core to the other was observed. Alternative mechanisms to induce optical switching that have been proposed for such fibers are electrostatic actuation [18] and optomechanical forces [2]. For many applications, forms of electrical actuation are preferred [12,13,14,15,18], but operation at low voltages/low powers would often only be achieved by placing electrically conducting electrodes close to the optically guiding cores, *i.e.*, by embedding metals into the fiber.

Here we report the fabrication of nanomechanical optical fibers with integrated electrodes by a multi-material co-draw technique: a dual suspended-core structure fabricated in lead-silicate glass (Schott F2) at the center of the fiber is surrounded by four tin electrodes. Using these fibers, we then demonstrate a novel electrically-induced actuation mechanism whereby electric current in the electrodes leads to Joule heating of the fiber by a few degrees Celsius, sufficient to modify the transmission properties of fibers of approximately 0.5 m length.

This design opens new possibilities for applications of optical fibers as MEMS [16]. Such MEMS-type fibers are transmitting light in the same way as standard fibers, but simultaneously they can perform some active signal processing functionality, such as switching or modulating, thereby avoiding e.g., the use of chip-based micromirrors and microlenses. Moreover, because of the long propagation length of fibers compared to integrated devices, already nanometer-scale actuation is sufficient to achieve observable effects [17,18].

2. FABRICATION OF NANOMECHANICAL OPTICAL FIBERS WITH EMBEDDED METAL ELECTRODES

The electrode-embedded nanomechanical optical fiber was fabricated in several steps:

- A dual-core glass preform was fabricated in Schott F2 glass, which has a refractive index of n_0 = 1.595 at an operating wavelength of 1550 nm leading to tighter mode confinement than silica [19]. Because of the low softening temperature of F2 glass (592 °C), glass extrusion with a stainless steel die could be used for preform fabrication, similar to the previous fabrication of other high-precision specialty fiber preforms based on compounds glasses including lead-silicate glasses [20], tellurite glasses [21] and chalcogenide glasses [22,23].

- The preform was then drawn into a smaller dimension cane of 2 mm diameter; see Figure 1a.

- In the next step, a multi-material preform was then assembled: The dual-core cane was placed in the middle of a F2 jacket tube (10/8 mm OD/ID) and surrounded by four glass tubes filled with tin rods (melting temperature 232 °C), as shown in Figure 1b. Four extra F2 rods were used to fill the gaps between the jacket and the tubes and to secure the position of the cane.

- Under nitrogen purging conditions this preform was then pre-heated to a temperature above the melting temperature of tin for 60 min in order to melt the tin and condense it down to the bottom of the preform. This step was found to be critical to minimize air bubbles between the metal and the glass, thus avoiding discontinuities of the tin electrodes along the fiber.

- Finally, the preform was drawn into fiber using a drawing temperature of 730 °C [24]. Two different fibers from different preforms were fabricated with this method. Figure 1c shows a scanning electron microscopy (SEM) picture of one of the fibers with the two suspended cores clearly seen in the center; a magnified picture is shown in Figure 1d. We note that the fiber is not perfectly symmetric due to a small asymmetry of the stacked preform and slightly insufficient pressure being applied to the core structure during fiber drawing. Table 1 summarizes the dimensions of the two fibers.

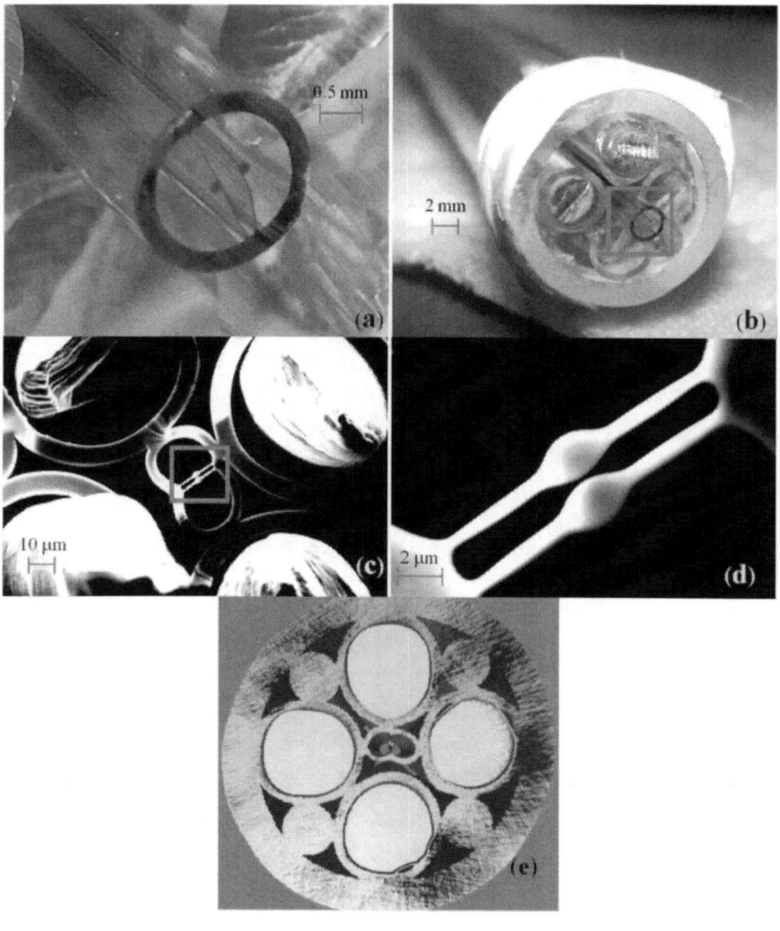

Figure 1. (**a**) Dual suspended-core cane; (**b**) Stacked preform with the cane in the center and four tin rods in glass tubes; (**c**) SEM picture of the dual-core fiber 1 with embedded metal electrodes after drawing. The length of the membranes is 12 μm, and the gap between the two cores is 200 nm; (**d**) Detail of the cores in the blue square of (**c**); (**e**) Advanced X-ray computed tomography image of the fiber, electrodes in yellow and glass in blue.

Table 1. Dimensions of the fibers used for the optical experiments.

Fiber	Core size (μm)	Gap (μm)	Membrane thickness × Length (μm)
Fiber 1	2.0 × 3.1	0.2	0.5 × 12
Fiber 2	2.0 × 3.0	0.5	0.6 × 15

For both fibers the average diameter of the four tin electrodes is 50 μm. It was confirmed that the electrodes are continuous over tens of meters by measuring the electrical resistance along the fiber. Images produced by X-ray computed tomography also showed that the tin electrodes filled the glass tubes without air bubbles, see Figure 1e. The optical loss of fiber 1 was measured for TE and TM polarizations (parallel and orthogonal to the membranes, respectively) as 2.5 and 4.8 dB/m. The loss mechanisms are expected to be the intrinsic loss of lead-silicate glasses (1 dB/m at 1.55 μm) and scattering due to surface roughness. The latter affects the TM mode more since this mode has larger optical fields at the core surfaces, leading to larger losses than for the TE mode.

3. OPTICAL CHARACTERIZATION

For the optical characterization of the fiber under electrical actuation, both ends of the fiber were carefully cleaved, removing the glass part but leaving the four electrodes for connecting purposes. The ends of the fiber were placed on two metallic V-grooves and fixed with magnets in such a way that the fiber was not bent or twisted along its length.

A diagram of the experimental setup for the optical characterization is shown in Figure 2. Linearly polarized light from a 1550 nm wavelength pigtailed laser diode (2 mW) was focused into one of the cores with a microscope objective (L1) with 40× (or 60×) magnification and numerical aperture NA = 0.65 (or 0.75). At the output, a 100× long working distance objective (L2) with NA = 0.85 was used to image the output beam onto a CCD camera (SCOR-20-1550, Ophir-Spiricon Inc., Logan, UT, USA).

The four electrodes at each end were connected to the V-groove using silver paint and two copper wires coming from the power supply were soldered to each V-groove. In this way, one side of the metal electrodes was connected to the positive side of a power supply, and the other side of the metal electrodes was connected to the negative.

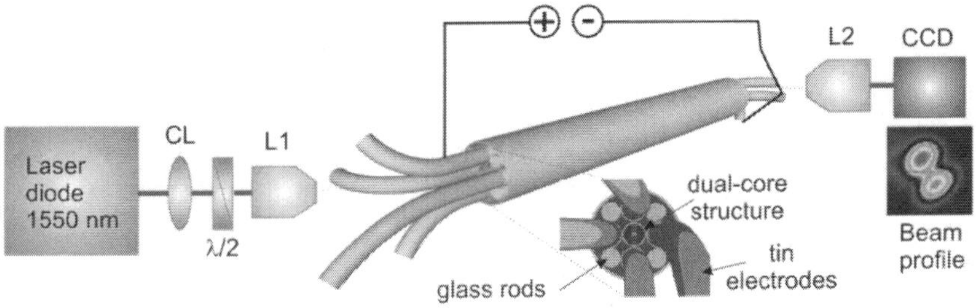

Figure 2. Setup for optical characterization of the dual-core fiber. CL: collimating lens; λ/2: half-wave plate; L1, L2: microscope objective lenses.

3.1. Experiments with Fiber 1

We first performed experiments with a piece of fiber 1 in Table 1 of 65 cm length. When the voltage of the source was varied, the current showed a linear Ohmic response with a total resistance of 11 Ω. Light was coupled into the fiber using the 40× objective as lens L1 (Figure 2). The polarization state was chosen normal to the membranes (TM polarization) because of the slightly stronger core-to-core coupling and thus higher switching contrast of this mode compared to that of TE polarization.

At room temperature, the guided light pattern shown on the CCD camera was stable. As the electric power increased and the fiber was therefore slowly heating up, a noticeable switching of light between the cores was observed, see Figure 3a. A sequence of light patterns was recorded consisting of 16 frames at 7.5 frames per second for each electric power level. The maximum intensity of light in each core was extracted from the 16 Gaussian profiles, averaged and normalized to the sum of intensities in both cores. The starting contrast was 0.55/0.45 (upper/lower core). As the electric power increased, a change in intensity of the light in each core was observed, see Figure 3a, reaching a maximum contrast of 0.38/0.62 at 0.4 W of electric power. The maximum applied electric power was 0.9 W where instabilities in the output light pattern and light coupling to higher order modes were observed. As shown in Figure 3a, an increase of power of 0.7 W was required to switch between the two points where both cores show the same output intensity. This behavior was repeatable, and the same results were observed when ramping the power up or down; no hysteresis effect was observed.

The experiment was repeated with a different section of fiber 1. This piece was 70 cm long with 13.4 Ω resistance and showed optical switching performance very similar to the previous piece of fiber. Again, the instability of the recorded output field pattern was observed at higher electrical powers (>0.6 W). However, we found experimentally that by decreasing the incident optical power four times with respect to the previous experiment (0.5 mW instead of 2 mW), the beam profile remained stable and we were able to increase the applied electric power up to 1.8 W, as can be seen in Figure 3b. The reason for this dependence of mode field stability on optical power is not exactly known at this time, yet we note that losses of 1 mW of optical power by material absorption in the fiber cores correspond to approximately 0.1 mW per μm²of cross section, which is comparable to the electric heating power distributed over the much larger cross section of the electrodes. It is thus conceivable that laser light absorption, even at such low power levels, leads to small localized heating effects in the fiber cores which in turn modify the optical coupling length between cores and thus result in unstable, nonlinear back action on the light field distribution.

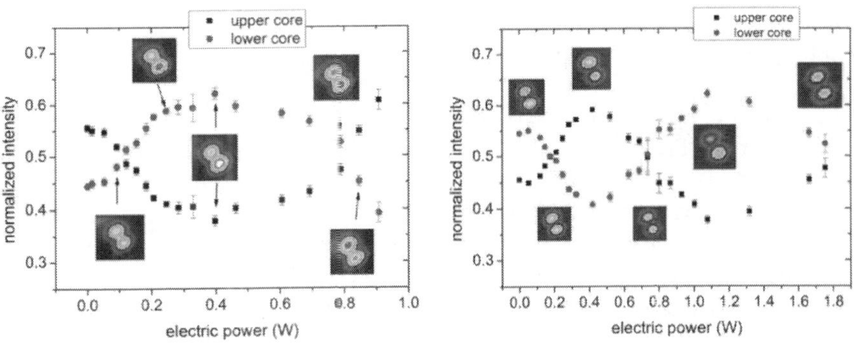

(a) (b)

Figure 3. Normalized output intensity in the upper (black squares) and lower (red dots) cores as function of electric power for two different pieces of fiber 1 with fiber lengths of (**a**) 65 cm and (**b**) 70 cm. The intensity profiles shown were recorded with the CCD camera.

With this reduced optical power, the fiber showed continuous switching with electrical power up to 1.8 W with a maximum contrast of 0.38/0.62, Figure 3b. The switching shown in Figure 3b is not linear with power. The reason for this behavior is not fully understood, but it could be caused by a small temperature dependence of the launching conditions and/or by interference of higher order modes. However, both pieces of fiber 1 exhibited the same behavior from 0 to 0.9 W indicating that the results are repeatable with different fibers of similar dimensions.

3.2. Experiments with Fiber 2

As discussed in more detail in Section 4, we explain the observed optical switching by the thermal expansion of the fiber structure due to Joule heating by the electric current. In order to confirm this thermal mechanism, we carried out another set of experiments to compare the optical switching achieved by electric current applied to the electrodes and by external heating of the fiber. Here we used a 68-cm length of fiber 2 with a total resistance of 14 Ω and the 60× objective was used as lens L1 to couple TM-polarized light into the fiber.

A cylindrical heater was designed such that a fiber placed at the axis of the heater could be uniformly heated. The heater is 45 cm long and 8 mm in diameter, and it can be controlled in steps of 0.1 °C up to a maximum of 50 °C. A thermocouple measures the temperature in the middle of the heater. The two experiments, heating the fiber via the electrodes and externally, were carried out one after the other with the fiber placed inside the heater. First, the fiber was heated by applying electric current and then, after cooling, the temperature of the heater was increased.

The resulting output intensities from the two cores for the current heating are shown in Figure 4a. As can be seen inFigure 4a, the starting output contrast is nearly 0.50/0.50. The initial temperature for this experiment was 22.5 °C. At low electric power levels (<0.4 W) small changes of the intensity in the two cores are recorded, which increase to the largest observed contrast of 0.65/0.35 at 0.7 W electric power. The electric power needed to switch from one core to the other is 0.427 W.

The measurement results for switching by external heating are shown in Figure 4b. The room temperature had increased to 25.8 °C compared to the previous experiment, leading to a shift of the curves in Figure 4b with respect to those of Figure 4a, but otherwise the behavior observed in the two experiments is very similar. The largest contrast was found at 32 °C with external heating, *i.e.*, a temperature change of 5.6 °C was needed to switch light between the two fiber cores. The equivalence of the two experimental results confirms that the change in temperature is in fact the switching mechanism for the electrically driven actuation of our dual-core nanomechanical fiber. It also provides an indirect measurement of the fiber core temperature increase with Joule heating in the electrodes (5.6 °C for 0.427 W).

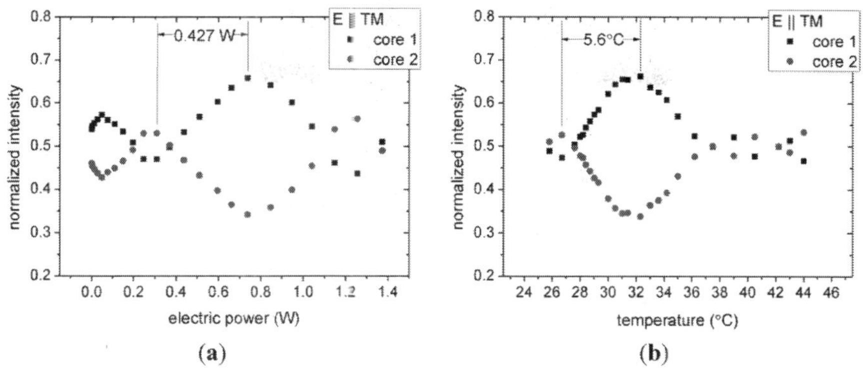

Figure 4. Normalized output intensity in core 1 (black squares) and core 2 (red dots) of fiber 2 as a function of (**a**) electric power; and (**b**) temperature.

4. MODELLING OF THERMALLY INDUCED SWITCHING

The mechanism of optical switching in the dual-core fiber can be interpreted as follows. Being closely placed to each other, the optical cores are coupled through their evanescent fields, forming symmetric and antisymmetric superposition mode pairs for each polarization (TE and TM supermodes). Each of these modes is characterized by an effective mode index (n_s and n_a denote symmetric and antisymmetric supermode index, respectively), which depends on the optical coupling strength between the two cores. If light is launched into a single core, both supermodes are excited and an interference pattern is formed along the length of the fiber, which is observed as light periodically switching between the cores. The beat length, L_b, the propagation distance at which the light transfers from one core to the other and back, depends on the mode index difference and the wavelength λ as $L_b = \lambda/(n_s - n_a)$. The fiber output thus depends on the total phase difference occurring between the supermodes along the fiber:

$$\delta = 2\pi(n_s - n_a)L/\lambda \tag{1}$$

where L is the fiber length. Modifying the fiber properties by some form of actuation leads to a change of δ, *i.e.*, a shift of the interference pattern of the supermodes that is seen as a continuous switching of the light between the cores at the output. One switch (half of a duty cycle) corresponds to a phase difference variation of π.

By running an electric current through the electrodes, Joule heating raises the temperature of the fiber. Unlike the previous study where switching was induced by optical core deformations caused by the metallic electrode expansion [13], in our fiber the influence of the electrode expansion is suppressed by the air gaps between the electrodes and the optical cores. The electrodes rather act as an internal heater. Having four electrodes connected to the power supply ensures uniform temperature of the internal part of the fiber eliminating stress due to temperature gradient. Thus, we identify three other contributions to the variation of δ: A. Thermal elongation of the fiber leads to a larger propagation distance and hence increases the phase difference (assuming that the mode indices do not change); B. The increase of the material refractive index with temperature leads to stronger localization of the optical fields inside the cores and thus reduces core-to-core coupling, which in turn decreases the phase difference between the supermodes; C. Uniform thermal expansion in the transverse direction of the inner fiber structure increases both the core size and the air gap between the cores, again reducing the core-to-core coupling efficiency. As a result, the change of the phase difference $d\delta$ between the supermodes induced by a temperature variation dT can be summarized as:

$$d\delta = 2\pi L\left(\alpha(n_s - n_a)dT + \left.\frac{\partial(n_s - n_a)}{\partial T}\right|_{\substack{\text{material}\\\text{index}}} dT + \left.\frac{\partial(n_s - n_a)}{\partial T}\right|_{\substack{\text{transverse}\\\text{expansion}}} dT \right)/\lambda \qquad (2)$$

where α is the thermal expansion coefficient. The first term is positive, while the other two contribute with negative sign.

To quantify the impact of these three effects in the dual-core fiber, we calculated effective mode indices for the fundamental symmetric and antisymmetric modes and variation of the mode indices induced by the effects A, B and C using a finite element method (Comsol Multiphysics [25]). The optical and thermal properties of F2 glass used in the calculations (refractive index n = 1.595 at 1.55 μm, thermal expansion coefficient $\alpha = 9.2 \times 10^{-6}\ K^{-1}$, and thermo-optic coefficient $dn/dT = 2.7 \times 10^{-6}\ K^{-1}$) were taken from the data sheet of the glass manufacturer [26]. The simulations were performed for a range of core dimensions (core width varied from 1 to 2 μm, keeping the length to width ratio fixed at 1.5) and air gaps (varied from 0 to 500 nm).

For this range of fiber dimensions the two cores are strongly coupled and the beat length was found to vary from a few tens of μm (for small cores and small gap) up to 1 mm (for large cores and large gap). The change of the phase difference$d\delta/dT$ between the TM supermodes induced by each of the three effects separately in a 50-cm long piece of fiber is shown inFigure 5a–c, while Figure 5d shows the total phase difference change caused by all of the effects. Similar results were also obtained for the TE supermodes. The figures show that the fiber sensitivity to temperature changes strongly depends on the core size and the air gap between the cores. Smaller cores and smaller core separations lead to stronger optical coupling between the cores and to larger sensitivity to temperature variations. As discussed above, effect A leads to an increase of δ with increasing temperature, while effects B and C decrease δ and the total phase difference change has a negative value. The contribution of the thermal expansion in the transverse direction (effect C) is approximately 2.5 times larger than the impact of the thermal elongation (effect A) and almost an order of magnitude larger than the effect of the material index change (effect B).

Calculations of the temperature of the cores required for switching the light from one core to the other for the TM and TE modes in a 50-cm long piece of fiber with a core size of 2 μm × 3 μm are shown in Figure 5e. The switching temperature varies from 5 to 25 K depending on the air gap between cores. This is of the same order of magnitude as the experimental switching temperature, but slightly larger. Note that following Equation (2) the phase change scales linearly with the fiber length and thus the required temperature change for optical switching is inversely proportional to the fiber length.

5. DISCUSSION AND CONCLUSIONS

In this work we reported the successful fabrication of a new kind of multi-material MOF based on lead silicate glass with integrated tin electrodes by a co-draw technique. The fiber has a dual-core structure, with two elliptical cores being independently suspended by thin membranes. The optical coupling between the two cores is extremely sensitive to any structural changes, and thus the fiber can be used as a sensor or an integrated all-fiber switch. The embedded conducting electrodes allow for novel electric actuation mechanisms operating at low voltages and/or low power due to their close vicinity to the optically guiding cores.

In previous work, we discussed two different methods of dual-core fiber actuation, mechanical and electrostatic. In the mechanical method one suspended core was moved by 8 nm by applying air pressure of ~100 mbar to one of the air holes through a slit in the cladding [17] leading to one optical switching cycle in a 40-cm long fiber. However, the mechanical method is slow and requires placing the fiber in a bulky pressure chamber. Alternatively, electrostatic MEMS-type actuation by induced electric dipole forces was proposed to allow for fast switching on a microsecond time scale [18]. This method relies on applying electric voltages to electrodes embedded in the fiber jacket, but because of the weak induced forces at realistic voltages the method also requires very thin and long membranes (e.g., 0.1 μm × 25 μm) supporting the fiber cores, making this an extremely challenging fiber design for fabrication. The fibers presented in this paper exhibit the required metal electrodes close to the cores for this actuation mechanism, but the supporting membranes are too stiff given the dimensions of Table 1. Our simulations suggest that voltages higher than ±200 V would have to be applied to the electrodes to observe optical switching in a 50-cm long fiber, generating an electric field sufficiently strong to lead to dielectric breakdown of air in the fiber holes [27].

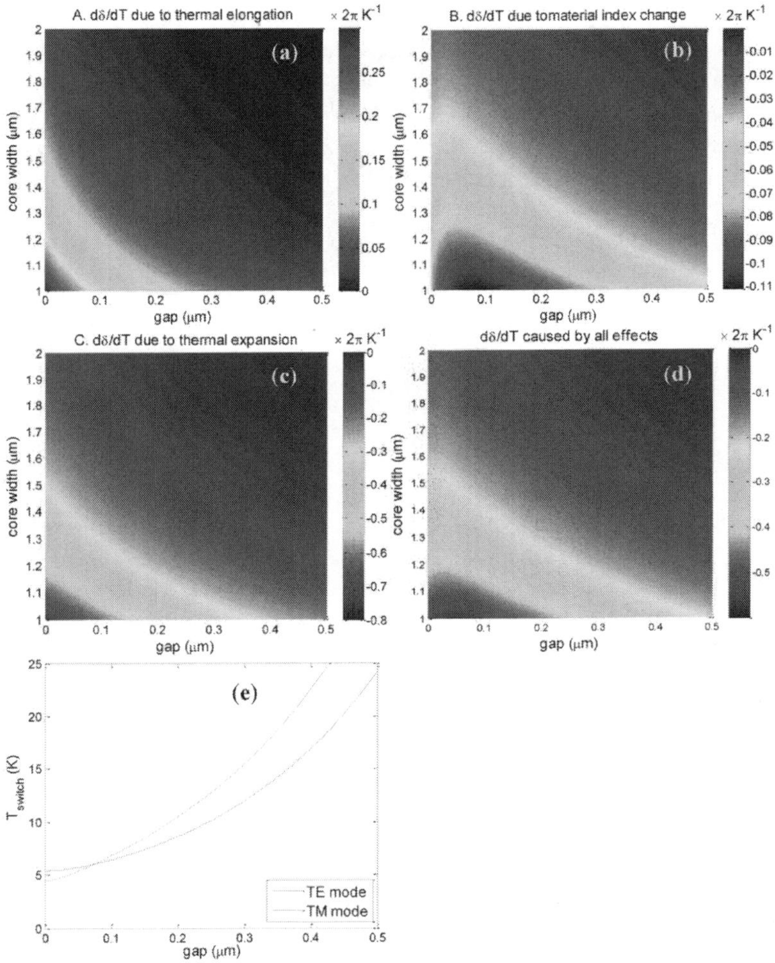

Figure 5. (**a–d**) Variation of the phase difference $d\delta/dT$ for TM polarization with core size and air gap size in a 50 cm piece of fiber induced by thermal effects A, B and C, and by the sum of these effects; (**e**) Temperature required to observe an optical switch in a 50-cm piece of fiber with core size 2 μm × 3 μm.

Here we demonstrated another electrically induced actuation method of the dual-core fiber based on Joule heating: by passing electric current through the embedded metal electrodes, the temperature of the fiber increases slightly, changing the coupling length between the fiber cores. Optical switching in ~70 cm of fiber was observed by applying ~0.4 W of electric power. In a reference experiment using an external heater we found that a temperature difference of 5.6 °C is required to produce one optical switch. Numerical simulations further confirm the thermal origin of the actuation mechanism. The calculations suggest that in the range of the studied fiber dimensions the effect of thermal expansion in the transverse directions dominates over the change of material index with temperature and elongation of the fiber. This novel actuation mechanism has the advantage of working at relatively robust fiber geometries with thick membranes supporting the dual cores, but it requires Watt-level actuation powers and the switching time will be limited to the millisecond range by the thermal response time of the fibers. Further improvements of the fabrication technique will therefore be required to demonstrate dual-core optical fibers with longer and more flexible membranes for fast and low-power optical switching by electrostatic actuation.

ACKNOWLEDGMENTS

We gratefully acknowledge support for this work from the Engineering and Physical Sciences Research Council (EPSRC) through funding for the EPSRC Centre for Innovative Manufacturing in Photonics, the "Photonic Hyperhighway" Programme Grant, and the "Tunable Optical Buffer" grant (EP/J012874/1). We also thank Wei H. Loh for his invaluable contribution to this work.

AUTHOR CONTRIBUTIONS

Zhenggang Lian fabricated and characterized the optical fibers; Martha Segura performed the switching experiments; and Nina Podoliak performed the numerical simulations; Zhenggang Lian, Martha Segura, and Nina Podoliak wrote the paper; Xian Feng contributed to the discussion of the experimental results; Nicholas White contributed to design and building of the setup; and Peter Horak led the interpretation of the results and edited the paper.

CONFLICTS OF INTEREST

The authors declare no conflict of interest.

REFERENCES

1. Monro, T.M.; Warren-Smith, S.; Schartner, E.P.; Franois, A.; Heng, S.; Ebendorff-Heidepriem, H.; Afshar, S. Sensing with suspended-core optical fibers. *Opt. Fiber Technol.* **2010**, *16*, 343–356.

2. Butsch, A.; Kang, M.S.; Euser, T.G.; Koehler, J.R.; Rammler, S.; Keding, R.; Russell, P.S. Optomechanical nonlinearity in dual-nanoweb structure suspended inside capillary fiber. *Phys. Rev. Lett.* **2012**, *109*.

3. Russell, P. Photonic crystal fibre. *Science* **2003**, *299*, 358–362.

4. Fini, J.; Nicholson, J.; Windeler, R.; Monberg, E.; Meng, L.; Mangan, B.; DeSantolo, A.; DiMarcello, F. Low-loss hollow-core fibers with improved single-modedness. *Opt. Express* **2013**, *21*, 6233–6242.

5. Aref, S.; Amezcua-Correa, R.; Carvalho, J.; Frazão, O.; Caldas, P.; Santos, J.; Araújo, F.; Latifi, H.; Farahi, F.; Ferreira, L.; *et al.* Modal interferometer based on hollow-core photonic crystal fiber for strain and temperature measurement.*Opt. Express* **2009**, *17*, 18669–18675.

6. Liu, S.; Wang, Y.; Hou, M.; Guo, J.; Li, Z.; Lu, P. Anti-resonant reflecting guidance in alcohol-filled hollow core photonic crystal fiber for sensing applications. *Opt. Express* **2013**, *21*, 31690–31697.

7. Kornaszewski, L.; Gayraud, N.; Stone, J.M.; MacPherson, W.N.; George, A.K.; Knight, J.C.; Hand, D.P.; Reid, D.T. Mid-infrared methane detection in a photonic bandgap fiber using a broadband optical parametric oscillator. *Opt. Express***2007**, *15*, 11219–11224.

8. Nisoli, M.; de Silvestri, S.; Svelto, O. Generation of high energy 10 fs pulses by a new pulse compression technique.*Appl. Phys. Lett.* **1996**, *68*, 2793–2795.

9. Anderson, P.N.; Horak, P.; Frey, J.G.; Brocklesby, W.S. High-energy laser-pulse self-compression in short gas-filled fibers. *Phys. Rev. A* **2014**, *89*.

10. Travers, J.C.; Chang, W.; Nold, J.; Joly, N.Y.; Russell, P.S. Ultrafast nonlinear optics in gas-filled hollow-core photonic crystal fibers. *J. Opt. Soc. Am. B* **2011**, *28*, A11–A26.

11. Li, C.; Rishad, K.P.M.; Horak, P.; Matsuura, Y.; Faccio, D. Spectral broadening and temporal compression of ~100 fs pulses in air-filled hollow core capillary fibers. *Opt. Express* **2014**, *22*, 1143–1151.

12. Sazio, P.J.A.; Amezcua-Correa, A.; Finlayson, C.E.; Hayes, J.R.; Scheidemantel, T.J.; Baril, N.F.; Jackson, B.R.; Won, D.-J.; Zhang, F.; Margine, E.R.; *et al.* Microstructured optical fibers as high-pressure microfluidic reactors. *Science* **2006**,*311*, 1583–1586.

13. Chesini, G.; Cordeiro, C.M.B.; de Matos, C.J.S.; Fokine, M.; Carvalho, I.C.S.; Knight, J.C. All-fiber devices based on photonic crystal fibers with integrated electrodes. *Opt. Express* **2009**, *17*, 1660–1665.

14. Zhang, W.Q.; Manning, S.; Ebendorff-Heidepriem, H.; Monro, T.M. Lead silicate microstructured optical fibres for electro-optical applications. *Opt. Express* **2012**, *21*, 31309–31317.

15. Larsen, T.T.; Bjarklev, A.; Hermann, D.S.; Broeng, J. Optical devices based on liquid crystal photonic bandgap fibres.*Opt. Express* **2003**, *11*, 2589–2596.

16. Craighead, H.G. Nanoelectromechanical systems. *Science* **2000**, *290*, 1532–1535.

17. Lian, Z.; Horak, P.; Feng, X.; Xiao, L.; Frampton, K.; White, N.; Tucknott, J.; Rutt, H.; Payne, D.N.; Stewart, W.; *et al.* Nanomechanical optical fiber. *Opt. Express* **2012**, *20*, 29386–29394.

18. Podoliak, N.; Lian, Z.; Loh, W.H.; Horak, P. Design of dual-core optical fibers with NEMS functionality. *Opt. Express***2014**, *22*, 1065–1076.

19. Ebendorff-Heidepriem, H.; Warren-Smith, S.; Monro, T. Suspended nanowires: Fabrication, design and characterization of fibers with nanoscale cores. *Opt. Express* **2009**, *17*, 2646–2657.

20. Mairaj, A.K.; Feng, X.; Shepherd, D.P.; Hewak, D.W. Channel waveguide lasers in a lead silicate glass fashioned using the extrusion technique. *Appl. Phys. Lett.* **2004**, *85*, 2727–2729.

21. Ebendorff-Heidepriem, H.; Kuan, K.; Oermann, M.; Knight, K.; Monro, T. Extruded tellurite glass and fibers with low OH content for mid-infrared applications. *Opt. Mater. Express* **2012**, *2*, 432–442.

22. Savage, S.D.; Miller, C.A.; Furniss, D.; Seddon, A.B. Extrusion of chalcogenide glass preforms and drawing to multimode optical fibres. *J. Non Cryst. Solids* **2008**, *354*, 3418–3427.

23. Lian, Z.; Li, Q.Q.; Furniss, D.; Benson, T.M.; Seddon, A.B. Solid Microstructured chalcogenide glass optical fibers for the near- and mid-infrared spectral regions. *IEEE Photonics Technol. Lett.* **2009**, *21*, 1804–1806.

24. Bayindir, M.; Sorin, F.; Abouraddy, A.F.; Viens, J.; Hart, S.D.; Joannopoulos, J.D.; Fink, Y. Metal-insulator-semiconductor optoelectronic fibres. *Nature* **2004**, *431*, 826–829.

25. Comsol Multiphysics 4.3a. Available online: http://www.comsol.com/ (accessed on 25 July 2014).

26. Schott Optical Glass Datasheet (2011). Available online: http://www.schott.com/ (accessed on 25 April 2013).

27. Podoliak, N.; Lian, Z.; Segura, M.; Loh, W.H.; Horak, P. Electrostatic actuation of nanomechanical optical fibers with integrated electrodes. *SPIE Proc.* **2014**, *9128*.

4

MODE-LOCKED FIBER LASERS WITH SIGNIFICANT VARIABILITY OF GENERATION REGIMES

S. Kobtsev[a], S. Smirnov[a], S. Kukarin[a], S. Turitsyn[a, b]

a Division of Laser Physics and Innovation Technologies, Novosibirsk State University, 630090 Novosibirsk, Russia

b Aston Institute of Photonic Technologies, Aston University, Birmingham, B4 7ET Birmingham, UK

HIGHLIGHTS

- Polarisation controller settings vary pulse/train spectrum width more than tenfold.
- Single/double-scale pulses of similar width lead to different NL-converted radiation.
- Each start of mode-locked regime requires control of major pulse/train parameters.

ABSTRACT

We demonstrate a great variability of single-pulse (with only one pulse/wave-packet traveling along the cavity) generation regimes in fiber lasers passively mode-locked by non-linear polarization evolution (NPE) effect. Combining extensive numerical modeling and experimental studies, we identify multiple very distinct lasing regimes with a rich variety of dynamic behavior and a remarkably broad spread of key parameters (by an order of magnitude and more) of the generated pulses. Such a broad range of variability of possible lasing regimes necessitates developing techniques for control/adjustment of such key pulse parameters as duration, radiation spectrum, and the shape of the auto-correlation function. From a practical view point, availability of pulses/wave-packets with such different characteristics from the same laser makes it imperative to develop variability-aware designs with control techniques and methods to select appropriate application-oriented regimes.

KEYWORDS

Fiber laser; Passively mode-locked laser; Ultrafast laser; Non-linear polarization evolution

1. INTRODUCTION

Mode-locked fiber lasers are widely used as sources of ultra-short optical pulses [1], [2],[3], [4], [5] and [6]. The growing use of fiber lasers for this purpose is supported by customer demand for reliable femto- and pico-second lasers with a long (or, better, unlimited) operation lifetime and without any need for manual adjustments. Fiber lasers mode-locked due to non-linear polarization evolution (NPE) may satisfy these requirements [7] and [8] as they make direct use of their resonators' non-linear properties and are able to generate pulses with durations as short as 50 fs and even less[9] and [10]. Being at the same time a promising tool and a unique platform for observing spectacular non-linear optical interactions, NPE mode-locked fiber lasers attract much attention from both scientists and laser engineers. Such lasers support a large variety of generation regimes [11], including pulses of different shapes [12], [13], [14] and [15], multi-pulse lasing at different fundamental repetition frequency multiples [16], as well as complex soliton structures [14] and [16] and soliton rains [17]. In the present work, we limit our analysis to single-pulse lasing, that is, when only one laser pulse/wave-packet is present over a single round trip of the cavity. However, as our results show, even within such a limited case, a large variety of lasing regimes is found, which may considerably (up to an order of magnitude or even more) differ from each other in pulse energy, duration, bandwidth, etc., and may dramatically differ from each other in pulse structure being either "conventional" (single-scale) smooth pulses or double-scale pulses (wave-packets) with complex internal structure and phase fluctuations [18]. This has an important practical impact, calling for a variability-aware design of laser systems with embedded control techniques and methods to select appropriate output pulses.

2. EXPERIMENTAL SETUP

The setup used in our experiments is similar to those described in our previous studies[18], [19] and [20], see Fig. 1. The fiber laser has a ring cavity with a wave-division-multiplexing (WDM) used to couple in the pump radiation. Either 2 m-long Er- or 8 m-long Yb-doped optical fiber is used in different experiments as the active medium, both of which produce qualitatively similar results at net-normal and all-normal cavity dispersion respectively. Either SMF-28 or normal-dispersion fiber (NDF) is used to elongate the cavity and thus increase pulse energy. Output laser radiation is extracted through a fiber polarization beam splitter (FPBS) or through an additional coupler inserted into the cavity. Mode-locked operation was achieved by adjusting fiber polarization controllers PC1, 2. Operation regimes were studied with the help of a fast oscilloscope, an optical spectrum analyzer (OSA), and an optical pulse auto-correlator.

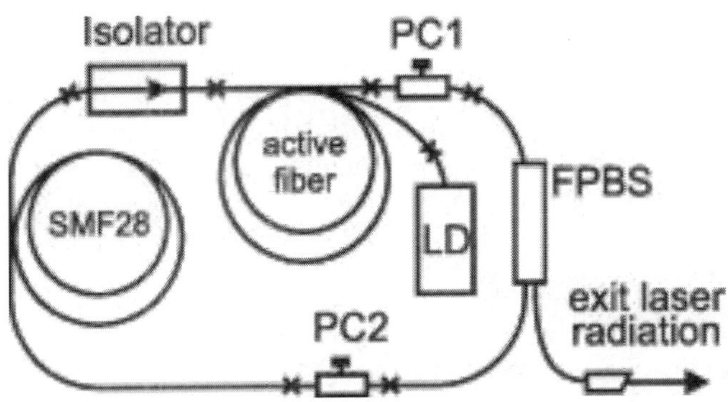

Figure. 1. Laser layout used in experiments. LD—laser diode, FPBS—fiber polarization beam splitter, PC1, 2—fiber polarization controllers.

3. NUMERICAL MODEL

In order to investigate the variety of possible single-pulse lasing regimes and their properties, we used a well-established numerical model based on a set of modified, non-linear Schrödinger equations for orthogonal polarization components of the field envelope [21]:

$$\frac{\partial A_x}{\partial z} = i\gamma\left\{|A_x|^2 A_x + \frac{2}{3}|A_y|^2 A_x + \frac{1}{3}A_y^2 A_x^*\right\} + \frac{g_0/2}{1 + E/(P_{sat} \cdot \tau)} A_x - \frac{i}{2}\beta_2 \cdot \frac{\partial^2 A_x}{\partial t^2} \tag{1}$$

$$\frac{\partial A_y}{\partial z} = i\gamma\left\{|A_y|^2 A_y + \frac{2}{3}|A_x|^2 A_y + \frac{1}{3}A_x^2 A_y^*\right\} + \frac{g_0/2}{1 + E/(P_{sat} \cdot \tau)} A_y - \frac{i}{2}\beta_2 \cdot \frac{\partial^2 A_y}{\partial t^2} \tag{2}$$

where A_x, and A_y are the orthogonal components of the field envelope, z is the longitudinal coordinate, t – time, $\gamma = 4.7 \times 10^{-5}$ (cm W)$^{-1}$, $\beta_2 = 23$ ps^2/km – non-linear and dispersion coefficients, $g_0 = 540$ dB/km – unsaturated gain coefficient,$P_{sat} = 52$ mW – saturation power for the active fiber. In order to accelerate simulations and spare computational resources, we reduced cavity length to 10 m, which corresponds to cavity round-trip time $\tau = 48$ ns. Similarly to [18] and [19], we integrated Eqs. (1) and (2) over 10^4 round-trips, using white noise as the initial condition for the first iteration. The term proportional to g_0 describes saturated optical gain and therefore was omitted while simulating pulse propagation in a passive fiber. Polarization controllers were taken into account through applying unitary matrices. Eqs. (1) and (2) were integrated numerically using step-split Fourier method. We checked carefully that the main results of this paper were not influenced by variation of numerical integration step, step number, mesh width, and number of mesh points, provided that these parameters were chosen in proper limits.

4. RESULTS AND DISCUSSION

The laser layout depicted in Fig. 1 includes two fiber polarization controllers, which provide several degrees of freedom, including the ability to trigger mode-locked operation and then switch between different generation regimes or adjust laser parameters. Since investigating lasing properties in multi-dimensional space is quite complicated, we confined ourselves to statistical study of different generation regimes. The approach consisted of choosing the polarization controller settings (tilt/slew angles) in a random way and integrating Eqs. (1) and (2) with white noise as the initial conditions. After a certain number (typically 10^3...10^4) of round-trips, we may obtain one of the following solutions: (i) quasi-CW laser operation (no mode-lock reached; thus Eqs.(1) and (2) and their numerical solution are not valid); (ii) multi-pulse mode-locked operation (that is, two or more pulses co-exist in the cavity); and (iii) single-pulse, mode-locked operation. If a single-pulse, mode-locked operation is reached after the fixed amount of cavity round-trips, the program saves the results (tilt angles of PC1, 2, pulse duration, energy, optical spectrum, temporal intensity distribution, etc.). Otherwise, no information is kept about this program run. In either case, the program selects a new combination of random PC settings and proceeds with the next run, thus accumulating information about single-pulse operation regimes.

The lasing regimes found in simulation demonstrate both quantitative and qualitative differences. Qualitatively, one can distinguish two main types of single-pulse generation regimes: fully coherent, single-scale, and double-scale pulses with complex inner structure [19]. Temporal distribution of radiation intensity for fully coherent pulses features a smooth envelope and can be described by a single parameter, that is, the envelope width (single-scale or "conventional" pulses). In contrast, temporal intensity distribution of the radiation of partially coherent pulses is stochastic: inside wave-packets with overall duration of several picoseconds to several nanoseconds; there are fast stochastic variations of radiation intensity with typical time scale of a hundred to several hundred femtoseconds.

Correspondingly, temporal distribution of radiation intensity for such pulses (wave-packets or double-scale pulses) is defined by two temporal parameters: the pulse-train envelope width and the typical intensity fluctuation time inside the train. Overall parameters of the entire wave packet, such as bandwidth, energy, and duration, fluctuate around their average values within a big range from one packet to another. Intensity fluctuations inside single wave-packets may also vary from relatively small values up to peak wave-packet intensity. Lasing regimes with strong intensity fluctuations are usually classified as noise-like generation.

Double-scale pulses have not yet received a universal designation. Different terms are found across the available literature: noise-like pulses [22] and [23], double-scale lumps[19], femtosecond clusters [24], etc. This type of pulse can be easily identified by a singular auto-correlation function shape featuring a narrow (100–200 fs) peak on a broader picosecond pedestal. Significant attention to these pulses is predominantly due to the presence of femtosecond components with high peak power, while the cavities of fiber lasers generating them may have relatively large dispersion.

There is also a series of intermediate possibilities between "conventional" laser pulses and noise-like generation which manifests a relatively small and variable fraction of intensity noise and phase fluctuations on the background of single-scale laser pulses[18]. Remarkably, even inside a single type of lasing (for example, single-scale or double-scale pulses), the studied laser supports a large variety of sub-regimes that correspond to different settings of PC1, 2 and thus differ vis-à-vis energy, duration, bandwidth, etc. Generated pulse parameters may vary by an order of magnitude or even more, depending on PC settings. Probability density functions (PDF) for rms-bandwidth (seeFig. 2(a) and (b)) and rms-duration (see Fig. 2(c) and (d)) obtained by randomly changing simulation PC settings are shown in Fig. 2. These PDFs reveal the extent of parameter variability in different realizations of two main single-pulse lasing regimes, namely conventional lasing (Fig. 2(a) and (c)) and double-scale pulse generation (Fig. 2(b) and (d)). For instance, rms-bandwidth varied in random simulation runs from 0.2 up to 3.9 nm for single-scale pulses and from 0.3 up to 7.4 nm for double-scale pulses being a function of PC settings. (Note that given values are pulse rms-bandwidth, which is usually several times as narrow as the spectrum's full width at half-maximum, FWHM. As an example, for Π-shaped spectrum, the ratio between spectral FWHM and rms-bandwidth is 3.5. For differently shaped spectra, this ratio may vary).

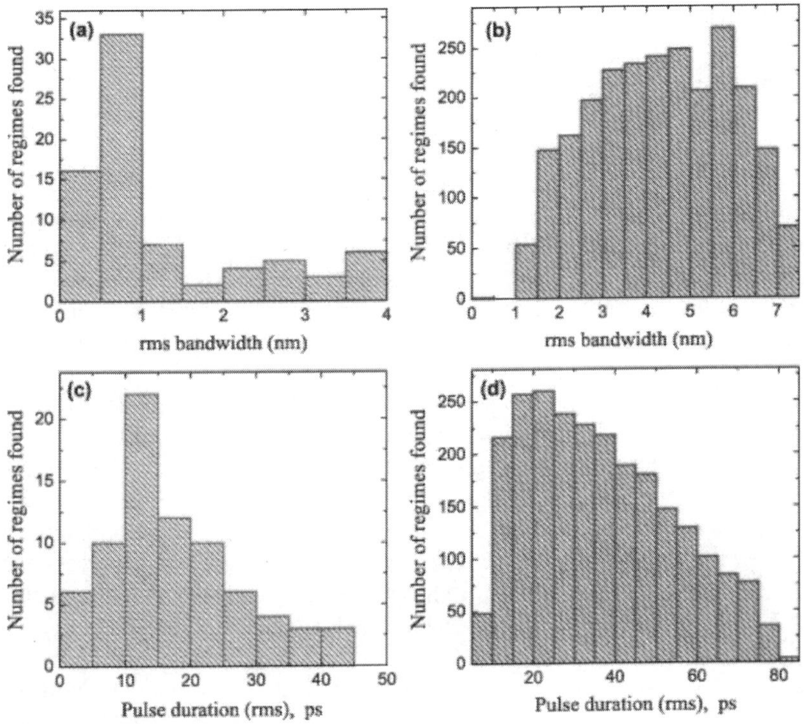

Figure. 2. Rms bandwidth (a, b) and rms-pulse duration (c, d) variability in simulated single-scale (a, c) and double-scale (b, d) single-pulse generation regimes.

Another important difference between lasing regimes attainable at different angular settings of intra-cavity PCs is related to efficiency of non-linear laser pulse conversion. Very recently it was found that efficiency of second harmonic generation (SHG) may vary by an order of magnitude or more depending, along with other laser pulse parameters, on PC settings (see [20] and Fig. 3(a and b)). In order to generalize our conclusions, we study dimension-relative SHG efficiency, which does not depend on pump power and thickness of the thin non-linear crystal, but is sensitive to mode correlations and fluctuations, thus allowing us to easily compare different lasing regimes from the viewpoint of efficiency of non-linear frequency conversion. It is defined [20] as the ratio of two SH powers, of which the first is obtained when the non-linear crystal is pumped by a laser pulse and the second is generated with single-mode monochromatic pumping of the same power. As our simulation shows, double-scale pulses have comparable or higher SHG-relative efficiency compared with that of single-scale laser pulses of the same duration, see Fig. 3(c).

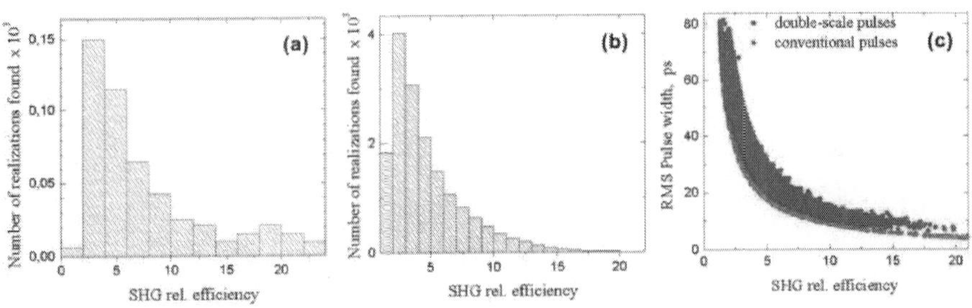

Figure. 3. Relative SHG efficiency in single-scale (a) and double-scale (b) single-pulse generation regimes and its correlation with rms pulse duration, ps (c).

We also conducted a comparative study of Raman scattering spectra generated by single-scale and double-scale laser pulses in a long stretch of extra-cavity fiber. As is illustrated below, notwithstanding similar duration of single-scale and double-scale pulse envelopes, Raman spectra resulting from these pulses are substantially different.

Shown in Fig. 4 and Fig. 5 are the radiation spectra and auto-correlation functions of single-scale and double-scale pulses studied in our experiment. Pulses of both types were generated in the laser cavity of Fig. 1 at different polarization controller positions. Auto-correlation functions were registered using commercial scanning auto-correlator "Tekhnoscan FS-PS-Auto" with working range of 5 fs to 35 ps. Certain noisiness of the recorded auto-correlation functions is due to a rather small fraction of the output radiation used for their measurement. Optical spectra were similarly registered by using a small amount of output power guided into an optical spectrum analyzer Advantest Q8381A. Both auto-correlation function and spectra of the generated pulses were measured in real time in order to know with certainty, which particular type of pulses is generated at a given moment. Mode locking was triggered by mutual adjustment of the polarisation controllers and pump radiation power within 1.3–1.5 W. Single-scale pulses have a П-shaped optical spectrum and quasi bell-shaped auto-correlation function, whereas spectrum of double-scale pulses is much smoother and bell-shaped, and its ACF contains a narrow 180-fs long peak on a wide 12-ps long pedestal.

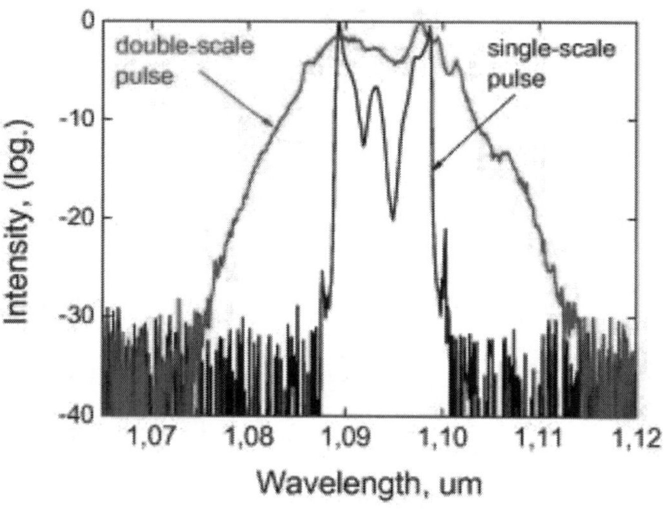

Figure. 4. Radiation spectra of single-scale (black curve) and double-scale (red curve) pulses used in the experiment. (For interpretation of the references to colour in this figure legend, the reader is referred to the web version of this article.)

Fig. 6 shows Raman scattering spectra of single-scale (red curves) and double-scale (black curves) pulses propagating along a 1.2-km long phosphosilicate fiber. In order to produce sufficient spectral conversion, pulses from the reference fiber oscillator could be passed through a fiber optic amplifier to reach the average output power of 840 mW. This figure presents broader Raman scattering spectra of double-scale pulses in comparison to analogous spectra for single-scale pulses. Most likely, this is a result of femtosecond components present in the structure of double-scale pulses, which have higher non-linear conversion efficiency.

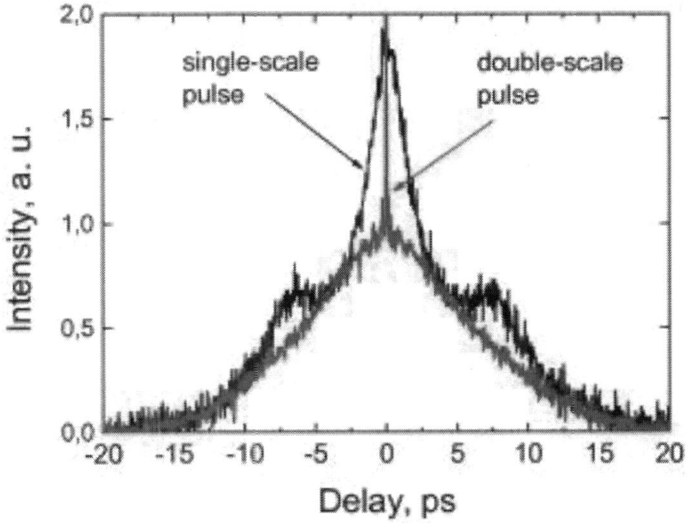

Figure. 5. Auto-correlation functions of single-scale (black curve) and double-scale (red curve) pulses used in the experiment. (For interpretation of the references to colour in this figure legend, the reader is referred to the web version of this article.)

5. ULTRA-LONG MODE-LOCKED FIBER LASERS: DISCUSSION

In the preceding sections, we have shown a great variety of generation regimes in NPE mode-locked fiber lasers instantiated by different settings of intra-cavity polarization controllers. By adjusting these settings, either manually or through automated actuators, one can achieve switching between generation regimes and, consequently, significant changes in the generated laser pulse parameters. However, fiber lasers offer another important degree of freedom that can be used for designing novel generation regimes: resonator length. As was demonstrated by multiple groups [25], [26], [27], [28], [29], [30],[31], [32] and [33], the cavity length has a substantial impact on the output pulse parameters in mode-locked fiber lasers. Specifically, it is possible to control the pulse energy and repetition rate by changing the laser cavity. Indeed, the pulse repetition rateν_{rep} is inversely proportional to the laser cavity length L, $\nu_{rep} \sim c/(n \times L)$, where c/n is the speed of light inside the fiber cavity. The energy of a generated single pulse E at constant average output power P_{av} is inversely proportional to the pulse repetition rate, and, therefore, directly proportional to the cavity length: $E \sim (P_{av}/\nu_{rep}) \sim (P_{av} \times L \times n/c)$. Hence, it is possible to proportionally increase the pulse energy and reduce the repetition rate by laser cavity elongation.

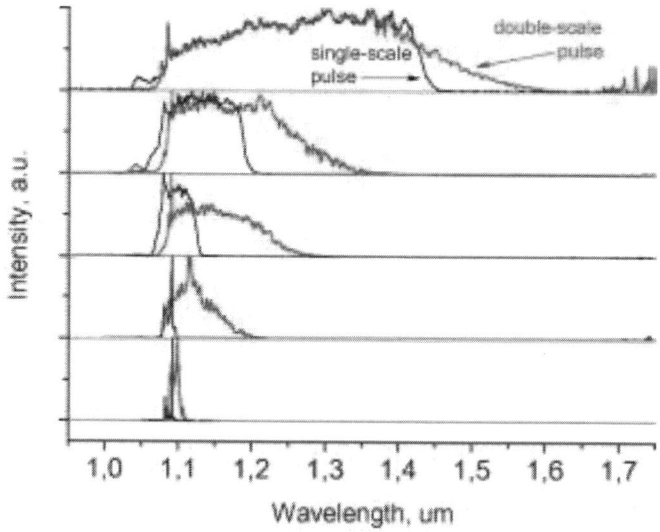

Figure. 6. Raman radiation spectra from 1.2-km long phosphosilicate fiber pumped with single-scale (black curves) and double-scale (red curves) pulses. (For interpretation of the references to colour in this figure legend, the reader is referred to the web version of this article.)

It would be pertinent to note that this approach was earlier used in solid-state lasers [34]. However, for an open cavity containing discrete elements, the practical possibility of elongation is limited to approximately 100 m. Fiber lasers present a remarkable technical opportunity to increase length of resonant cavities. For instance, earlier publications reported passive mode-locking in 400 m long fiber lasers [35] and [36]. An important breakthrough in this field dates back to 2008 when mode-locked operation of a fiber laser with a 3.8 km cavity was achieved due to the effect of non-linear polarization evolution of radiation [37]. At that time, the record high for mode-locked master oscillators pulse energy of 3.9 μJ was reported at a record low for this type of laser's pulse repetition rate of 77 kHz. The latter parameter was further reduced to 37 kHz [38] and to ~8 kHz [39], even though no improvement in pulse energy was achieved. Notwithstanding the respective elongation of the laser cavity to 8 and 25 km, the pulse energy remained unchanged at around 4 μJ. The existence of some upper limit to the pulse energy when the cavity of a mode-locked master oscillator is increased is the subject of the active current research [31] and is one of the field's important research challenges.

From the first experimental demonstration of high-energy pulse generation in ultra-long NPE mode-locked master oscillators [37], many research groups studied mode-locked fiber lasers with long and ultra-long resonators of different configurations. They demonstrated the possibility of passive mode locking in long cavities due to saturable absorbers, such as single-wall carbon nanotubes [25], [40] and [41], semiconductor saturable absorber mirrors [42], [43] and [44], and non-linear optical loop mirrors [45]. It was experimentally and theoretically confirmed that passively mode-locked fiber master oscillators give rise to a broad variety of generation regimes in different cavity configurations [46], [47], [48], [49], [50], [51], [52] and [53], including generation of single and multiple pulses over a cavity round-trip, pulses with various duration, envelope shape, and spectrum width. The results of Refs [30] and [52] show that in ultra-long lasers, there is high probability of generating double-scale partially coherent pulses, whereas "normal" (single-scale) pulses with a smooth temporal optical phase shape and large chirp become difficult to obtain. This indicates a step difference from the traditional mode-locking theory and may be considered as a partial mode-locking, when phase locking is not between all resonator modes, but rather between groups/clusters of modes. Nevertheless, experimentally generating strongly chirped pulses in a 1 km laser cavity with mode-locking due to a carbon nanotube saturable absorber was reported in[25] and [40]. The paper [28] reports lasing in 1.4 km long Yb-doped fiber laser with steep spectral edges, which are attributed to a "conventional", stable, single-pulse generation regime [18] and [19].

Therefore, in principle, mode-locked fiber master oscillators with long and ultra-long cavities may be used to generate pulses with relatively high energies without any additional amplification or Q-switching, at the same time as they exhibit a broader variety of operation regimes compared to lasers with "short" resonators (on the order of several meters). In regimes with one pulse over the cavity round-trip, mode-locked fiber master oscillators can produce both fully and partially coherent pulses (double-scale wave-packets). Such systems require further analysis and numerical modeling.

The variety of generation regimes in mode-locked fiber master oscillators with long and ultra-long cavities, as well as significant variability of parameters of single-pulses/wave-packets produced therein, are of great interest both for fundamental research in laser physics and for practical applications of laser pulses with new unique parameters.

6. CONCLUSIONS

Our simulation and experiments demonstrate considerable spread of pulse parameters generated by fiber lasers mode-locked due to NPE. This finding should encourage users to carefully control at least the duration and optical spectrum of laser pulses, since the pulse parameters may change considerably, for example, due to relaxation of stress introduced by fiber polarization controllers, equivalent to a change in PC settings. This is of primary importance in the case of long and ultra-long fiber lasers, which are characterized by substantial length of non-linear interactions, which lead to considerable amplification of perturbations and make such lasers highly prone to undesired irregular regime hops. Note also that NPE mode-locked fiber lasers are relatively sensitive to ambient conditions because fiber birefringence depends on temperature, mechanical stress, and so on. Uncontrollable fiber birefringence changes make it necessary to adjust polarization controllers in order to trigger mode-locked operation or to stabilize it. Adjustment, however, may lead to the start of a mode-locked regime substantially different from the one before the adjustment.

Another important point of our study is that various lasing regimes should be carefully chosen in order to optimally meet the requirements of a particular laser application. Using second harmonic generation and Raman frequency conversion as a test bed, we have shown that different lasing regimes' efficiency may vary considerably even if laser pulses generated in different regimes have similar energies and duration.

The technological remedy (for example, in the form of control and adjustment elements) for the discussed output pulse characteristic variability should be incorporated into laser system designs.

ACKNOWLEDGMENTS

This work was supported by the Marie Curie International Exchange Scheme, Research Executive Agency Grant "TelaSens" No 269271; ERC project ULTRALASER; Grants of Ministry of Education and Science of the Russian Federation (agreement No. 14.B25.31.0003; basic research order No. ZN-06-14 (2419); order No. 3.162.2014/K); Russian President Grant

MK-4683.2013.2; Grant of SKOLKOVO Foundation (G-13-131); Council of the Russian President for the Leading Research Groups (project No. NSh-4447.2014.2).

REFERENCES

1. L.E. Nelson, D.J. Jones, K. Tamura, H.A. Haus, E.P. Ippen, Ultrashort-pulse fiber ring lasers, Appl. Phys. B 65 (1997) 277–294.

2. A. Tünnermann, J. Limpert, S. Nolte, Ultrashort pulse fiber lasers and amplifiers, in: F. Dausinger, F. Lichtner, H. Lubatschowski (Eds.), Femtosecond Technology for Technical and Medical Applications, Topics in Applied Physics, vol. 96, Springer-Verlag, Berlin, Heidelberg, 2004, pp. 35–54.

3. F.W. Wise, A. Chong, W.H. Renninger, High-energy femtosecond fiber lasers based on pulse propagation at normal dispersion, Laser Photonics Rev. 2 (2008) 58–73.

4. J. Hecht, Fiber lasers bring femtoseconds to the masses, Laser Focus 1, 2010, http://www.laserfocusworld.com/articles/2010/01/fiber-lasers-bring.html.

5. O.G. Okhotnikov (Ed.), Fiber Lasers, Wiley-VCH Verlag, Weinheim, 2012.

6. M.E. Fermann, Mode locked fiber lasers, past, present and future. CLEO Technical Digest, OSA, 2012, CTu1I.5.pdf http://wr.lib. tsinghua. edu.cn/sites/default/files/CTu1I.5.pdf.

7. V.J. Matsas, T.P. Newson, D.J. Richardson, D.N. Payne, Self-starting, passively mode-locked fiber ring soliton laser exploiting non-linear polarization rotation, Electron. Lett. 28 (1992) 1391–1393.

8. M. Hofer, M.H. Ober, F. Haberl, M.E. Fermann, Characterization of ultrashort pulse formation in passively mode-locked fiber lasers, IEEE J. Quantum Electron. 28 (1992) 720–728.

9. F.Ö. Ilday, J.R. Buckley, H. Lim, F.W. Wise, W.G. Clark, Generation of 50-fs, 5-nJ pulses at 1.03 lm from a wave-breaking-free fiber laser, Opt. Lett. 28 (2003) 1365–1367.

10. F. Ilday, J. Buckley, L. Kuznetsova, F. Wise, Generation of 36-femtosecond pulses from a ytterbium fiber laser, Opt. Express 11 (2003) 3550–3554.

11. P. Grelu, N. Akhmediev, Dissipative solitons for mode-locked lasers, Nat. Photonics 6 (2012) 84–92.

12. L.M. Zhao, D.Y. Tang, T.H. Cheng, C. Lu, Nanosecond square pulse generation in fiber lasers with normal dispersion, Opt. Commun. 272 (2007) 431–434.

13. B. Ortaç, A. Hideur, M. Brunel, C. Chédot, J. Limpert, A. Tünnermann, F.Ö. Ilday, Generation of parabolic bound pulses from a Yb-fiber laser, Opt. Express 14 (2006) 6075–6083.

14. J.M. Soto-Crespo, P. Grelu, N. Akhmediev, N. Devine, Soliton complexes in dissipative systems: vibrating, shaking, and mixed soliton pairs, Phys. Rev. E 75 (2007) 016613.

15. W.H. Renninger, A. Chong, F.W. Wise, Pulse shaping and evolution in normaldispersion mode-locked fiber lasers, IEEE J. Sel. Top. Quantum Electron. 18 (2012) 389–398.

16. Ph. Grelu, J.M. Soto-Crespo, Temporal soliton "molecules" in mode-locked lasers: collisions, pulsations, and vibrations, Lect. Notes Phys. 751 (2008) 137–173.

17. S. Chouli, P. Grelu, Rains of solitons in a fiber laser, Opt. Express 17 (2009) 11776–11781.

18. S. Smirnov, S. Kobtsev, S. Kukarin, A. Ivanenko, Three key regimes of single pulse generation per round trip of all-normal-dispersion fiber lasers mode-locked with nonlinear polarization rotation, Opt. Express 20 (2012) 27447–27453.

19. S. Kobtsev, S. Kukarin, S. Smirnov, S. Turitsyn, A. Latkin, Generation of doublescale femto/pico-second optical lumps in mode-locked fiber lasers, Opt. Express 17 (2009) 20707–20713.

20. S. Smirnov, S. Kobtsev, S. Kukarin, Efficiency of non-linear frequency conversion of double-scale pico-femtosecond pulses of passively modelocked fiber laser, Opt. Express 22 (2014) 1058–1064.

21. G.P. Agrawal, Nonlinear fiber optics, third ed., Academic Press, 2001.

22. M. Horowitz, Y. Barad, Y. Silberberg, Noise-like pulses with a broadband spectrum generated from an erbium-doped fiber laser, Opt. Lett. 22 (1997) 799–801.

23. O. Pottiez, R. Grajales-Coutiño, B. Ibarra-Escamilla, E.A. Kuzin, J.C. Hernández-García, Adjustable noise-like pulses from a figure-eight fiber laser, Appl. Opt. 50 (2011) E24–E31.

24. B. Nie, G. Parker, V.V. Lozovoy, M. Dantus, Energy scaling of Yb fiber oscillator producing clusters of femtosecond pulses, Opt. Eng. 53 (2014) 051505.

25. E.J.R. Kelleher, J.C. Travers, E.P. Ippen, Z. Sun, A.C. Ferrari, S.V. Popov, J.R. Taylor, Generation and direct measurement of giant chirp in a passively mode-locked laser, Opt. Lett. 34 (2009) 3526–3528.

26. S.V. Smirnov, S.M. Kobtsev, S.V. Kukarin, S.K. Turitsyn, Mode-locked fiber lasers with high-energy pulses, in: K. Jakubczak (Ed.), Laser Systems for Applications, InTech, Rijeka, Croatia, 2011, pp. 39–58.

27. S. Kobtsev et al. / Optical Fiber Technology 20 (2014) 615–620 619

28. X. Li, X. Liu, X. Hu, L. Wang, H. Lu, Y. Wang, W. Zhao, Long-cavity passively mode-locked fiber ring laser with high-energy rectangular-shape pulses in anomalous dispersion regime, Opt. Lett. 35 (2010) 3249–3251.

29. L.J. Kong, X.S. Xiao, C.X. Yang, Low-repetition-rate all-fiber all-normaldispersion Yb-doped mode-locked fiber laser, Laser Phys. Lett. 7 (2010) 359.

30. N. Li, J. Xue, C. Ouyang, K. Wu, J.H. Wong, S. Aditya, P.P. Shum, Cavity-length optimization for high energy pulse generation in a long cavity passively modelocked all-fiber ring laser, Appl. Opt. 51 (2012) 3726–3730.

31. I.A. Yarutkina, O.V. Shtyrina, M.P. Fedoruk, S.K. Turitsyn, Numerical modeling of fiber lasers with long and ultra-long ring cavity, Opt. Express 21 (2013) 12942–12950.

32. C. Aguergaray, A. Runge, M. Erkintalo, N.G. Broderick, Raman-driven destabilization of mode-locked long cavity fiber lasers: fundamental limitations to energy scalability, Opt. Lett. 38 (2013) 2644–2646.

33. T. Liu, D. Jia, J. Yang, J. Chen, Z. Wang, T. Yang, An ultra-long cavity passively mode-locked fiber laser based on nonlinear polarization rotation in a semiconductor optical amplifier, Laser Phys. 23 (2013) 095005.

34. E.J.R. Kelleher, J.C. Travers, Chirped pulse formation dynamics in ultra-long mode-locked fiber lasers, Opt. Lett. 39 (2014) 1398–1401.

35. V.Z. Kolev, M.J. Lederer, B. Luther-Davies, A.V. Rode, Passive mode locking of a Nd:YVO4 laser with an extra-long optical resonator, Opt. Lett. 28 (2003) 1275–1277.

36. J.U. Kang Jr., R. Posey, Demonstration of supercontinuum generation in a longcavity fiber ring laser, Opt. Lett. 23 (1998) 1375–1377.

37. K.H. Fong, S.Y. Kim, K. Kazuro, H. Yaguchi, S.Y. Set, Generation of low-repetition rate high-energy picosecond pulses from a single-wall carbon nanotube modelocked fiber laser, Optical Amplifiers and their Applications Conference (OAA 2006), Whistler, British Columbia, Canada, OMD4, June 2006.

38. S. Kobtsev, S. Kukarin, Y. Fedotov, Ultra-low repetition rate mode-locked fiber laser with high-energy pulses, Opt. Express 16 (2008) 21936–21941.

39. S.M. Kobtsev, S.V. Kukarin, S.V. Smirnov, Y.S. Fedotov, High-energy modelocked all-fiber laser with ultralong resonator, Laser Phys. 20 (2010) 351–356.

40. A. Ivanenko, S. Turitsyn, S. Kobsev, M. Dubov, Mode-locking in 25-km fiber zlaser, European Conference on Optical Communication (ECOC), 1–3, (2010).

41. J.R. Kelleher, J.C. Travers, Z. Sun, A.G. Rozhin, A.C. Ferrari, S.V. Popov, J.R. Taylor, Nanosecond-pulse fiber lasers mode-locked with nanotubes, Appl. Phys. Lett. 95 (2009) 111108.

42. Y. Senoo, N. Nishizawa, Y. Sakakibara, K. Sumimura, E. Itoga, H. Kataura, K. Itoh, Ultralow-repetition-rate, high-energy, polarization-maintaining, Er-doped, ultrashort-pulse fiber laser using single-wall-carbon-nanotube saturable absorber, Opt. Express 18 (2010) 20673–20680.

43. L. Chen, M. Zhang, C. Zhou, Y. Cai, L. Ren, Z. Zhang, Ultra-low repetition rate linear-cavity erbium-doped fiber laser modelocked with semiconductor saturable absorber mirror, El. Lett. 45 (2009) 731–733.

44. X. Tian, M. Tang, P.P. Shum, Y. Gong, C. Lin, S. Fu, T. Zhang, High-energy laser pulse with a submegahertz repetition rate from a passively mode-locked fiber laser, Opt. Lett. 34 (2009) 1432–1434.

45. M. Zhang, L.L. Chen, C. Zhou, Y. Cai, L. Ren, Z.G. Zhang, Mode-locked ytterbiumdoped linear-cavity fiber laser operated at low repetition rate, Laser Phys. Lett. 6 (2009) 657.

46. F. Ai, Z. Cao, X. Zhang, C. Zhang, B. Zhang, B. Yu, Passively mode-locked fiber laser with kilohertz magnitude repetition rate and tunable pulse width, Opt. Laser Technol. 43 (2011) 501–505.

47. A. Zaviyalov, O. Egorov, R. Iliew, F. Lederer, Rogue waves in mode-locked fiber lasers, Phys. Rev. A 85 (2012) 013828.

48. M. Baumgartl, J. Abreu-Afonso, A. Díez, M. Rothhardt, J. Limpert, A. Tünnermann, Environmentally stable picosecond Yb fiber laser with low repetition rate, Appl. Phys. B 111 (2013) 39–43.

49. B.N. Nyushkov, V.I. Denisov, S.M. Kobtsev, V.S. Pivtsov, N.A. Kolyada, A.V. Ivanenko, S.K. Turitsyn, Generation of 1.7-uJ pulses at 1.55 um by a self-modelocked all-fiber laser with a kilometers-long linear-ring cavity, Laser Phys. Lett. 7 (2010) 661–665.

50. B.N. Nyushkov, A.V. Ivanenko, S.M. Kobtsev, S.K. Turitsyn, C. Mou, L. Zhang, V.I. Denisov, V.S. Pivtsov, Gamma-shaped long-cavity normal-dispersion modelocked Er-fiber laser for sub-nanosecond high-energy pulsed generation, Laser Phys. Lett. 9 (2012) 59–67.

51. J.L. Dong, W.C. Xu, Z.C. Luo, A.P. Luo, H.Y. Wang, W.J. Cao, L.Y. Wang, Tunable and switchable dual-wavelength passively mode-locked fiber ring laser with high-energy pulses at a sub-100 kHz repetition rate, Opt. Commun. 284 (2011) 5719–5722.

52. P. Grelu, W. Chang, A. Ankiewicz, J.M. Soto-Crespo, N. Akhmediev, Dissipative soliton resonance as a guideline for high-energy pulse laser oscillators, J. Opt. Soc. Am. B 27 (2010) 2336–2341.

53. S.M. Kobtsev, S.V. Smirnov, Fiber lasers mode-locked due to nonlinear polarization evolution: golden mean of cavity length, Laser Phys. 21 (2011) 272–276.

54. S.K. Turitsyn, Theory of energy evolution in laser resonators with saturated gain and non-saturated loss, Opt. Express 17 (2009) 11898.

5

MODIFIED POWDER-IN-TUBE TECHNIQUE BASED ON THE CONSOLIDATION PROCESSING OF POWDER MATERIALS FOR FABRICATING SPECIALTY OPTICAL FIBERS

Jean-Louis Auguste [1,*], Georges Humbert [1], Stéphanie Leparmentier[1], Maryna Kudinova [1], Pierre-Olivier Martin [1], Gaëlle Delaizir [2], Kay Schuster [3] and Doris Litzkendorf [3]

[1] Xlim Research Institute, UMR 7252 CNRS/University of Limoges, 123 avenue Albert THOMAS, 87060 LIMOGES, France

[2] SPCTS, UMR 7315 CNRS/University of Limoges, European Ceramic Center, 12 rue Atlantis, 87068 LIMOGES, France

[3] Leibnitz Institute of Photonic Technology IPHT, Albert-Einstein-Str. 9, 07745 Jena, Germany

ABSTRACT

The objective of this paper is to demonstrate the interest of a consolidation process associated with the powder-in-tube technique in order to fabricate a long length of specialty optical fibers. This so-called Modified Powder-in-Tube (MPIT) process is very flexible and paves the way to multimaterial optical fiber fabrications with different core and cladding glassy materials. Another feature of this technique lies in the sintering of the preform under reducing or oxidizing atmosphere. The fabrication of such optical fibers implies different constraints that we have to deal with, namely chemical species diffusion or mechanical stress due to the mismatches between thermal expansion coefficients and working temperatures of the fiber materials. This paper focuses on preliminary results obtained with a lanthano-aluminosilicate glass used as the core material for the fabrication of all-glass fibers or specialty Photonic Crystal Fibers (PCFs). To complete the panel of original microstructures now available by the MPIT technique, we also present several optical fibers in which metallic particles or microwires are included into a silica-based matrix.

KEYWORDS

powder in tube; process; glass; metal; oxidation state; multimaterial optical fiber; species diffusion; specialty optical fiber

1. INTRODUCTION

Standard silica singlemode optical fibers are widely used in many applications, such as optical components, fiber sensors, non-linear optical devices, and of course, for long-haul telecommunications.In the 1970s, research efforts were mainly focused on the development of efficient processes for elaborating highly transparent materials and for fabricating optical fibers with attenuation coefficient as low as possible in the telecom wavelength bands.

Chemical Vapor Deposition (CVD) techniques and synthesized-silica glass became the main actors of optical fiber success, with an attenuation coefficient record as low as 0.15 dB/km at the 1550 nm wavelength (PureBand® Sumitomo, Japan). In the 1990s, rare-earth doped silica based optical fiber was an important breakthrough, offering the ability to amplify light without any optic/electronic converter and to develop optical fiber lasers.

Another breakthrough occurred in 1996 [1,2,3] with the introduction of the concept of Photonic Crystal Fiber, which is composed of a silica-core surrounded by a two dimensional (2D) periodic pattern of air-holes (running along the fiber length). The following decade was mainly devoted to the development of original PCF topologies for specific applications. For example, highly dispersive PCF at the telecom band (with chromatic dispersion lower than −1000 ps/(nm.km) at 1550 nm) [4] were presented as corrective optical devices for long-length data transmission applications. All-solid or hollow-core PCFs were also studied as active media for generating non-linear effects or as passive components for gas sensors, for example.

The slump in the demand on the telecommunication market in the last 15 years has initiated the use of optical fibers in other application domains such as optical sensors, optical fiber lasers, nonlinear devices, and high power delivery cables, requiring shorter fiber lengths (often limited to a few meters). The constraint on the optical losses has then been released, providing access to the use of other optical glasses with specific properties (e.g., higher linear and nonlinear refractive indices) but with a larger attenuation coefficient (than a pure silica one). New fiber fabrication techniques have also been proposed or adapted from previous ones. For example, the core suction method that consists in sucking up molten lead-SF6 glass (Schott) into a silica tube for realizing the fiber preform enables the fabrication of short length of optical fiber with soft-glass materials (glasses with lower transition point temperature (T_G) than silica) [5]. A similar process was developed for inserting (along few centimeters) soft-glass or metal into an optical fiber by applying an external pressure on molten materials for filling the air holes of a PCF [6].

Others groups in the academic world have also worked on optical fibers composed of multimaterials. Ballato and colleagues (at Clemson University, United States) have reused the rod-in-tube technique for fabricating specialty optical fibers from a rod of a specific material (such as Silicon) filled in a silica tube [7]. They have also used the powder-in-tube method that consists in filling a tube with a powder material and drawing directly this preform down to an optical fiber [8]. This method, developed in the 1970s [9], appears to be a nice and simple way to produce optical fibers composed of multimaterials. Recently, Silitec Fibers SA proposed and patented [10] an extension of this concept for producing single-mode optical fibers by replacing the thick and expensive overcladding silica tube of the fiber preform (fabricated by MCVD) with synthetic pure silica powder, leading to a large reduction of the fabrication costs.

This industrial development was achieved by adding an intermediate process stage that consists in consolidating the preform filled with powder, for enabling stable and long fiber length productions with high reproducibility. A few years later (in 2007), we started to work at the Xlim research institute on the development of multimaterials optical fibers. In this context, in strong collaboration with Silitec Fibers SA, we have developed an original consolidation unit for improving the fabrication of multimaterials optical fibers with the powder-in-tube technique. The principles of operations of this Modified Powder-in-Tube (MPIT) technique, its advantages and fabrication constraints, are presented in Section 2. Examples of specialty optical fibers fabricated with MPIT technique are presented in Section 3 in order to demonstrate its potentialities for developing optical fibers composed of different glasses or of silica and metal.

2. THE POWDER CONSOLIDATION PROCESS

2.1. Description of the Fabrication Process

In the last 20 years, the use of powders in optical fiber preforms has demonstrated interesting results for drawing state-of-the-art specialty optical fibers. Glass powders can be used to form the core and/or the optical cladding of the preform. Powder based processes present several advantages compared to conventional techniques like MCVD, extrusion or rod-in-tube techniques. It offers large freedoms in the fiber designs allowing the realizations of symmetric or asymmetric fiber topologies, composed of two or more different material compositions, each material inserted somewhere into the fiber structure. Furthermore, it can be easily combined with the stack-and-draw process [1] enabling the association of the properties of powder materials with PCF ones. Most of all, it allows for the fabrication of optical fibers from a large variety of optical materials (glasses, ceramics, metals) reduced to powders, in contrast with CVD techniques that are limited by the availability of gas precursors and to their low doping concentrations into silica glass.

Even though this fabrication process can be easily described—basically, it consists in drawing down to fiber (with a diameter of few hundred of micrometers) a glass tube filled with powder(s)—all the phenomena involved during the fabrication are complex and difficult to anticipate theoretically. An important phenomenon to manage during the drawing stage is the air trapped between the powder particles. During the high temperature processing like the preform drawing stage, air could expand suddenly causing severe perturbations such as formation of uncontrolled air holes into the fiber core or cladding, leading to large propagation losses, or as in the worst case, mechanical break of the fiber during its fabrication.

The size distribution and packing density (solid filling ratio in the porous material) of powder particles are parameters of great importance that determine the air fraction in the powdered preform. In order to reduce the porosity of the powder-based preforms, we propose a compaction heat-treatment of the preform under low pressure at a temperature above the transition temperature (T_G) of the glass powder material.

- For $T_G < T < T_{SOFT}$ (with T_{SOFT} the glass softening temperature), the glass grains initiate a molten state and are able to deform. Primary vacuum ensures slight air elimination and granulates tend to agglomerate. After cooling, the final aspect of the preform is a spongious-like solid rod in the surrounding glass tube. This process is called preform consolidation.

- For $T_{SOFT} < T$, the glass powder material is softening and air can evacuate by natural bubbles collapsing that is enhanced by vacuum pumping. After cooling, the preform is a dense glassy rod. This process is called preform vitrification (see Figure 1b,c).

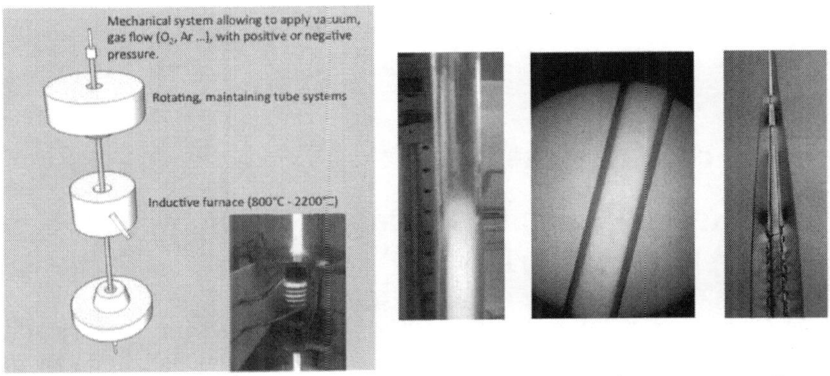

(a) (b) (c) (d)

Figure 1. (**a**) Scheme of the Consolidation Unit; (**b**) photograph picture of a pure silica powder vitrified only in the half top. Pictures of pure silica powder based preform under observation with a polariscope; (**c**) after vitrification process showing absence of large defects and mechanical stress; (**d**) without consolidation (remaining part after the fiber drawing stage) showing cracks and large air inclusions in the core.

In this way, we are able to produce all-solid preforms from powdered raw materials without formation of air bubbles or cracks during the drawing stage.

Such processes are realized by using specific equipment named "Consolidation Unit", developed at the XLIM Research Institute in collaboration with the Silitec Fibers SA company, Boudry, Switzerland (Figure 1a). In this equipment, both ends of the preform are hermetically sealed and connected to a vacuum pump or gas mass-flows for pressurizing it with helium or oxygen gas. Gases pressures or depressions are either controlled from the top or the bottom of the preform. The preform is held vertically with rotary mandrels. Preform rotation is applied to large diameter preforms for keeping homogeneous consolidation processing in the whole preform cross section. An inductive furnace (with a heating zone of 7 cm) is moved up and down along the preform. All parameters of the consolidation process are controlled by dedicated software.

In conclusion, the powder-in-tube technique could be significantly improved by a consolidation processing since it enables (i) a vitrification of the glass powder for removing air trapped between particles and for allowing stable, reproducible fiber drawings; (ii) the use of specific gases or gas mixtures during the preform heat-treatment for cleaning or for oxidizing or reducing materials into the perform; (iii) the realization of new fiber designs by implementing rods and tubes anywhere in the material powder (with or without a triangular arrangement).

2.2. Processing Features

2.2.1. Thermo-Mechanical Property Mismatches between Fiber Materials

Optical fibers are made of at least with two different core/cladding materials. That the thermo-mechanical properties are mismatched between these two materials is an important characteristic to take into account for the consolidation process and the fiber drawing stage. Indeed, it may involve formation of large mechanical stresses at the fiber core/cladding interface leading to large optical losses, uncontrolled birefringence and mechanical fiber brittleness. This phenomenon is lowered by using materials with a similar thermal expansion coefficient (α) and liquidus (for crystals) or transition (for glasses) temperatures.

The thermo-mechanical stress (σ) induced in an optical fiber composed of disparate core and cladding glass materials could be estimated at a temperature T with the following Equation [11]:

$$\sigma = \frac{E\,(T - T_G)(\alpha_1 - \alpha_2)}{1 - \mu}\left(\frac{r}{R}\right)^2$$

where the indices 1 and 2 correspond respectively to the fiber core and cladding, μ and E are the Poisson ratio and Young's modulus of the glass in the core, and r/R is the core/fiber radius ratio. The ultimate compressive strength of pure silica is 1.1 GPa, *i.e.*, silica breaks when a force higher than this value is applied on it. Additional stresses on the fiber structure may also appear during the drawing process, due to technological defects or glass quenching.

Therefore, the consolidation process of glass powder preforms (before the drawing stage) is suitable for fabricating optical fibers composed of multimaterials with different thermo-mechanical properties:

- In the case of a low T_G mismatch between the glass core and pure silica cladding ($T_G(SiO_2)$ ~ 1200 °C), the consolidation of the powder leads to rearrangements of macroscopic defects (reduction of the air fraction, annealing of glass powder materials...).

- In the case of a large T_G mismatch between the glass core and pure silica cladding, the consolidation process leads to an all-solid preform by softening only the core glass material.

To test and demonstrate the interest of the consolidation process we have fabricated several multimaterial optical fibers with different thermo-mechanical properties. These fibers were fabricated by filling a silica tube with different glass powders.

A lanthano-aluminosilicate (SAL) glass developed at the IPHT Institute in Jena, (Germany), with the following composition: 82% SiO_2-11.6% Al_2O_3-6.4% La_2O_3 was first used as the core material [12]. This glass presents a T_G ~ 900 °C and $\alpha \sim 2.8 \times 10^{-6}$ K^{-1} (7 times higher than pure silica). Tens of meters of fiber long length have been successfully fabricated (Figure 2a). An optical fiber with a core composed of a lead-silicate optical glass, for which T_G ~ 600 °C (half compared to silica T_G) and $\alpha \sim 10^{-5}$ K^{-1} (20 times higher than pure silica) has also been fabricated (Figure 2b). Figure 2c shows the cross section of fiber composed of borosilicate core (T_g~ 560 °C) and a silica cladding.

To emphasis further the suitability of the MPIT technology for fabricating specialty optical fibers composed of glasses with different thermo-mechanical properties, we have realized a fiber composed of a tellurite glass core and a borosilicate cladding. Borosilicate is used to reduce the thermo-mechanical mismatch induced by the tellurite glass T_g (350 °C). Using borosilicate tubes for the preform preparation is a suitable way for fabricating optical fibers in which the core material has an important optical interest but presents a very low T_G compared to pure silica.

2.2.2. Thermal Consolidation under Controlled Reducing or Oxidizing Atmospheres

The consolidation process enables to monitor air or gas pressure during the process. This is also useful to initiate redox chemical reactions into the preform structure. The oxidation state of materials can be controlled and changed during the heat-treatment process.

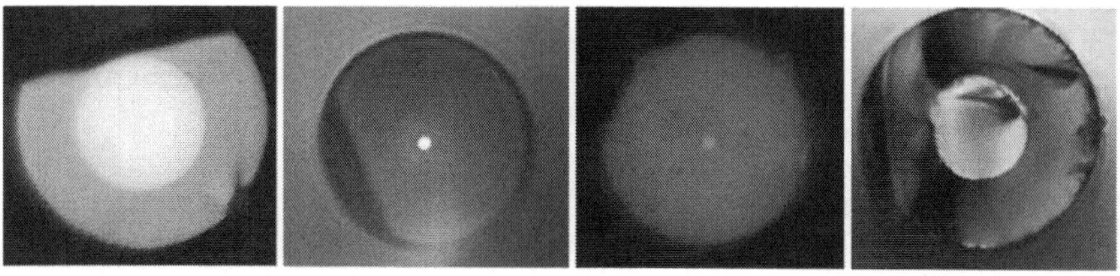

Figure 2. Microscope pictures of multimaterial optical fibers realized by the Modified Powder-in-Tube (MPIT) technique with (**a**) large SAL glass core and pure silica cladding; (**b**) lead-silicate glass core and pure silica cladding; (**c**) a borosilicate glass core surrounded by pure silica cladding; (**d**) Tellurite core material and borosilicate cladding.

We have exploited this functionality for realizing specialty optical fibers with metallic palladium (Pd) particles embedded into the silica glass cladding [13]. Such optical fibers are difficult to fabricate with more conventional techniques such as rod-in-tube or CVD methods, since they are limited by the metal concentration range and the materials or precursors availability.

Preforms were realized by using pure SiO_2 and PdO powders mixture for the cladding material. The vitrification stage was performed under primary vacuum at a temperature close to the silica softening temperature. Scanning Electron Microscopy (SEM) observations (*Cf.* Figure 3a) coupled with EDS elemental analysis (Energy Dispersive X-ray Spectroscopy) and XRD (X-Ray Diffraction) measurements (BRUKER D8 equipment, Bruker, Rheinstetten, Germany) (*Cf.* Figure 3b) on fabricated samples demonstrate total reduction of the palladium oxide (Pd^{2+}) to metallic palladium particles (Pd^0) into the silica cladding material during the preform consolidation process.

Palladium aggregates formations are observed on SEM pictures in the silica cladding. Nevertheless, the mean metallic particles diameter was estimated (from the XRD diagram by using the Debye-Scherrer approximation) to be 60 and 40 nm (with some aggregates as shown in Figure 3a) for initial PdO concentration of 1% mol or 0.5% mol in the silica powder, respectively. The size of the metal particles can be influenced by the initial PdO concentration as well as the dwell time and temperature of the thermal reduction process.

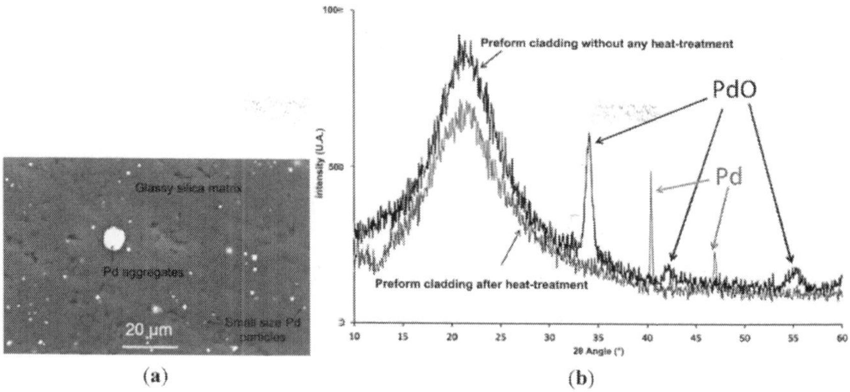

Figure 3. (**a**) SEM observation of the cross-section of a preform after heat-treatment, observation in the SiO_2-Pd cladding with initially 1% mol PdO; (**b**) comparison of XRD diagrams for non-heat-treated and heat-treated 99% SiO_2 + 1% PdO (% mol) composite materials (the large pic centered to $2\theta = 20°–22°$ is due to the silica amorphous network and is typical to all glass structures).

2.2.3. Chemical Diffusion between the Core/Cladding Materials

An important effect that can occur during the preform consolidation and the fiber drawing is the diffusion of species.

As an example, hydrogen diffusion through silica has been intensively investigated due to its negative impact on optical losses in the telecom bands or its positive effect for increasing the germanium-doped core sensitivity to UV laser excitation (for Bragg grating inscription) [14]. Diffusion effects have also been observed in optical fiber with high GeO_2 doped silica core [15,16]. As species diffusion occurs along their concentration gradient, generating in the molten zone of the preform during the drawing process diffusion, from the core to the cladding, while oxygen and silicon atoms migrate from the cladding to the core. Spontaneous diffusion of phosphorus from the cladding to the core has also been demonstrated, even during the preform preparation by the MCVD method.

Diffusion of species has also been reported in optical fibers fabricated by the rod-in-tube techniques, *i.e.*, in fibers with much larger concentration gradients. As an example, during the realization of crystalline-core/glass-cladding optical fibers [17], the crystallinity of pure or doped YAG (Yttrium Aluminium Garnet) rods was used as the core material vanished during the fiber drawing stage. The authors have also noticed the influence of the fiber core diameter on the diffusion magnitude, demonstrating that smaller core diameter leads to larger diffusion of silicon (*i.e.*, silica) from the cladding to the core.

It is therefore essential to study chemical diffusion phenomenon in multimaterial optical fibers fabricated by the MPIT technique. Optical fibers composed of a SAL core and silica cladding have been fabricated with different core diameters for investigating species diffusion.

The refractive index profiles (RIP) of the fabricated optical fibers were measured by using an EXFO NR-9200 Optical Fiber Analyzer (Quebec, QC, Canada, based on the refractive near field technique) working at the 667.94 nm wavelength, with a spatial resolution of 0.5 μm and a measurement accuracy of 10^{-4}.

For each fiber, the refractive index profile of the core presents a gradient shape, although the preform has a step-index shape indicating that migration of species has occurred during the drawing stage. One preform was drawn down to different diameters for investigating the core diameter influence on the diffusion phenomenon. As expected, the amplitude of the refractive index gradient decreases with the fiber core diameter, as shown in Figure 4a,b illustrates the maximum of the core/cladding refractive index difference value *versus* the core radius for all the optical fibers fabricated from different preforms. These results emphasize and confirm that smaller core leads to lower amplitude of the core/cladding refractive index gradient.

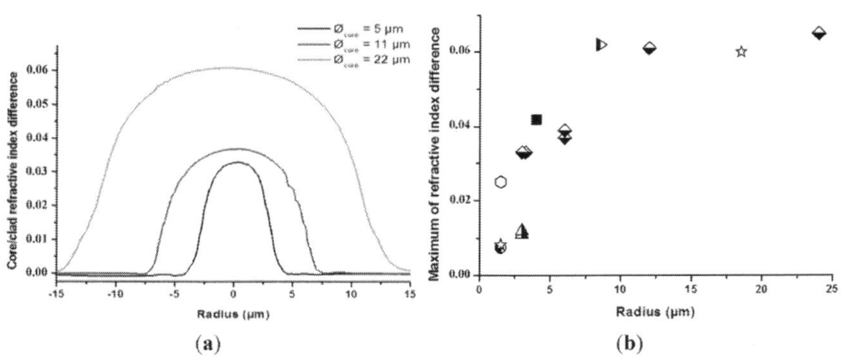

Figure 4. (**a**) Refractive index profiles for some SAL/silica fibers with different core diameters but realized during the same drawing run; (**b**) influence of the SAL core radius on the maximum of the core/cladding refractive index difference. Different forms of points correspond to different drawing campaigns.

In order to understand the origin of the gradient index shape of the SAL/silica fibers RIP, atomic concentrations of silicon, aluminum and lanthanum were investigated by EDS elemental analysis measurements on several fibers (with a core diameter of 6 μm or smaller). Measurements have been realized at different locations on the cross section of the fiber core.Figure 5 presents the atomic concentrations plotted *versus* the distance (l) from the cladding/core interface, *i.e.*, l = 0 at the core/cladding interface and l = 3 μm at the fiber core center. These measurements clearly demonstrate the evolution of the Si, Al and La atomic concentrations in agreement with the gradient shape of the RIP of the optical fiber. Lanthanum and aluminum concentrations are larger at the core center than at the core-cladding interface leading to a higher refractive index value.

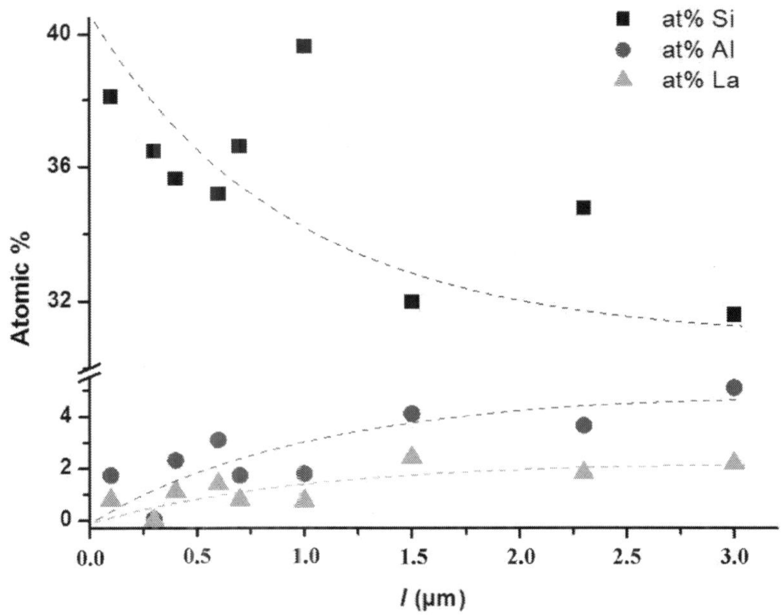

Figure 5. Atomic concentration of silicon, aluminum and lanthanum measured in a SAL core *versus* the distance ($l = r_{core} - r$) from the silica-cladding/SAL-core interface.

Furthermore, the initial atomic concentrations of the Si, Al and La species in the SAL glass used are 27.33%, 4.64% and 2.56%, respectively. It is worth noting that measured atomic concentration of Si is much larger than the theoretical one. This value is around 39% at the core-cladding interface and decreases to 32% in the core center. This behavior could be explained by the diffusion of silicon from the silica cladding to the SAL glass core. This is in agreement with reported observations on optical fibers fabricated by the rod-in-tube technique presented in [17]. In contrast, diffusion phenomena from the core to the cladding material are observed for lanthanum and aluminum by the reduction of their atomic concentration (but below theoretical values) at positions closer to the core-cladding interface. These behaviors are induced by the concentration gradient between the SAL glass core and the pure silica cladding. They also explained the shape of the RIP and the reduction of refractive index magnitude.

This study demonstrates that the diffusion of species in the fiber core and cladding materials are important phenomena which ones have to be considered for designing and fabricating specialty optical fibers by the MPIT technique.

3. EXAMPLES OF SPECIALTY OPTICAL FIBERS FABRICATED WITH THE MODIFIED POWDER IN TUBE TECHNIQUE

3.1. Lanthano-Aluminosilicate Glass Core/Silica Cladding Optical Fibers

This process is extremely well adapted for fabricating optical fibers composed of a core and a cladding with different optical materials. Typically, the cladding is formed by the tube (silica or borosilicate glass) and the core is realized by filling the tube with the selected optical material under the powder form.

Researchers at IPHT Institute in Jena (Germany) have intensively developed compositions in the SiO_2-Al_2O_3-La_2O_3 glass system (*i.e.*, SAL glasses) [12]. Lanthana is an interesting component for increasing linear and nonlinear refractive indices of glasses. Alumina is used to ensure lanthanum solubility into the vitreous matrix. Among several glass compositions studied, the composition 82%SiO_2-11.6%Al_2O_3-6.4%La_2O_3 has been used for fabricating optical fibers composed of a SAL core and a silica cladding. The linear and nonlinear refractive indices of this SAL glass are respectively n = 1.545 (at the 580 nm wavelength) and $n_2 = 5 \times 10^{-20}$ m^2/W (at the 1064 nm wavelength) allowing large core/cladding refractive index difference ($\Delta n = +8.6 \times 10^{-2}$) and enhanced nonlinear effects compared to pure silica.

Several optical fibers have been fabricated with this SAL glass. After cleaning SAL glass powder and the surrounding silica tube, the powder is inserted into the tube and the preform is consolidated by heating it at a temperature between SAL glass T_G and T_{SOFT}. Silica tubes with different inner diameters have been used to fabricate optical fibers with different core diameters enabling multimode or singlemode guidance in the near infrared spectrum. Cross section pictures of optical fibers with large and small SAL glass cores are shown in Figure 6a.

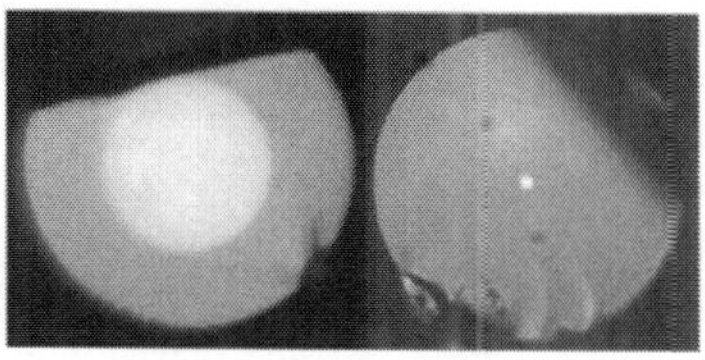

(a)

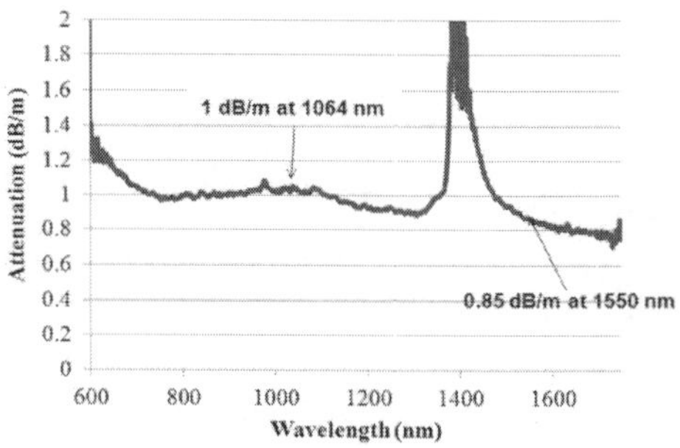

(b)

Figure 6. (**a**) Microscope pictures of cross section of two of SAL-core/silica-cladding optical fibers realized by the MPIT process ((**left**): core/fiber diameters = 50/95 μm; (**right**): 2.5/110 μm); (**b**) Attenuation coefficient *versus* wavelength of a SAL-core/silica-cladding optical fiber with a core diameter of 2.5 μm (measured by the cut-back method with fiber sample lengths of 28 and 5 m).

It is worth noting that the bulk material absorption of SAL around 0.6 dB/m at the 1.2 μm wavelength (measured at the IPHT Institute) leads to an attenuation coefficient of the fabricated optical fiber around 1 dB/m in the near infrared spectrum (1000–1600 nm), as shown in Figure 6b. Even if this attenuation coefficient is large, the larger linear refractive index ($+8.6 \times 10^{-2}$) and nonlinear refractive index compared to silica ones are interesting properties for developing functionalized or component optical fibers. For example, we have fabricated a specific step-index SAL optical fiber for generating nonlinear effects with a laser that emits femtosecond pulses around the wavelength of 1.55 μm (*Cf.* Figure 7a). The fiber was fabricated with a small core diameter (3.7 μm) in order to take advantage of the large core-cladding step index for tailoring the chromatic dispersion curve of the fundamental mode and for enhancing nonlinearities (*i.e.*, increasing the nonlinear coefficient). As shown on the measured refractive index profile of the fiber plotted in Figure 7b, the step-index difference is around 3.5×10^{-2}.

This value is lower than predicted ($\Delta n = +8.6 \times 10^{-2}$) due to silica diffusion from the cladding to the fiber core (*Cf.* Section 2.2). The chromatic dispersion of this SAL glass composition is equal to zero around 1791 nm. The diffusion of silica associated with the strong confining of the light within a small core allow to shift this zero dispersion wavelength (ZDW) down to 1447 nm (*Cf.* Figure 7c), enabling generation of nonlinear effects in anomalous dispersion regime by using pump laser operating around 1.55 µm. This example demonstrates that even simple fiber design could lead to interesting results when the core is made of a specific material.

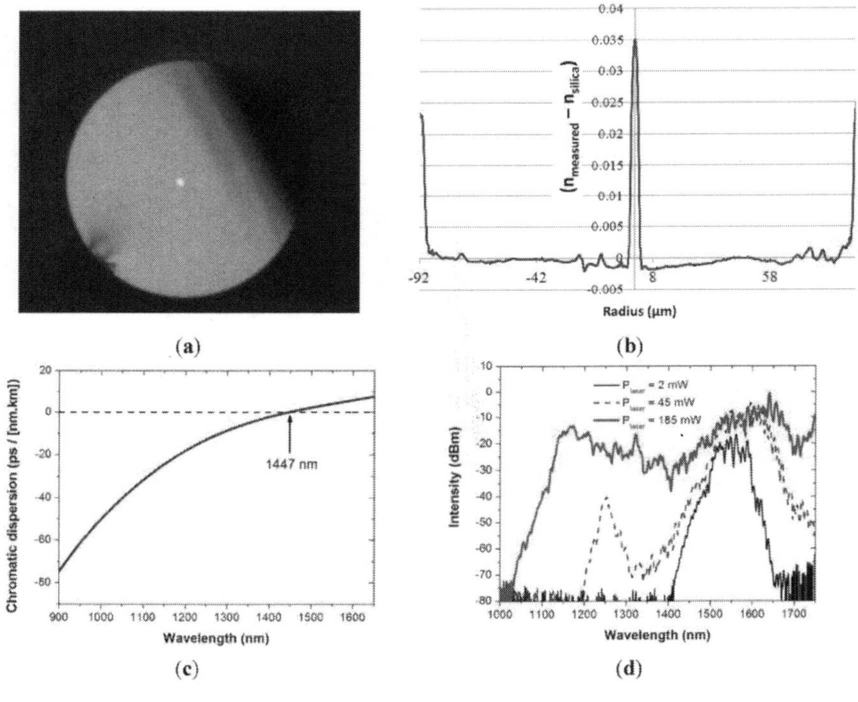

Figure 7. (**a**) Microscope picture of the cross section of an optical fiber composed of a SAL core of 3.7 µm and a silica cladding (with an external diameter of 210 µm); (**b**) RIP of the fiber (measured at 66,794 nm with EXFO NR9200); (**c**) chromatic dispersion*versus* wavelength of the fundamental mode guided in the fiber core. Measurements were done on a fiber sample of 40 cm long length (with a low-coherence interferometric setup in Mach-Zehnder configuration); (**d**) transmitted spectra of femtosecond pulses (centered around 1.55 µm) propagated in the fiber (sample length = 3 m) with different input powers.

In this prospect, a 3 m long length sample of this optical fiber was pumped with a laser emitting femtosecond pulses around 1.55 μm (with pulse width <100 fs, average power ~200 mW and repetition rate = 110 MHz). The generated nonlinear effects lead to a large spectral broadening of the laser output spectrum as shown in Figure 7d. More details on this experiment can be found in [12].

3.2. Multi Glasses Microstructured Optical Fibers

This process is also extremely well adapted for fabricating microstructured optical fibers, *i.e.*, fibers with more complex topologies by associating it with others fabrication processes such as the stack-and-draw technique. This method is the most used way for fabricating PCF. It consists in stacking glass capillaries and/or rods into a silica tube that is drawn down to structured glass rods (canes) of few millimeter diameters. The PCF is finally fabricated by drawing down a glass tube filled with a cane to a fiber of few hundred micrometers diameter. This technique offers much more flexibility for the fiber design, providing access to new properties and functionalities. The salient properties of PCF are endless single mode propagation, tailored chromatic dispersion curve, extremely high nonlinear coefficient and/or high birefringence.

To demonstrate the possibility of associating both fabrication processes for taking advantages of PCF technology and material optical properties, we have realized a PCF composed of a SAL glass core. The fiber was simply performed by replacing in the stack, the silica rod used to form the core by a fabricated rod composed of SAL glass surrounded by a thin silica layer (result of the silica tube used in the MPIT process). A picture of the fiber cross section is shown in Figure 8a. The core with a diameter of 5.6 μm is composed of a 1.25 μm SAL glass core surrounded by a silica layer. The microstructured cladding of air holes induces strong light confinement leading to a large nonlinear coefficient and to a shift of the ZDW down to shorter wavelength enabling generation of nonlinear effects as reported [18] in similar SAL glass based PCF fabricated with an alternative process (the SAL glass core rod was machined after the glass casting and fine cooling into a parallelepiped mold).

The stack-and-draw technique is also used for fabricating another class of PCF that guides light into a hollow-core or solid-core by photonic bandgap effects. Solid-core Photonic BandGap (PBG) fibers are composed of a silica core surrounded by a 2D crystal of rods with a higher refractive index [19]. The light guidance is driven by this crystal cladding which one exhibits bandgaps in specific guiding conditions, leading to transmission windows with unusual optical properties.

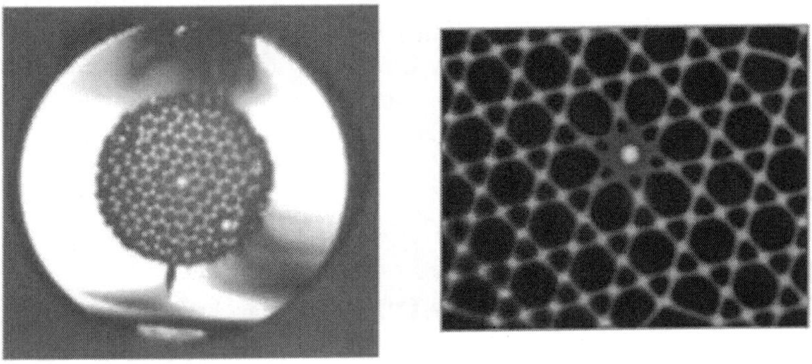

(a) (b)

Figure 8. Microscope pictures of a PCF composed of a SAL core with a diameter of 1.25 μm (**a**) whole fiber cross-section (**b**) zoom in the core.

In order to demonstrate the potential of the consolidation process of powder materials as an alternative solution for developing special fibers, we have fabricated a solid-core PBG fiber composed of a 2D crystal composed of SAL glass rods surrounding a silica core. The fiber was fabricated by filling the air holes of a PCF cane (composed of a silica core as shown in Figure 9a) with SAL-core/silica-cladding optical fibers (*cf.* Figure 9b). The PBG fiber is then fabricated by drawing down a silica tube filled with the cane. A cross section picture of the solid-core PBG is shown in Figure 9c. The fiber is composed of a 2D crystal made of two rings of SAL glass rods with a diameter of 2 μm spaced by 7.5 μm.

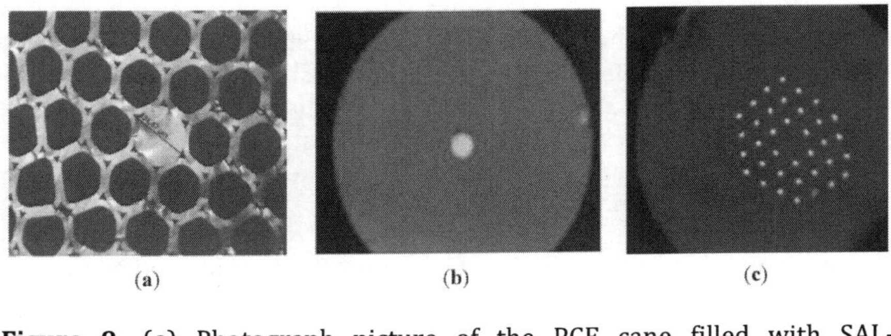

(a) (b) (c)

Figure 9. (**a**) Photograph picture of the PCF cane filled with SAL-core/silica-cladding optical fibers; (**b**) microscope picture of the cross section of the SAL-core/silica-cladding optical fiber (external diameter = 150 μm); (**c**) microscope picture of the fabricated fiber composed of a silica core surrounded by a 2D pattern of SAL glass inclusions.

It is worth noting that this fiber could also be realized by drawing (in two steps) a stack of SAL glass and silica rods. This fabrication demonstrates the interest of the MPIT process combined to the stack-and-draw technique for developing solid-core PBG fibers or related ones. In contrast with usual solid-core PBG fibers that are based on Ge-doped rods fabricated by MCVD process, this original technological route offers more flexibility in the choice of the rod materials and as a consequence on the conception of the fiber properties. For example, using rods with a larger refractive index compared to SAL glass or silica ones may increase the light confinement into the core leading to lower bend loss sensitivity and to a modification of the chromatic dispersion curve.

3.3. Metal–Silica Optical Fibers

The flexibility of the MPIT process allows us to use other materials than pure glass compositions. Metals could also offer interesting properties and functionalities to optical fibers. To demonstrate the potentiality of the MPIT technique for fabricating new specialty fibers composed of metals and glasses, we have investigated and tested the realization of three classes of fibers composed of (i) a silica core including copper particles; (ii) a copper wire surrounded by a silica cladding; (iii) a PCF with copper wires into the air/silica microstructure.

3.3.1. Optical Fibers with Metallic Ions and Particles Embedded into the Core

Metallic nanoparticles (e.g., Au, Cu, Ag) have unique optical and electrical properties that differ, especially optical absorption, drastically from bulk material. The physical origin of the light absorption by the metallic nanoparticles is the excitation of coherent oscillations of the conduction band electrons by light that is known as surface plasmon resonance [20]. This phenomenon has been employed to optical switches, waveguides, spectroscopy, chemical and biological sensors, *etc.*, [21,22,23,24]. As a consequence, metallic nanoparticles embedded and stabilized into optical fibers pave the way to many types of applications with such a simple device.

Optical fibers with copper ions into the core present interesting properties, as extremely high resonant nonlinearity coefficient exhibited by Cu^{2+} (*i.e.*, CuO) ions for example [25]. Moreover, modification of the copper ions oxidation states from Cu^{2+} (*i.e.*, CuO) to Cu^{+} (*i.e.*, Cu_2O) under gamma radiations are measurable regarding the luminescence spectrum of the copper, allowing the development of gamma ray dosimeter [26].

Such optical fibers could be prepared by inserting metallic oxide during the realization of the preform with CVD techniques for example. Flash vaporization process could also be joined to the MCVD method to allow silica layers doping with many heavy metal or rare earth-ions that do not have precursor materials in gaseous form compatible with CVD techniques [27]. However, it is worth noting that these processes enable only the insertion of metallic ions through metallic oxide precursors, since to the best of our knowledge, common CVD processes do not allow for working on reducing atmosphere. Furthermore, these processes require preform collapsing operated at very high temperature and under depression. This fabrication stage is critical for particles embedded in the latest deposited layers which ones can be removed during this operation.

Others techniques, such as the simple powder-in-tube technique are also willing to produce metallic particles embedded into optical fiber. First trials have been performed using a silica cladding and a core made of 70% SiO_2-20% Al_2O_3-10% La_2O_3 (SAL') + 0.5% CuO (mol%) (*Cf.* Figure 10a). A post drawing heat-treatment at 850 °C (>T_G) under a reducing Ar-H_2(5%) atmosphere leads to the presence of a highly red reflective ring at the interface core-cladding as shown in Figure 10b. XRD pattern of the core fiber (*Cf.* Figure 10c) was obtained using a Single Crystal X-Ray Diffraction System Kappa-CCD (Kα-Mo = 0.71 Å). We clearly show a broad peak centered at 2θ = 10° corresponding to the glassy core and a small peak at 2θ = 19.5° highlighted by a green line corresponding to nanometric copper particles Cu^0. The formation of this Cu^0 ring at the interface core/cladding results from the reduction of Cu^{2+} and Cu^+ ions due to the Ar-H_2 atmosphere in the fiber after drawing. To balance the depleted charge that results from the reduction reaction at the interface core/cladding, Cu^+ and Cu^{2+} ions diffuse from the core center to the interface and are reduced also leading to the ring. It is noteworthy that the heat-treatment near T_G (850 °C) allows ions to become mobile.

As demonstrated in Section 2.2.2, the MPIT process can be directly used to reduce in one step Pd^{2+} into Pd^0 under vacuumed conditions. By regarding the Ellingham's diagram of some metal/metal oxide redox couples, this technique could be extended also to other metallic particles than palladium, including copper. Different trials are still in progress.

3.3.2. Long Metallic Wires Distributed in Optical Fibers

Another possibility offered by the MPIT process is the fabrication of optical fibers including metallic wires. To explore this potentiality, we have started by fabricating a fiber composed of a copper wire surrounded by a silica cladding. A silica tube filled with pure copper powder was consolidated and then drawn down to fiber with an external diameter of 200 μm. As shown in Figure 11b, the fiber is composed of a copper wire with a diameter of 50 μm surrounded by a silica cladding. It is worth noting that, even if there is a large difference between the softening temperature of silica (1665 °C) and the fusion temperature of copper (1080 °C), we were able to draw hundreds of meters of such fibers.

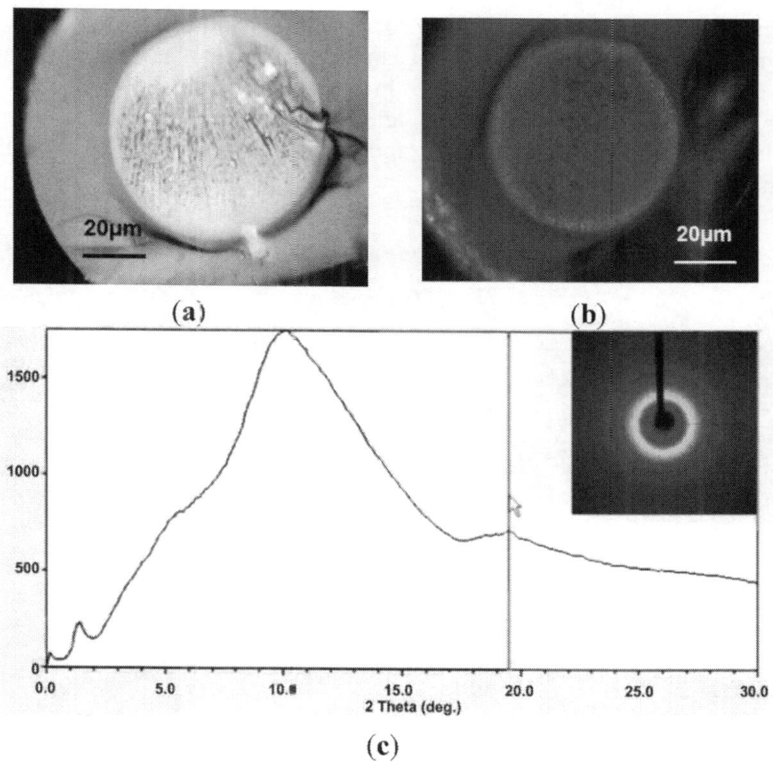

Figure 10. Microscope picture of cross section of a SAL glass +0.5% CuO core optical fiber **(a)** before and **(b)** after a post-drawing heat-treatment at 850 °C under a reducing Ar-H_2 (5%) atmosphere **(c)** X-Ray Diffraction (XRD) pattern of the SAL-Cu optical fiber (Kα-Mo, Kappa CCD).

In order to quantify the quality of this copper wire, its resistivity was measured on a fiber sample of two meters long. The evaluated resistivity is 1.7×10^{-8} Ω·m, which is very close to the theoretical one (1.68×10^{-8} Ω·m).

This realization demonstrates the interest of the MPIT technique for developing high quality electric micro-wires (in diameters) for specific electrical or RF applications. It also gives us basic skills and expertise for developing specialty optical fibers that associate optical and metallic wire properties.

In this respect, we fabricated a PCF with copper wires inserted into the air/silica microstructured cladding. To fabricate it, we inserted two larger capillaries in the PCF stack that were filled with consolidated copper rods. This stack was drawn down to cane of few millimeter diameters (*Cf.* Figure 12a) and then a silica tube filled with a cane was drawn to a fiber of 250 μm diameter. A picture of the fiber cross section composed of a silica core surrounded by a photonic crystal cladding of air holes and two copper wires of 20 μm diameters, is shown in Figure 12b. Light propagation in the core (along few centimeter long length fiber sample) was successfully measured with a CCD camera, as shown by the near-field picture of the fundamental mode in Figure 12c.

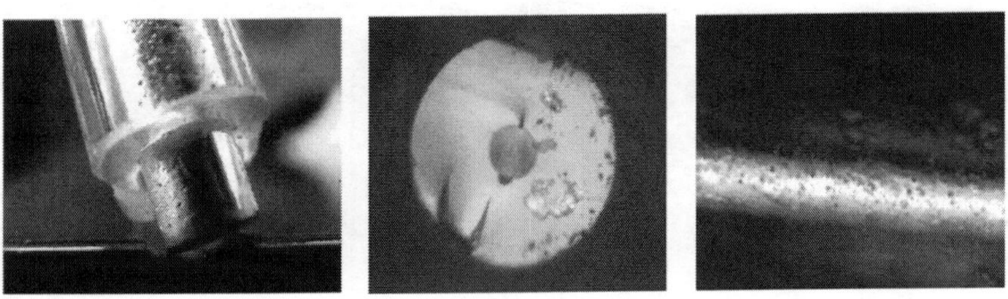

(a) (b) (c)

Figure 11. (**a**) Photograph picture of a silica preform with a pure copper core. Microscope pictures of the fabricated fiber with a copper-core diameter of 50 μm; **(b)**cross section; (**c**) longitudinal picture.

Even if, the drawing process could be improved this first fabrication result demonstrates the high potential of the MPIT technique for developing specialty optical fibers with added functionalities.

4. CONCLUSIONS

We have presented in this paper an original route to produce specialty optical fibers, based on the powder-in-tube technique combined with a consolidation process realized under specific oxidizing or reducing atmosphere.

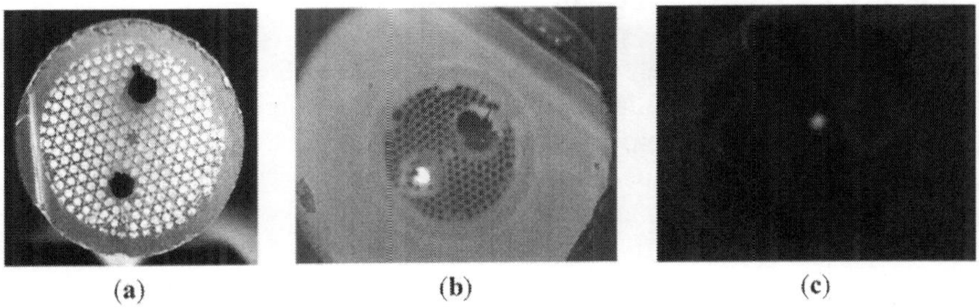

Figure 12. (**a**) Photograph picture of a fabricated PCF cane with two copper rods; (**b**) microscope picture of the fabricated PCF with two copper wires of 20 μm diameters; (**c**) measured near-field optical pattern of the light guided in the fiber core.

After the presentation of the principle of operation of this process for consolidating or vitrifying a powder based optical preform, we demonstrated its flexibility for drawing materials with large thermo-mechanical different properties, for controlling and modifying oxidation states of powder materials inside the preform. We also investigated experimentally the chemical diffusions of materials in the preform (during the fiber fabrication) and demonstrated their impacts on the fiber RIP and optical properties.

We presented preliminary fabrication results of optical fibers composed silica and a lanthano-aluminosilicate glass used as the core material in step-index fiber or in specialty PCFs. To complete the panel of original microstructured fibers now available by the MPIT technique, we also presented several optical fibers in which metallic particles or microwires are included in a silica-based matrix.

We believe that we have demonstrated in this paper the flexibility and simplicity of the MPIT technique compared to classical manufacturing processes, in order to develop new class of specialty optical fibers.

ACKNOWLEDGMENTS

Authors would like to thank the COST Action TD1001 Novel and Reliable Optical Fiber Sensor Systems for Future Security and Safety Applications (OFSeSa) for support and helpful discussions around fiber sensors.

We would like to sincerely thank the Silitec Fibers SA team (Switzerland) and more particularly Philippe Ribaux, Carlos Pedrido and Frédéric Sandoz for all the discussions helping for the design and the improvement of the consolidation unit developed at the XLIM Institute.

Authors would like to thank the funding provided by the Thuringian Ministry of Economics, Labor, and Technology (TMWAT) under contract FKZ B715-07036 (TechFas), TNA I-1/2010 (NEODIN) and 2012 VF0020 (LASIL) with financial support from the European Social Fund (ESF) and European Regional Development Fund (EFRE) is gratefully acknowledged.

Thanks are also to the Région LIMOUSIN (Vortex Project 2012) and European Regional Development Fund (EFRE) for his financial support.

CONFLICTS OF INTEREST

The authors declare no conflict of interest.

REFERENCES

1. Knight, J.C.; Birks, T.A.; Russell, P.; Atkin, D.M. All-silica single-mode fiber with photonic crystal cladding. *Opt. Lett.* **1997**, *22*, 484–485.

2. Russell, P. Photonic crystal fibers. *Science* **2003**, *299*, 358–362.

3. Russell, P. Photonic-crystal fibers. *J. Lightwave Technol.* **2006**, *24*, 4729–4749.

4. Gérôme, F.; Auguste, J.L.; Maury, J.; Blondy, J.M.; Marcou, J. Theoretical and experimental analysis of a chromatic dispersion compensating module using a dual concentric core fiber. *J. Lightwave Technol.* **2006**, *24*, 442–448.

5. Goel, N.K.; Stolen, R.H.; Morgan, S.; Kim, J.K.; Kominsky, D.; Pickrell, G. Core-suction technique for the fabrication of optical fiber preforms. *Opt. Lett.* **2006**, *31*, 438–440.

6. Schmidt, M.; Wondraczek, L.; Russell, P. Pressure-assisted flow of glass melts in narrow capillaries. In Proceedings of the GOMD Conference, Corning, NY, USA, 1–4 October 2010. SII-035-2010.

7. Ballato, J.; Hawkins, T.; Foy, P.; Stolen, R.; Kokuoz, B.; Ellison, M.; McMillen, C.; Reppert, J.; Rao, A.M.; Daw, M.; *et al.* Silicon optical fiber. *Opt. Express* **2008**, *16*, 18675–18683.

8. Ballato, J.; Snitzer, E. Fabrication of fibers with high rare-earth concentrations for Faraday isolator applications. *Appl. Opt.* **1995**, *34*, 6848–6854.

9. Van Uitert, L.G.; Pinnow, D.A.; Williams, J.C.; Rich, T.C.; Jaeger, R.E.; Grodkiewicz, W.H. Borosilicate glasses for fiber optical waveguide. *Mater. Results Bull.* **1973**, *3*, 469–476.

10. Pedrido, C. Method for Fabricating an Optical Fiber, Preform for Fabricating an Optical Fiber, Optical Fiber and Apparatus. Dätwyler Fiber Optics SA. Patent WO 2005/102946 A1 and WO 2005/102947 A, 3 November 2005.

11. Clare, A.G.; Parker, J.M. The effect of refractive index modifiers on the thermal expansion coefficient of fluoride glasses. *Phys. Chem. Glasses* **1989**, *30*, 205–210.

12. Litzkendorf, D.; Grimm, S.; Schuster, K.; Kobelke, J.; Schwuchow, A.; Ludwig, A.; Kirchhof, J.; Leich, M.; Jetschke, S.; Dellith, J.; *et al.* Study of lanthanum aluminum silicate glasses for passive and active optical fibers. *Int. J. Appl. Glass Sci.* **2012**, *3*, 321–331.

13. Leparmentier, S.; Auguste, J.L.; Humbert, G.; Delaizir, G.; Delepine Lesoille, S.; Bertrand, J.; Buschaert, S.; Perisse, J.; Mace, J.R. Palladium particles embedded into silica optical fibers for hydrogen gas detection. In Proceedings of the SPIE Photonics Europe, Bruxelles, Belgium, 2 May 2014; Volume 9128.

14. Lemaire, P.J. Reliability of optical fibers exposed to hydrogen: Prediction of long-term loss increases. *Opt. Eng.* **1991**,*30*, 780–790.

15. Lee, K.G.; Oh, C.H.; Ryu, K.S.; Kang, H.J. Mass-production of zero water peak fiber by VAD process. In Proceedings of the Asia-Pacific Optical Communications Conference, Shanghai, China, 7–10 November 2005.

16. Dianov, E.M.; Mashinsky, V.M.; Neustruev, V.B.; Sazhin, O.D. Origin of excess loss in singlemode optical fibers with high GeO_2 doped silica core. *Opt. Fiber Technol.* **1997**, *3*, 77–86.

17. Ballato, J.; Hawkins, T.; Foy, P.; Kokuoz, B.; Stolen, R.; McMillen, C.; Daw, M.; Su, Z.; Tritt, T.M.; Dubinskii, M.; *et al.* On the fabrication of all-glass optical fibers from crystals. *J. Appl. Phys.* **2009**, *105*, 053110:1–053110:9.

18. Kobelke, J.; Schuster, K.; Litzkendorf, D.; Schwuchow, A.; Kirchhof, J.; Bartelt, H.; Tombelaine, V.; Leproux, P.; Couderc, V.; Labruyère, A. Microstructured fibers with high lanthanum oxide glass core for nonlinear applications.*SPIE* **2009**, *7357*, 1–12.

19. Luan, F.; George, A.K.; Hedley, T.D.; Pearce, G.J.; Bird, D.M.; Knight, J.C.; Russell, P. All-solid photonic bandgap fiber.*Opt. Lett.* **2004**, *29*, 2369–2371.

20. Kelly, K.L.; Coronado, E.; Zhao, L.L.; Schatz, G.C. The optical properties of metal nanoparticles: The influence of size, shape, and dielectric environment.*J. Phys. Chem. B* **2003**, *107*, 668–677.

21. Nikolajsen, T.; Leosson, K.; Bozhevolnyi, S.I. Surface plasmon polariton based modulators and switches operating at telecom wavelengths. *Appl. Phys. Lett.* **2004**, *85*, 5833–5835.

22. Hochberg, M.; Baehr-Jones, T.; Walker, C.; Scherer, A. Integrated plasmon and dielectric waveguides. *Opt. Express***2004**, *12*, 5481–5486.

23. Zia, R.; Selker, M.D.; Brongersma, M.L. Leaky and bound modes of surface plasmon waveguides. *Phys. Rev. B* **2005**,*71*, 165431:1–165431:9.

24. Krasteva, N.; Besnard, I.; Guse, B.; Bauer, R.E.; Mullen, K.; Yasuda, A.; Vossmeyer, T. Composite films for vapor sensing applications. *Nano Lett.* **2002**, *2*, 551–555.

25. Lin, A.; Kim, B.H.; Moon, D.S.; Chung, Y.; Han, W.-T. Cu^{2+} doped germano-silicate glass fiber with high resonant nonlinearity. *Opt. Express* **2007**, *15*, 3665–3672.

26. Huston, A.L.; Justus, B.L.; Falkenstein, P.L.; Miller, R.W.; Ning, H.; Altemus, R. Remote optical fibre dosimetry. *Nucl. Instrum. Methods Phys. Res. B* **2001**, *182*, 55–67.

27. Lenardič, B.; Kveder, M.; Lisjak, D.; Guillon, H.; Bonnafous, S. Novel method for fabrication of metal- or oxide-nanoparticle doped silica-based specialty optical fibers. *SPIE* 2011, *7934*.

6

OPTICAL FIBER NANOTIPS COATED WITH MOLECULAR BEACONS FOR DNA DETECTION

Ambra Giannetti [1,*], Andrea Barucci [1], Franco Cosi [1], Stefano Pelli [1,2], Sara Tombelli [1], Cosimo Trono [1] and Francesco Baldini [1]

[1] CNR—Institute of Applied Physics "Nello Carrara", Via Madonna del Piano 10, 50019 Sesto Fiorentino (FI), Italy

[2] Museo Storico della Fisica e Centro Studi e Ricerche Enrico Fermi, Piazza del Viminale 1, 00184 Rome, Italy

ABSTRACT

Optical fiber sensors, thanks to their compactness, fast response and real-time measurements, have a large impact in the fields of life science research, drug discovery and medical diagnostics. In recent years, advances in nanotechnology have resulted in the development of nanotools, capable of entering the single cell, resulting in new nanobiosensors useful for the detection of biomolecules inside living cells. In this paper, we provide an application of a nanotip coupled with molecular beacons (MBs) for the detection of DNA. The MBs were characterized by hybridization studies with a complementary target to prove their functionality both free in solution and immobilized onto a solid support. The solid support chosen as substrate for the immobilization of the MBs was a 30 nm tapered tip of an optical fiber, fabricated by chemical etching. With this set-up promising results were obtained and a limit of detection (LOD) of 0.57 nM was reached, opening up the possibility of using the proposed nanotip to detect mRNAs inside the cytoplasm of living cells.

KEYWORDS

molecular beacon; optical fiber nanotip; nanosensor; optical biosensors; fluorescence; mRNA; DNA; surviving

1. INTRODUCTION

Usually, information on basic cellular signaling processes associated with diseases is not directly deducible from measurements on an average of a pool of cells. In fact, new information can only be obtained by monitoring the cellular signaling pathways directly inside intact cells [1]. Traditional approaches for studying cells and molecular biology outside living cells (e.g., by using biomolecules purified from cells), while being highly productive, often lose valuable information about cellular mechanisms that can only be observed in their natural environment. Hence, even if challenging, this progress in cellular physiology study requires new detection strategies at the nanoscale level applied to individual cells with large temporal and spatial resolution. In this framework, nanotechnology is becoming a key tool, opening new ways to explore and understand cellular and subcellular mechanisms [2]. In recent years, the advances in nanotechnology have led to the development of nanotools, capable of entering the single cell and detecting target biomolecules inside the living cells if coupled to the specific receptors. Examples of these nanotools are: fluorescent quantum dots, used to visualize dynamic processes in living cells [3,4,5]; carbon nanotubes or polymeric nanoparticles, used to penetrate the cell membranes as carrier of drugs or for imaging purpose [6,7,8]; gold nanorods, used for photothermal therapy [9]; and nanoneedles and nanotips which can penetrate the cell membranes with minimal cell damage [10,11].

Among these nanotools, nanoprobes based on fiber-optical nanotips were conceived as integrated nanodevices consisting of a recognition molecule coupled to the optical transducing element (the optical fiber) interfaced to a photo- or spectrometric detection system, aimed to perform diagnosis within single living cells [12,13,14,15].

Chemical nanosensors were first developed for monitoring calcium and nitric oxide, and other physiochemical parameters in single cells [16], showing excellent detection limits, as well as photostability, reversibility, and millisecond response. McCulloch *et al.* [17] prepared nanometric optical fiber sensors for intracellular pH measurements, while other groups [18] developed an optical fiber sensor for the measurement of dissolved oxygen based on sol-gel immobilization technology. Vo-Dinh and coworkers have developed nanobiosensors to detect biochemical targets inside living single cells [19,20,21,22], while Zhang [23] reported about an integrated device for optical stimulation and spatiotemporal electrical recording of neural activity in light-sensitized brain tissue.

In the field of DNA/RNA analysis in cell, MBs were proposed by Tyagi and Kramer [24,25,26,27,28,29]. MBs are single-stranded DNA molecules that possess a stem-loop structure commonly named hairpin structure. The loop portion of the molecule can form a double-stranded DNA in the presence of a complementary sequence. MB can be labeled with a fluorophore and a quencher at the two side-ends of the stem, constituted by a more or less long chain of complementary bases, which keeps these two moieties in close proximity to each other. Since the fluorophore is characterized by an emission band, which overlaps the absorption band of the quencher, this proximity causes the fluorescence of the fluorophore to be quenched by energy transfer. The fluorophore fluorescence is restored upon the opening of the stem due to the hybridization of the MB with the target sequence.

A great many fiber nanotips are characterized by a metal coating which allows them to illuminate exactly only the fiber nanotip. Actually, it is very difficult to quantify the exact fraction of light that will reach the nanotip since, below a certain inner diameter, even the lowest guided mode runs into cutoff, where the wave vector becomes imaginary and thus the mode field decays exponentially [30]. However, although the transmitted light fraction can be extremely small at the apex of the nanotip, when applying the nanoprobe for intracellular analysis, the dimensions of a single cell which are in the range 1 μm–10 μm, lead to the fact that the examined part of the nanoprobe will be not only the apex of 30 nm and consequently a larger portion of hundreds of nanometers will be exposed to the intracellular content. The absence of the metal coating on the fiber nanotip implies the excitation of the MB immobilized along the whole fiber nanotip. On the other hand, only the portion of the MB inside the cell which interacts with the intracellular target will emit fluorescence. A partial loss of localization associated with this approach can be accepted for some intracellular application where there is not a strong necessity of localization, if this is accompanied with an increase of the fluorescence signal associated to the excitation of a much larger number of molecules of MB.

In this paper we present preliminary results concerning the use of an optical fiber nanotip coupled to a MB for DNA detection (Figure 1). The solid support chosen as substrate for the immobilization of the MB was a tip of an optical fiber tapered at nanoscale size, intended to be used in the future for mRNA detection inside the cytoplasm of living cells. The nanotip was fabricated by chemical etching, starting from 500 micron—diameter of the multimode optical fiber—down to 30 nm at the tip. Then, the fiber tip was silanized, and the MB was attached via a covalent-binding procedure.

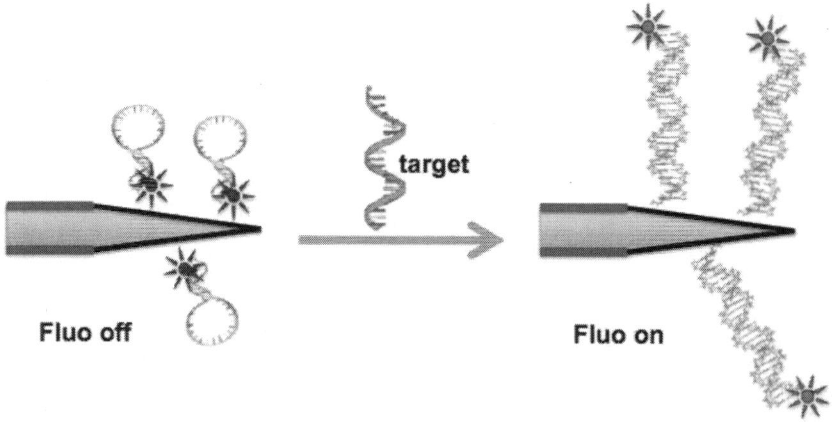

Figure 1. Scheme of the sensing mechanism.

In particular, the attention was focused on the mRNA for survivin, a protein highly expressed in most types of cancer. In this case, the MB could act not only as detector of the over-expression of the mRNA for the survivin (diagnosis) but also as the blocking agent of the synthesis of the protein itself (therapy), as we demonstrated with Real-time PCR and western blotting experiments which showed a time-dependent reduction of survivin mRNA and protein after 100 nM-MB treatment, respectively, with the molecular beacon transfected into A375 melanoma cells [31,32].

After the characterization of the sole MB in solution carried out to verify the reliability and the effectiveness of the fluorophore/quencher pair immobilized at the extremities of the single-stranded DNA and at the same time of the good interaction of the MB with its target, a thorough opto-chemical characterization was performed in order to evaluate the possibility of getting a reliable, reproducible and robust system consisting of the MB immobilized onto the fiber nanotip.

2. EXPERIMENTAL SECTION

2.1. Chemicals

All the reagents for the preparation of phosphate buffered saline (PBS), 40 mM, pH 7.4; tris hydrochloride (tris-HCl), magnesium chloride ($MgCl_2$) for the preparation of tris-buffer (10 mM tris-HCl, 10 mM $MgCl_2$, pH 8); hydrofluoric acid (HF); hydrochloric acid (HCl); sulfuric acid (H_2SO_4); hydrogen peroxide (H_2O_2); methanol (MeOH) were purchased from Sigma-Aldrich (Milan, Italy). (3-Aminopropyl)triethoxysilane (APTES) was purchased by abcr GmbH (Karlsruhe, Germany). N-succinimidyl 3-(2-pyridyldithio) propionate (SPDP) was purchased from Pierce (Thermofisher Scientific, Milan, Italy). The molecular beacon for survivin, its complementary target sequence, the non-specific random sequence and the labeled linear probe were purchased from IBA (Gottingen, Germany).

2.2. Molecular Beacon and Target Sequences

The sequence for the survivin molecular beacon [31] was chosen among several published sequences [33,34,35,36] for its better performances in terms of folding properties and greater reactivity with the target sequence. A different fluorophore-quencher pair, ATTO647N (λ_{abs} 644 nm, λ_{em} 669 nm) and BlackBerry® Quencher 650 (λ_{max} ~ 650 nm, useful absorbance between 550 and 750 nm), was chosen with respect to the one used for the published MBs in order to obtain a greater quantum yield (QY = 65 of the ATTO647N instead of 0.28 of the Cy5). The sequences of the different oligonucleotides (MB, specific target, random sequence and linear probe) are reported below:

- MB 5'-(ATTO647N)CGACGGAGAAAGGGCTGCC ACGXCG(BBQ)-3' X=C6-dT Thio
- Target 5'-CCCCTGCCTGGCAGCCCTTTCTC AAGGACC-3'
- Random sequence 5'-ATCGGTGCGCTTGTCG-3'
- Linear probe 5'-(ATTO647N)GAGAAAGGGC TGCCA(Thiol)-3'

All the measurement carried out with both the MB in solution and the MB immobilized on the fiber tip were carried out at room temperature.

2.3. Optical Fiber Nanotip Preparation

The optical fiber nanotip was manufactured by a dynamic chemical etching method [37,38], by mechanically rotating and dipping a 3 M silica optical fiber, (core diameter 480 micron) in a chemical etching solution (aqueous hydrofluoric acid) covered with a protection layer.

Using different dynamic regimes of the mechanical movements during the chemical etching process, it was possible to vary the cone angle, the shape and the roughness of the nanoprobes. It was found that the tip profiles were determined by the nonlinear dynamic evolution of the meniscus of the etching solution near the fiber. Different regimes can be generated by changing the ratio between the angular velocities and the ratio between the radii of optical fiber and vial, ranging from laminar flow to the onset of chaotic flow. The type of flow of the viscous HF solution caused by the rotation of the vial and the optical fiber can be described in the framework of the Taylor-Couette flow (TCF) theory [39,40].

Moreover, by an accurate control of the extraction speed of the fiber from the HF solution, the length and the angle of the nanotip were precisely controlled, and the capillarity effects at the nanoscale were minimized by limiting strong friction on the optical fiber nanotip in the final part of the etching process.

SEM pictures and AFM measurements have been performed on different nanotips showing our ability to reproduce typical tip features, such as short taper length (~200 μm), large cone angle (from 15° to 40°), small probe tip dimension (less than 30 nm), and roughness below 10 nm. The geometrical characterization demonstrated that, with this method, a high yield of reproducible nanotips can be obtained, overcoming some drawbacks of conventional etching techniques. The nanometric roughness is the key point to keep the scattering, which can perturb the measurements, at the lowest levels.Figure 2 shows an example of nanotip obtained using the fabrication method described above.

In particular, in the left top part a general view of the short taper length and the large cone angle (about 40° in this example) of the nanotip is shown, while in the left bottom an AFM image (10 μm × 10 μm) is reported, with an estimated roughness of the tip of about 10 nm. On the right panel is reported a zoom on the distal end, showing the apex angle (about 24°).

The glass nanotip was dipped in piranha solution (H_2SO_4:H_2O_2 7:3) for 30 min to carry out a strong cleaning and to activate the silanol groups present on the surface. Then the tapered fiber nanotip was functionalized by a 2-hour long silanization procedure using 5% APTES in MeOH/H_2O (1:1) in order to achieve the amino groups onto the surface of the nanotip. After several rinsing cycles in MeOH and deionized water, the nanotip was cured in oven at 100 °C for 1 h.

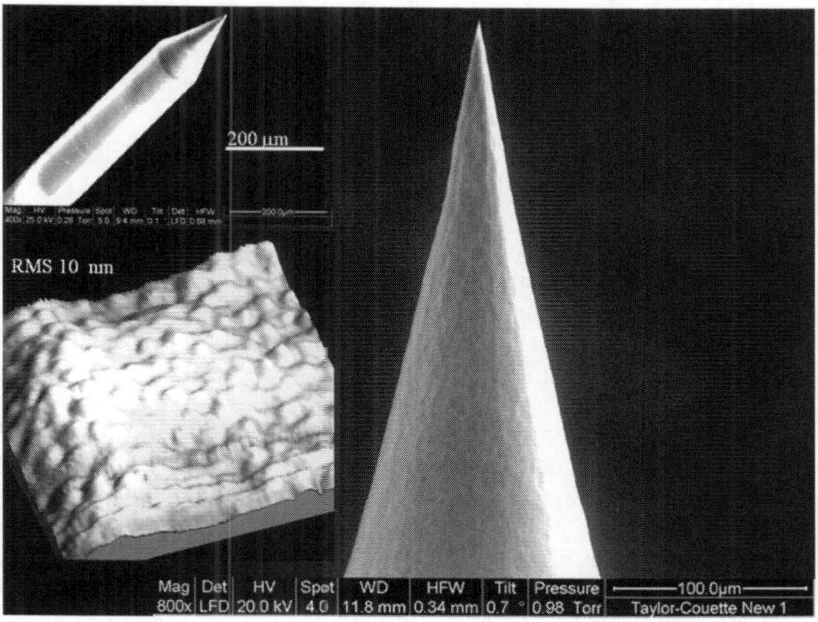

Figure 2. Example of a nanotip obtained using the fabrication method described inSection 2.3. In the left top a general view showing a taper length of about 200 μm and a large cone angle (about 40°), while in the left bottom an AFM image (10 μm × 10 μm) is reported, with an estimated roughness of about 10 nm. On the right panel is reported a zoom on the distal end, showing an apex angle of about 24°.

SPDP was used as crosslinker between the -NH_2 groups available onto the nanotip and the -SH groups of the linear probe or of the MB. After the treatment with a solution of SPDP 4 mM in PBS at pH 7.4 for 30 min, the nanotip was rinsed in PBS and then in tris-buffer. As already reported by Tombelli *et al* [41], the cross-linker SPDP could be used to form a cleavable disulphide bond to detach the MB from the delivery vehicle inside the cytoplasm. This could be an advantage if the aim is to release the MB inside the cells, and, consequently the nanotip is used simply as the delivery tool of the MB as therapeutic molecule inside the cell. On the other hand, if the probe is used as imaging tool, other cross-linkers, such as succinimidyl-4-(N-maleimidomethyl)cyclohexane-1-carboxylate (SMCC) should be considered in order to avoid any loss of signal caused by the MB detachment from the fiber nanotip.

Before its immobilization, the MB was heated for 5 min in water bath at about 70–80 °C (slightly higher than the melting temperature) and then left to slowly equilibrate down to room temperature in order to optimize the hairpin formation. The immobilization of 1 μM MB (in tris-buffer), or of 1 μM linear probe, onto the nanotip was performed over night at 4 °C.

2.4. The Optical Measurements

Scanning Electron Microscope (SEM, Quanta-200 FEI, Hillsboro, OR, USA) was used to characterize the geometrical features of the nanotips, whereas the surface roughness was checked by Atomic Force Microscopy (AFM, prototype developed by Pini *et al.* [42]).

An in-house optical set-up, illustrated in Figure 3, was used to characterize the fluorescent nanoprobes. Fluorescence measurements on fiber nanotips were carried out in a cuvette using, for the MB excitation, a LDH-P-C-635B laser diode (PicoQuant, Berlin, Germany) emitting at 635 nm filtered with a bandpass interference filter (FL635-10, ThorLabs, Newton, NJ, USA). The emitted fluorescence was collected by means of a GRIN lens coupled with a multimode optical fiber (diameter 200 μm), aligned with the fiber nanotip and then guided to an optical high-pass filter Thorlabs FEL0650, in order to filter the excitation light scattered out from the fiber tip, and finally acquired by a Shamrock 303i spectrograph (Andor, Belfast, United Kingdom). The cuvette was fixed inside a suitable holder onto a manual labjack (model 271 Labjack, Newport, Irvine, CA, USA) and the fiber was immersed into the cuvette from the top. In this way, it was possible, after the optical alignment, to move down the cuvette, to empty and fill it or to exchange the cuvette, and to move up again it to the exact same position, without changing the alignment of the nanotip and GRIN lens.

The experimental set-up for fluorescence measurement of the molecular beacon in solution was very similar to the one described in Figure 3, with the difference that the fluorescent solution into the cuvette was excited directly with the collimated laser beam, without the objective, and collected with the GRIN lens in normal direction with respect to the excitation laser beam direction. The measurements in solution were performed without the optical high-pass filter Thorlabs FEL0650, because of the better signal (fluorescence) to noise (scattered excitation light) ratio of this measurement configuration.

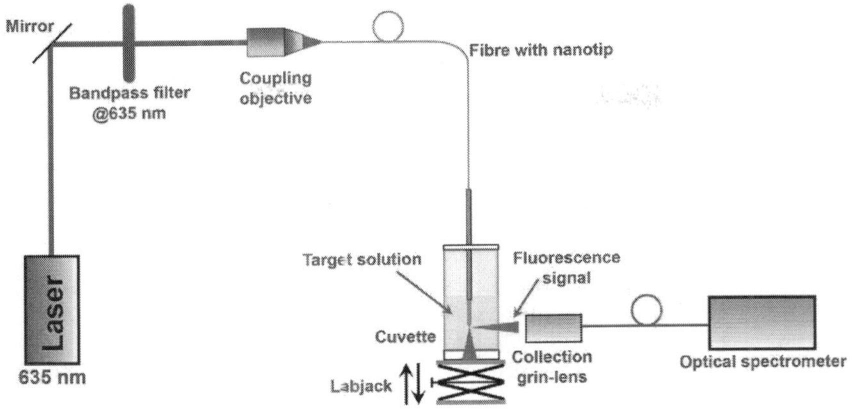

Figure 3. Optical experimental set-up.

3. RESULTS AND DISCUSSION

3.1. MB Characterization in Solution

The functionality of the MB was firstly investigated in buffer solution by hybridization with the specific target at different incubation times. Different buffers and different culture media were already tested and reported, such as Dulbecco's modified Eagle's Medium (DMEM) [43,44]. Even if tris-buffer is not mimicking the environmental condition within the cell, it is the one which provides the best performances of the molecular beacon in solution.

The MB was characterized by recording the fluorescence (λ_{ex} 635 nm) of the MB 100 nM in Tris buffer alone and after 1, 3 and 5 h of incubation with the specific target 0.2 μM (Figure 4). A final 10-times increase of the MB fluorescence in presence of the target was recorded with the signal increasing with time from 1 to 5 h. The characterization on such a long time, much longer than the interaction time between the MB and the target, was carried out in order to verify the stability of the molecular beacon and, consequently, of its emitted fluorescence also after that the interaction took place.

3.2. MB and Linear Probe Characterization onto the Nanotip

Before working with the molecular beacon, the fluorescent linear probe was immobilized onto the nanotip in order to evaluate the actual possibility of sensing the presence of a labeled oligonucleotide by direct excitation of the oligonucleotide fluorophore by means of the same fiber nanotip. After the probe immobilization, the fluorescence of the ATTO647N fluorophore labeling the probe, was measured (λ_{ex} 635 nm). Figure 5 shows the fluorescence spectrum of the linear probe immobilized onto the nanotip. The fluorescence peak shown in the Figure is split into two peaks, at 660 nm and 680 nm due to the not uniform transmission spectrum of the Thorlabs FEL0650 high-pass filter.

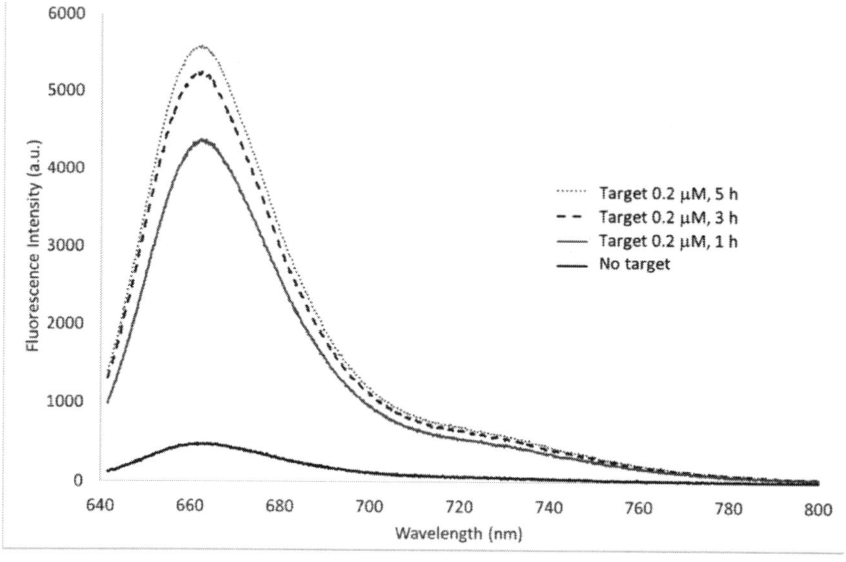

Figure 4. Fluorescence spectra of the molecular beacon (0.1 μM) after incubation for 1, 3 and 5 h with the specific target 0.2 μM. λ_{ex} 635 nm, exposure time 1 s.

The fluorescence signal of the probe was monitored for 30 min in order to evaluate the fluorescence stability: the average fluorescence intensity in 30 min was of 11,300 (a.u.) with a standard deviation of 436 (a.u.) leading to a signal variability of 3.8%.

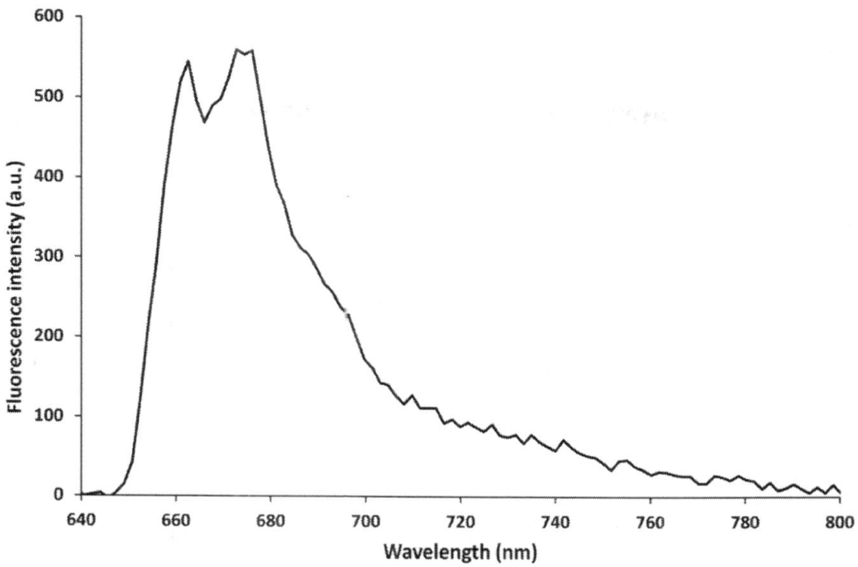

Figure 5. Fluorescence spectra of the linear probe immobilized onto the nanotip. λ_{ex}635 nm, exposure time 1 s.

In Figure 6 the fluorescence spectrum of the multimode fiber functionalized with the molecolar beacon is shown. In particular, the nanotip was immersed in pure buffer (Tris), in a solution of the random sequence and in two solutions of the specific target 1 and 10 μM and the fluorescence spectra were recorded after 30 min.

The performance of the fiber nanotip to act as sensor was evaluated by monitoring the fluorescence signal as a function of time, when exposed to different concentrations of the target. Its specificity when interacting with the random sequence was also investigated (Figure 7). In particular, after an initial stabilization of the MB fluorescence in buffer, the coated nanotip was first immersed in a solution 1 μM of the random sequence and the fluorescence was recorded for 30 min. Thereafter, the same nanotip was consecutively incubated for the same time in target solutions at different concentrations. As can be seen from the figure, only a 3.5% fluorescence increase was observed in presence of the random sequence, whereas the increase with only 0.01 μM of the target was of about 13%. Each point of the graph is evaluated as the sum of the optical intensity detected by the optical spectrograph between 650 and 720 nm.

The curve in Figure 7 well testifies for the feasibility of the MB-coated nanotip to act as a sensor, providing also information on the kinetics of the interaction between the MB and the target. The nanotip was exposed to the different solutions (random and target solutions) for 30 min in order to perform an opto-chemical characterization, even if the timing needed was showed to be minor than 10 min.

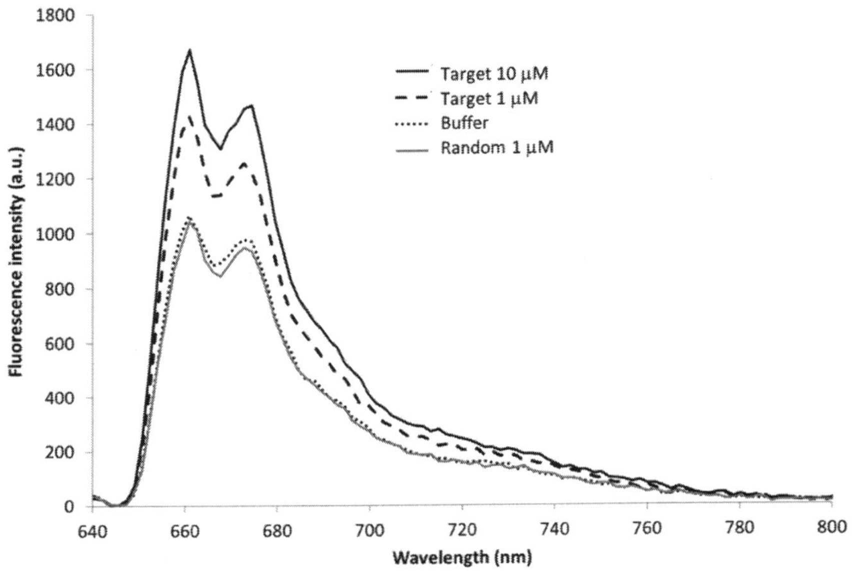

Figure 6. MB immobilized onto the nanotip, characterization in presence of the target (1 µM and 10 µM) and of the random sequence (1 µM). λ_{ex} 635 nm, exposure time 1 s.

Figure 8 shows the calibration curve obtained by averaging the experimental data at the plateau of the binding curve and calculating the related standard deviation for each target concentration. A Limit of Detection (LOD) of 0.57 nM for the proposed sensor was calculated from the calibration curve considering three times the greater difference between the experimental data and the fitting curve (maximum of residuals).

The limit of detection achieved opens up the possibility of using the proposed nanotip to detect mRNAs inside the cytoplasm of living cells. In fact, the concentration of specific mRNAs has been estimated to be in the nM range [10], although it is quite difficult to estimate the mRNA concentration inside the cytoplasm since it is not homogeneously distributed, but it has preferential locations. In the case of survivin mRNA, which is the molecule of interest, preferential localization is at the perinuclear position [34].

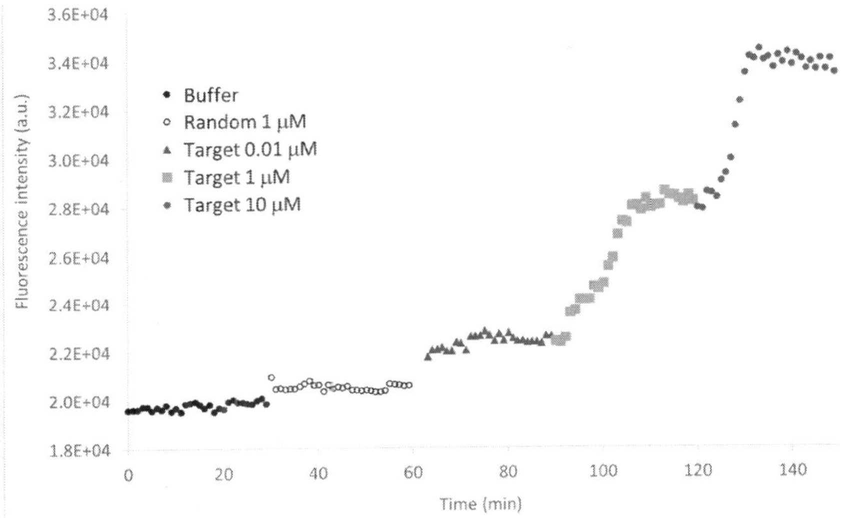

Figure 7. Fluorescence signal (*vs.* time) during the interaction of the MB immobilized onto the nanotip with the random sequence, and the target at different concentrations (0.01 µM, 1 µM and 10 µM).

By investigating the possibility of regenerating the nanotip for multiple use by the use of HCl, we discovered that it is possible to increase the performance of the sensor. Figure 9 shows the comparison between the response of the fiber nanotip to the 30 min exposure to 1 µM target solution (grey tracing) shown in Figure 7 (green tracing) and the response to the same target concentration after having exposed the fiber nanotip to HCl 2 mM for 1 min (black tracing). In this graph, the baselines of the two curves were both moved to the value of zero for a better comparison. After the HCl treatment, the nanotip was thoroughly washed in tris-buffer and left to equilibrate in this buffer until a stable fluorescence signal was reached, before exposing it to the target solution.

An 8-fold increase in the fluorescence signal due to the interaction of the MB with the target can be observed. This behaviour can be ascribed to the unfolding effect of HCl which is able to open the hairpin structure of the MB and the subsequent refolding of the beacon hairpin structure in a more efficient manner with respect to the untreated beacon which could be also folded around the nanotip and consequently hindered towards the hybridisation with the target. This fact can provide the explanation of the faster kinetics which is observed after the HCl treatment, with the plateau reached after only 4 min, instead of 15 min. The use of HCl 1–2 mM, among others regenerating solutions, is quite common in DNA biosensor regeneration [45,46,47]. The demonstration that the MB is not degraded by this treatment is given by the high and fast fluorescent increase from PBS (target zero) to the target 1 μM, as shown in Figure 8. The HCl treatment is therefore essential in order to achieve a correct and ready-to-interact MB layer on the fiber nanotip immediately after its immobilization which can occur also with secondary structures hampered towards the proper interaction with the target. The same treatment could also be used in the regeneration of the sensor after the interaction with the target in order to remove the target from the MB and make it fold again in its closed structure ready for a new hybridisation cycle.

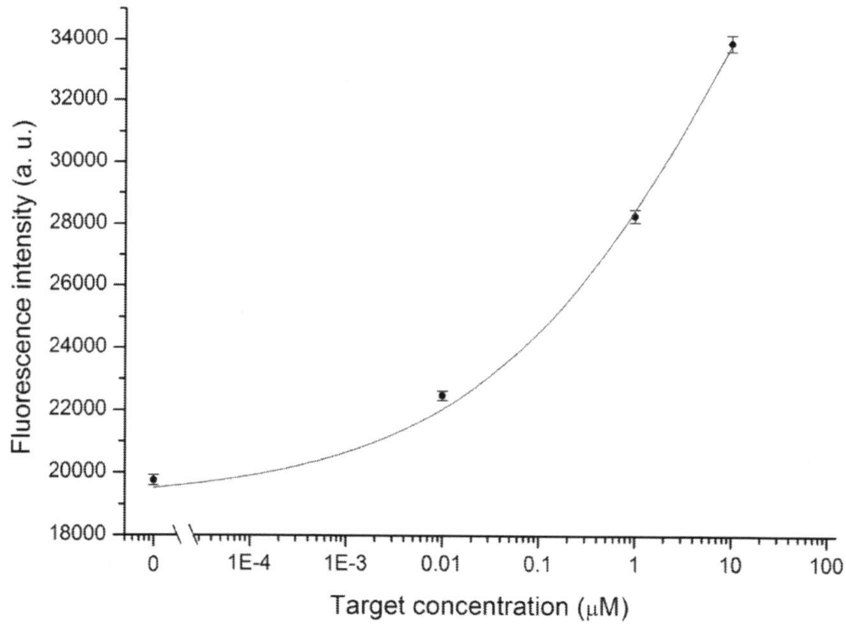

Figure 8. Calibration curve obtained from the interaction of the MB-coated nanotip with increasing concentrations of target.

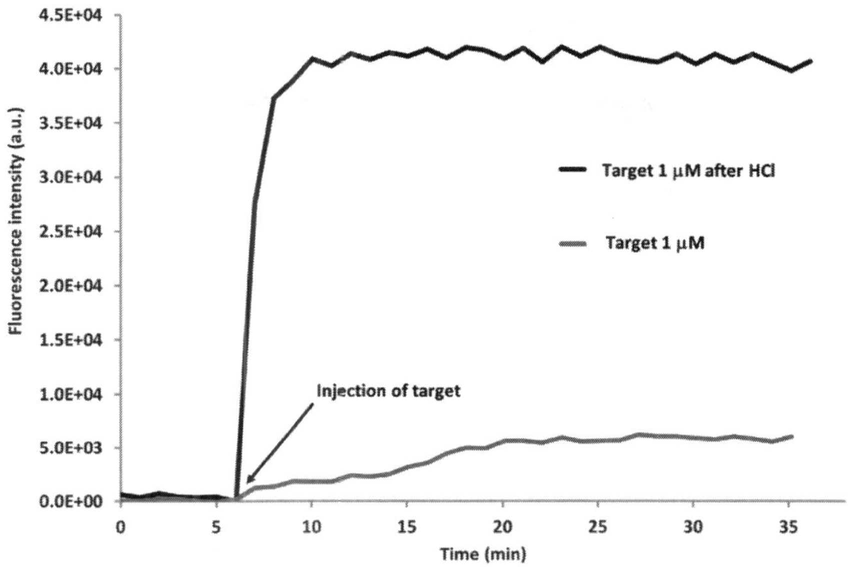

Figure 9. Kinetics of the fluorescence signal of the MB onto the nanotip after incubation with the target 1 μM without any previous treatment and after the treatment with HCl 2 mM.

4. CONCLUSIONS

The present paper describes the manufacture and optical characterization of an optical fiber nanotip coated with molecular beacons potentially capable to perform intracellular measurements. The optical characterization showed the good performance of the obtained nanotip in terms of selectivity and sensitivity, and a limit of detection of 0.57 nM was achieved. The treatment with HCl, besides allowing a regeneration of the nanotip, is able to provide a better performance of the molecular beacon with a further decrease of the limit of detection.

ACKNOWLEDGMENTS

The authors wish to thank Bruno Tiribilli for the acquisition of the AFM images, Giancarlo C. Righini for the fruitful discussion during the whole course of this study and the project FIRB ITALNANONET (RBPR05JH2P) for financial support.

AUTHOR CONTRIBUTIONS

Ambra Giannetti proposed the work, designed and supervised the experiments; Ambra Giannetti and Sara Tombelli performed the chemical/biochemical experiments; Andrea Barucci and Franco Cosi performed the fabrication of the nanotip; Cosimo Trono constructed the optical set-up and conduced the optical measurements; all authors wrote the paper and provided comments on it.

CONFLICTS OF INTEREST

The authors declare no conflict of interest.

REFERENCES

1. Vo-Dinh, T. Nanosensing at the single cell level. Spectrochim. Acta Part B At. Spectrosc. 2008, 63, 95–103.

2. Zheng, X.T.; Hu, W.; Wang, H.; Yang, H.; Zhou, W.; Li, C.M. Bifunctional electro-optical nanoprobe to real-time detect local biochemical processes in single cells. Biosens. Bioelectron. 2011, 26, 4484–4490.

3. Dahan, M.; Lévi, S.; Luccardini, C.; Rostaing, P.; Riveau, B.; Triller, A. Diffusion dynamics of glycine receptors revealed by single-quantum dot tracking. Science 2003, 302, 442–445.

4. Pathak, S.; Cao, E.; Davidson, M.C.; Jin, S.; Silva, G.A. Quantum Dot Applications to Neuroscience: New Tools for Probing Neurons and Glia. J. Neurosci. 2006, 26, 1893–1895.

5. Ma, Q.; Lin, Z.; Yang, N.; Li, Y.; Su, X.-G. A novel carboxymethyl chitosan–quantum dot-based intracellular probe for Zn2+ ion sensing in prostate cancer cells. Acta Biomater. 2014, 10, 868–874.

6. Duchi, S.; Sotgiu, G.; Lucarelli, E.; Ballestri, M.; Dozza, B.; Santi, S.; Guerrini, A.; Dambruoso, P.; Giannini, S.; Donati, D.; et al. Mesenchymal stem cells as delivery vehicle of porphyrin loaded nanoparticles: Effective photoinduced in vitro killing of osteosarcoma. J. Control Release 2013, 168, 225–237.

7. Giannetti, A.; Tombelli, S.; Baldini, F. Oligonucleotide optical switches for intracellular sensing. Anal. Bioanal. Chem.2013, 405, 6181–6196.

8. Sokolova, V.; Epple, M. Inorganic Nanoparticles as Carriers of Nucleic Acids into Cells. Angew. Chem. Int. Ed. 2008, 47, 1382–1395.

9. Ratto, F.; Matteini, P.; Centi, S.; Rossi, F.; Pini, R. Gold nanorods as new nanochromophores for photothermal therapies. J. Biophoton. 2010, 4, 64–73.

10. Kihara, T.; Yoshida, N.; Kitagawa, T.; Nakamura, C.; Nakamura, N.; Miyakea, J. Development of a novel method to detect intrinsic mRNA in a living cell by using a molecular beacon-immobilized nanoneedle. Biosens. Bioelectron. 2010,26, 1449–1454.

11. Yum, K.; Wang, N.; Yu, M.-F. Nanoneedle: A multifunctional tool for biological studies in living cells. Nanoscale 2010,2, 363–372.

12. Dufrêne, Y.; Garcia-Parajo, M.F. Recent progress in cell surface nanoscopy: Light and force in the near-field. Nano Today 2012, 7, 390–403.

13. Zheng, X.T.; Li, C.M. Single living cell detection of telomerase over-expression for cancer detection by an optical fibre nanobiosensor. Biosen. Bioelectron. 2010, 25, 1548–1552.

14. Zheng, X.T.; Yang, H.B.; Li, C.M. Optical Detection of Single Cell Lactate Release for Cancer Metabolic Analysis. Anal. Chem. 2010, 82, 5082–5087.

15. Vitol, E.A.; Orynbayeva, Z.; Friedman, G.; Gogotsi, Y. Nanoprobes for intracellular and single cell surface-enhanced Raman spectroscopy (SERS). J. Raman Spectrosc. 2012, 43, 817–827.

16. Brasuel, M.; Kopelman, R.; Kasman, I.; Miller, T.J.; Philbert, M.A. Ion Concentrations in Live Cells from Highly Selective Ion Correlations Fluorescent Nano-Sensors for Sodium. IEEE Proc. 2002, 1, 288–292.

17. McCulloch, S.; Uttamchandani, D. Development of a fibre optic micro-optrode for intracellular pH measurements. IEE Proc. Optoelectron. 1997, 144, 162–167.

18. Xu, H.; Aylott, J.W.; Kopelman, R.; Miller, T.J.; Philbert, M.A. A Real-Time Ratiometric Method for the Determination of Molecular Oxygen Inside Living Cells Using Sol–Gel-Based Spherical Optical Nanosensors with Applications to Rat C6 Glioma. Anal. Chem. 2001, 73, 4124–4133.

19. Vo-Dinh, T.; Alarie, J.P.; Cullum, B.M.; Griffin, G.D. Antibody-based nanoprobe for measurement of a fluorescent analyte in a single cell. Nat. Biotechnol. 2000, 18, 764–767.

20. Vo-Dinh, T.; Kasili, P. Fibre-optic nanosensors for single-cell monitoring. Anal. Bioanal. Chem. 2005, 382, 918–925.

21. Vo-Dinh, T.; Kasili, P.; Wabuyele, M. Nanoprobes and nanobiosensors for monitoring and imaging individual living cells. Nanomed. Nanotechnol. Biol. Med. 2006, 2, 22–30.

22. Zhang, Y.; Dhawan, A.; Vo-Dinh, T. Design and Fabrication of Fibre-Optic Nanoprobes for Optical Sensing. Nanoscale Res. Lett. 2011, 6.

23. Zhang, J.; Laiwalla, F.; Kim, J.A.; Urabe, H.; Van Wagenen, R.; Song, Y.-K.; Connors, B.W.; Zhang, F.; Deisseroth, K.; Nurmikko, A.V. Integrated device for optical stimulation and spatiotemporal electrical recording of neural activity in light-sensitized brain tissue. J. Neural Eng. 2009, 6.

24. Tyagi, S.; Kramer, F.R. Molecular beacons: Probes that fluoresce upon hybridization. Nat. Biotechnol. 1996, 14, 303–308.

25. Liu, X.; Farmerie, W.; Schuster, S.; Tan, W. Molecular Beacons for DNA Biosensors with Micrometer to Submicrometer Dimensions. Anal. Biochem. 2000, 283, 56–63.

26. Monroy-Contreras, R.; Vaca, L. Molecular Beacons: Powerful Tools for Imaging RNA in Living Cells. J. Nucleic Acids2011, 2011.

27. Santangelo, P.J. Molecular beacons and related probes for intracellular RNA imaging. Wiley Interdiscip. Rev. Nanomed. Nanobiotechnol. 2010, 2, 11–19.

28. Wang, Q.; Chen, L.; Long, Y.; Tian, H.; Wu, J. Molecular Beacons of Xeno-Nucleic Acid for Detecting Nucleic Acid.Theranostics 2013, 3, 395–408.

29. Boutorine, A.S.; Novopashina, D.S.; Krasheninina, O.A.; Nozeret, K.; Venyaminova, A.H. Fluorescent Probes for Nucleic Acid Visualization in Fixed and Live Cells. Molecules 2013, 18, 15357–15397.

30. Hecht, B.; Sick, B.; Wild, U.P.; Deckert, V.; Zenobi, R. Scanning near-field optical microscopy with aperture probes: Fundamentals and applications. J. Chem. Phys. 2000, 112, 7761–7774.

31. Carpi, C.; Fogli, S.; Giannetti, A.; Adinolfi, B.; Tombelli, S.; Da Pozzo, E.; Vanni, A.; Martinotti, E.; Martini, C.; Breschi, M.C.; et al. Theranostic

properties of a survivin-directed molecular beacon in human melanoma cells. PLoS ONE 2014,11.

32. Adinolfi, B.; Carpi, S.; Giannetti, A.; Nieri, P.; Pellegrino, M.; Sotgiu, G.; Tombelli, S.; Trono, C.; Varchi, G.; Baldini, F. Complex nanostructures based on oligonucleotide optical switches and nanoparticles for intracellular mRNA sensing and silencing. Proc. Eng. 2014, 87, 751–754.

33. Nitin, N.; Santangelo, P.J.; Kim, G.; Nie, S.; Bao, G. Peptide-linked molecular beacons for efficient delivery and rapid mRNA detection in living cells. Nucleic Acids Res. 2004, 32.

34. Santangelo, P.J.; Nix, B.; Tsourkas, A.; Bao, G. Dual FRET molecular beacons for mRNA detection in living cells.Nucleic Acids Res. 2004, 32.

35. Qiao, G.; Gao, Y.; Li, N.; Yu, Z.; Zhuo, L.; Tang, B. Simultaneous Detection of Intracellular Tumor mRNA with Bi-Color Imaging Based on a Gold Nanoparticle/Molecular Beacon. Chem. Eur. J. 2011, 17, 11210–11215.

36. Xue, Y.; An, R.; Zhang, D.; Zhao, J.; Wang, X.; Yang, L.; He, D. Detection of survivin expression in cervical cancer cells using molecular beacon imaging: New strategy for the diagnosis of cervical cancer. Eur. J. Obstet. Gynecol. Reprod. Biol.2011, 159, 204–208.

37. Barucci, A.; Cosi, F.; Pelli, S.; Soria, S.; Nunzi Conti, G.; Giannetti, A.; Righini, G. Method of Fabricating Structures, Starting from Material Rods. Patent Pending PCT/EP2014/071743, 10 October 2014.

38. Barucci, A.; Cosi, F.; Giannetti, A.; Pelli, S.; Griffini, D.; Insinna, M.; Salvadori, S.; Tiribilli, B.; Righini, G.C. Optical fibre nanotips fabricated by a dynamic chemical etching for sensing applications. J. Appl. Phys. 2015, 117.

39. Taylor, G.I. Stability of a Viscous Liquid Contained between Two Rotating Cylinders. Philosoph. Trans. R. Soc. London Ser. A Contain. Pap. Math. Phys. Charact. 1923, 223, 289–343.

40. Andereck, C.D.; Liu, S.S.; Swinney, H.L. Flow regimes in a circular Couette system with independently rotating cylinders. J. Fluid Mech. 1986, 164, 155–183.

41. Tombelli, S.; Ballestri, M.; Giambastiani, G.; Giannetti, A.; Guerrini, A.; Sotgiu, G.; Trono, C.; Tuci, G.; Varchi, G.; Baldini, F. Oligonucleotide switches and nanomaterials for intracellular mRNA sensing. Proc. SPIE 2013, 8798.

42. Pini, V.; Tiribilli, B.; Gambi, C.M.C.; Vassalli, M. Dynamical characterization of vibrating AFM cantilevers forced by photothermal excitation. Phys. Rev. B. 2010, 81.

43. Giannetti, A.; Baldini, F.; Ballestri, M.; Ghini, G.; Giambastiani, G.; Guerrini, A.; Sotgiu, G.; Tombelli, S.; Trono, C.; Tuci, G.; et al. Intracellular nanosensing and nanodelivery by PMMA nanoparticles. Lect. Notes Electri. Eng. 2014, 162, 69–75.

44. Giannetti, A.; Tombelli, S.; Trono, C.; Ballestri, M.; Giambastiani, G.; Guerrini, A.; Sotgiu, G.; Tuci, G.; Varchi, G.; Baldini, F. Intracellular delivery of molecular beacons by PMMA nanoparticles and carbon nanotubes for mRNA sensing. Proc. SPIE. 2013, 8596.

45. Lechuga, L.M. New frontiers in optical biosensing. In Proceedings of the 13th European Conference on Integrated Optics (ECIO 2007), Copenhagen, Denmark, 25–27 April 2007.

46. Mannelli, I.; Minunni, M.; Tombelli, S.; Wang, R.; Spiriti, M.M.; Mascini, M. Direct immobilization of DNA probes for the development of affinity biosensors. Bioelectrochemistry 2005, 66, 129–138.

47. Dos Santos Riccardi, C.; Yamanaka, H.; Josowicz, M.; Kowalik, J.; Mizaikoff, B.; Kranz, C. Label-free DNA detection based on modified conducting polypyrrole films at microelectrodes. Anal. Chem. 2006, 78, 1139–1145.

7

MATERIALS DEVELOPMENT FOR NEXT GENERATION OPTICAL FIBER

John Ballato [1,*] and Peter Dragic [2]

[1] The Center for Optical Materials Science and Engineering Technologies (COMSET) and the Department of Materials Science and Engineering, Clemson University, Clemson, SC 29634, USA

[2] Department of Electrical and Computer Engineering, University of Illinois at Urbana-Champaign, Urbana, IL 61801, USA

ABSTRACT

Optical fibers, the enablers of the Internet, are being used in an ever more diverse array of applications. Many of the rapidly growing deployments of fibers are in high-power and, particularly, high power-per-unit-bandwidth systems where well-known optical nonlinearities have historically not been especially consequential in limiting overall performance. Today, however, nominally weak effects, most notably stimulated Brillouin scattering (SBS) and stimulated Raman scattering (SRS) are among the principal phenomena restricting continued scaling to higher optical power levels. In order to address these limitations, the optical fiber community has focused dominantly on geometry-related solutions such as large mode area (LMA) designs. Since such scattering, and all other linear and nonlinear optical phenomena including higher order mode instability (HOMI), are fundamentally materials-based in origin, this paper unapologetically advocates material solutions to present and future performance limitations. As such, this paper represents a 'call to arms' for material scientists and engineers to engage in this opportunity to drive the future development of optical fibers that address many of the grand engineering challenges of our day.

KEYWORDS

optical fiber; high energy lasers; stimulated Brillouin scattering; stimulated Raman scattering

1. INTRODUCTION

Light today enables an annual global economic impact of about $7.5 trillion (USD) [1], which is approximately half of the gross domestic product (GPD) of the United States; and light's economic value is growing. The three largest market sectors relying on light-based products are those associated with telecommunications, transportation, and biotechnologies [2]. The telecommunications sector constitutes over 50% of the economic value. Core technologies enabled by light include e-commerce, Internet music and movies, and information routing and storage. The transportation sector includes laser-based manufacturing (e.g., precision cutting, drilling, and welding) of automobiles and airplanes. The biotechnological sector includes laser and optics-based diagnostic and surgical systems as well as Laser-Assisted in-situ Keratomileusis (LASIK) eye surgery, blood testing, and gene sequencing [2]. Smaller by market value but critical nonetheless are laser-based defense, security, and sensor systems, which will continue to drive technology development especially at "specialty" (high value and performance, low volume) levels.

Another way to view this cumulative effect on global goods and services is that light-based solutions exist (or could reasonably exist with further advancements) for many of the technology platforms identified as being most central to society over the next decade [3]. Despite this centrality of light to the many modern conveniences there has been decades-long erosion in the appreciation given to the materials as enablers of such technologies [4]. Modern and prototypic future systems that are prescribed as solutions to present performance limitations are strongly driven by 'structural' (geometric) as opposed to material considerations. Perhaps the best exemplars of these are photonic crystal fibers (PCFs) which are transversely periodic arrays of air holes surrounded by glass that run longitudinally down an optical fiber. Such designs range from the fairly straight-forward (note that the word "simple" was not used) to the intricately absurd; at least with an eye towards scalability and industrial, albeit low volume, manufacturing. Yet while PCFs can force light to do its bidding in a wide variety of useful ways, many phenomena, such as reduced nonlinearities, can be accomplished more simply through materials; or so this paper plans to promote and prove.

In reality and with practicality in mind, the true solution is both. It takes little more than opening a basic physics book to remember that properties are dependent upon both shape and substance; e.g., the capacitance of a capacitor—amongst the most elementary electronic devices—is dependent upon its geometry (area and thickness) and the material (specifically its permittivity) from which it is made. However, we, the materials community, are at a cross-road. A recent (albeit US-centric) study by Corning Incorporated [5] concludes by noting that "... less than one-quarter of students at US universities who are doing research in glass science are studying systems that would make them well prepared for a future career in the glass industry. In our experience ... students with expertise in glass families that are industrially relevant ... are more likely to be hired into a position in industry and also require less on the job technical training after being hired." The authors of that work go on to identify a very rich range of problems that materials scientists and engineers can work on (and get employed to solve!) that would have significant societal benefits.

Accordingly, this paper seeks to amplify further the recommendations of two recent works [4,5] by providing a review of materials-related solutions to power-scaling in high energy optical fiber laser-based systems. Given that the lowest threshold nonlinearities that presently limit performance are stimulated Brillouin scattering (SBS) and stimulated Raman scattering (SRS), these phenomena will be used as primary exemplars. Also to be discussed are higher order mode instabilities (HOMI) and management of the nonlinear refractive index. It is the authors' hope that this work, along with [4,5], will help to reinvigorate the global materials community to rejoin the development of next generation optical fibers.

2. RESULTS AND DISCUSSION

The intensity of light in both high-power continuous wave (CW) and high power-per-unit-bandwidth pulsed laser systems is sufficient to induce nonlinearities in the material response to the propagating electromagnetic field. The result is a number of nonlinear and parametric photonic processes that are widely used or alternatively considered to be parasitic to system performance [6], including Raman scattering (or stimulated Raman scattering, SRS), Brillouin scattering (or stimulated Brillouin scattering, SBS), Self-Phase Modulation (SPM), and Four-Wave Mixing (FWM). Each of these is described in greater detail in Section 2.1 as are their materials' origins.

Since these parasitic effects are intensity-dependent, one approach to their control is to spread the optical power out over a larger cross-sectional area thereby reducing the effective power-per-unit area. In practice, this has led to so-called "large mode area" (LMA) fibers where the dimension of the core in which the light is propagating is expanded relative to more conventional designs. However, such large core sizes, coupled with typical fiber numerical apertures (NAs) which together generally render such fibers 'few-moded' almost always promote multimode guidance such that the beam quality of the resultant laser light is deteriorated. Several methods to overcome this obstacle have been proposed and implemented. A few of these are outlined next.

In principle, restricting the light launch conditions such that only the fundamental optical mode (FOM) is excited is one way to achieve the highest beam quality [7] available from such fibers. However, mode coupling is brought about through fiber bends, packaging, and environment, leading to degradation in the quality (or M^2) of the beam in most practical situations. Therefore, an early approach to overcoming the inevitable multimode nature of the beam in these LMA fibers was the discovery that via careful bending of the fiber, higher order modes (HOMs) could be 'peeled away' from the core [8] leaving what effectively amounts to single mode operation. The technique calls for coiling the fiber into a preselected diameter such that the differential loss between the fundamental mode and HOMs is large. In this way, the fundamental mode dominates the laser characteristics and captures essentially all of the available gain. It too is possible to design a microstructured fiber for 'endlessly single mode operation' [9] with extremely low effective NA, but the intolerance of such fibers to bending has largely confined them to the rod-like regime, with the resultant lasers more closely resembling a cross-over of sorts between bulk and fiber lasers. In the restricted mode launch regime, it was subsequently found that excitation of a HOM in a few-moded fiber offered increased robustness of the mode propagation which was coupled with the added benefit that the HOMs have larger mode areas than the FOM [10].

Other "effectively" single mode LMA fibers were also subsequently developed to force what otherwise is a multimode design into behaving like a single mode fiber. As with the simple fiber coiling method described above, the goal is to provide some sort of differential loss mechanism such that the HOMs experience vastly greater propagation losses than the FOM. Hence the FOM will experience the highest gain and dominate the modal distribution at the laser output. A few recent fiber designs include the chirally-coupled core fiber (HOMs are coupled to and shed from one or more satellite waveguides that form a helix around the fiber core) [11], photonic bandgap Bragg fiber (where only the FOM propagates with low loss) [12], solid photonic bandgap fibers (where bending loss is exacerbated on HOMs relative to the FOM) [13], leakage channel fibers (wherein all modes are lossy, but the loss is minimized for the FOM) [14], and selectively-doped triple clad fibers (where the FOM has the highest overlap with the active region in the fiber) [15]. While there is clearly no shortage of methods utilized to improve the beam quality from an LMA fiber, such fibers generally are quite complex in their cross-sectional geometries. Furthermore, while effectively single mode operation does result, there is a power threshold above which single mode operation is unstable [16,17]. This 'modal instability' (MI) or 'higher order mode instability' (HOMI) yields a time varying output beam profile that negates many of the benefits of large mode areas fibers.

The authors have recently articulated a different approach to mitigating and, in some cases, completely removing from consideration, these parasitic effects based on the judicious choice of materials rather than on fiber geometry [4]. Specific examples are provided in Section 2.2 below. However, it is necessary first to describe the interaction of light with the optical material (fiber) through which they propagate and how such nonlinear and/or parasitic phenomena arise. Only then can materials-related solutions be more fully envisioned and appreciated.

2.1. The Materials Physics of Linear and Nonlinear Optical Phenomena

Depending on the application, nonlinear phenomena in optical fiber can be useful or deleterious. From the perspective of achieving high powers from active optical fiber systems, these processes are generally parasitic in nature. They give rise to significant limitations to power scaling and tremendous effort has been made in suppressing them (or in other words, elevating the threshold power at which these processes become significant). Since these processes are intensity-dependent, a larger mode diameter in an optical fiber usually implies a higher 'turn-on' threshold for these interactions. As described above, this has led to a number of fiber designs aimed at increasing the mode field diameter (MFD) in an optical fiber while maintaining robust single mode operation. While nonlinear effects can be controlled to some extent with proper waveguide design, each of these processes can be described mathematically via a combination of materials properties that cooperate to define a 'gain' or 'gain coefficient'. With careful selection of precursor materials, it is possible to tailor or design the composition such that these gain coefficients are reduced to acceptable levels for a given application.

The first nonlinear process investigated is stimulated Brillouin scattering (SBS). SBS is an interaction between hypersonic (thermally excited) acoustic waves and the optical signal in a fiber. Brillouin scattering begins as a spontaneous process with optical back-scattering most efficiently taking place from acoustic waves that are Bragg-matched to the optical wave. Simply put, the acoustic wave is a periodic longitudinal pressure, and therefore also density, variation. The spatially modulated density of the material then corresponds to a spatially modulated refractive index. Via electrostriction, the interference between the forward-propagating signal and back-scattered light feeds the acoustic (pressure) wave. This 'positive feedback' process increases in efficiency as the power is increased until 'threshold' is reached wherein the acoustic wave becomes a highly-efficient reflector to the optical signal. At this point Brillouin scattering is in the stimulated state. In general, SBS limits the amount of light per unit bandwidth that can be transmitted down or generated in an optical fiber. As such, it typically has the lowest threshold of all the nonlinear processes in narrow linewidth systems.

The Brillouin gain coefficient for a given optical wavelength (λ) is given to be $g_B = 2\pi n^7 p^2{}_{12}/c\lambda^2 \rho V_a \Delta\nu_B$ ([18], which has units of m/W), where n is the refractive index, p_{12} is a Pockels photoelastic constant, c is the speed of light, ρ is the mass density, V_a is the acoustic velocity, and $\Delta\nu_B$ is the Brillouin spectral width (which is inversely proportional to the acoustic damping time). Given the dependencies of the Brillouin gain on these physical parameters, beneficial materials properties for the suppression of SBS include: (1) a large acoustic velocity; (2) relatively low index; (3) relatively large mass density; (4) large Brillouin spectral width (*i.e.*, large acoustic attenuation coefficient); and (5) low p_{12}. Unfortunately, a large density typically suggests a relatively large refractive index, so the density and index tend to mitigate one another with respect tog_B. Regardless, a clear guideline with selection criteria can be identified for choosing materials for reduced Brillouin gain.

To model the system, an additive approach based on the work of Schott and Winkelmann [19,20] is utilized to calculate each of the terms in this equation for a mixed multicomponent glass (such as a germanosilicate glass). Specifically, these quantities are

(a) the acoustic velocity

$$V_a = \left(\sum_{i=1}^{N} (m_i - m_{i-1}) / V_{a,i}\right)^{-1};$$

(b) the mass density

$$\rho = \sum_{i=1}^{N} (m_i - m_{i-1}) \rho_i;$$

(c) the refractive index

$$n = \sum_{i=1}^{N} (m_i - m_{i-1}) n_i; \text{ and}$$

(d) the Brillouin spectral width ($\Delta\nu_B$),

$$\Delta\nu_B = (V_a/\pi)\sum_{i=1}^{N}(m_i - m_{i-1})(\nu_a/\nu_{\text{ref}})^2\alpha_i(\nu_{\text{ref}})$$

which is proportional to the acoustic attenuation coefficient α (units of m^{-1}). The scaling term $(\nu_a/\nu_{ref})^2$ accounts for the usual frequency-squared dependence of the intrinsic linewidth, while α is always specified at a reference acoustic frequency ν_{ref}. It is important to note that it is possible (but seemingly rare) for the scaling term to be a sub-square function [21], depending on the material and thermal history of the glass. For each calculation $(m_i - m_{i-1})$ is the volume of constituent i in the fiber. The Pockels' elasto-optic coefficient p_{12} is calculated utilizing a similar additive approach, but it is lumped in with the refractive index rather than calculated directly. For details, the reader is asked to refer to Reference [22]. The key point with regard to p_{12} is that for a material this value can be either positive or negative. Therefore a combination of materials with photoelastic constants of the former and the latter types can give rise to a composition wherein p_{12} is very close to zero. As a result, the possibility of a material with zero Brillouin gain exists. Such negation of certain glass coefficients in order to suppress or eliminate a nonlinear effect can also be applied to HOMI in high power optical fibers. It is well-known that restricted mode launch and careful modal control can give rise to effective single-mode operation in a few-moded fiber utilized in a fiber laser. However, at some sharp threshold power a dynamic randomization of the mode distribution at the laser output is observed (referred to as higher order mode instability, or HOMI), and is believed to be the result of a thermally-induced refractive index grating. The process is believed to be driven in part by dn/dT [16,17]; the grating forms via a longitudinally-periodic thermal variation in the fiber. It is interesting to point out that HOMI is not a process that can be reduced with a wider MFD, as it becomes more significant of a problem as the core diameter is increased [16,17].

While rendering a gain coefficient zero may be possible for some interactions, this may not be the case for others. An example is stimulated Raman scattering (SRS). Raman scattering is an interaction between an optical wave and optical phonons [18]. In the case of silica-dominated materials systems, the Raman spectrum has a peak gain positioned roughly 13.2 THz from the signal frequency. It can be considered to be a parasitic effect in high-peak-power fiber laser systems where wavelength control is mandatory [23]. The excitation of SRS from spontaneous scattering can lead to wavelength shifts and power instabilities that degrade laser performance, thus presenting the need for its suppression. However, the dependence of the Raman gain is on material properties (Raman cross section, refractive index) that do not appear to necessarily facilitate a zeroing of this value [23]. Instead, materials can be selected such that one or more of the following conditions are met [24]: (1) the material is highly disordered in order to broaden the Raman gain spectrum thereby reducing the peak value; (2) high concentrations of materials with low Raman gain are utilized; and (3) materials are utilized that have Raman spectra with minimal overlap. For (3), a mixture of two materials with similar-strength but not overlapping Raman gain spectra could cut the peak Raman gain coefficient in half.

Processes such as Self-Phase Modulation (SPM) and Four-Wave Mixing (FWM) are results of the dependence of the refractive index on the optical intensity [18]. SPM can be observed in pulsed (time-varying) systems, and it results from an intensity-induced time-varying refractive index (due to the time-varying nature of the signal itself). In multi-wavelength systems, FWM is the transfer of an amount of energy to a new wavelength by the amount that is (plus or minus) the difference in photon energy between two extant optical frequencies. In the case of non-degenerate FWM, if the existing optical frequencies are ν_1 and ν_2 (and $\nu_1 \neq \nu_2$) then the new frequencies will be at $2\nu_1 - \nu_2$ and $2\nu_2 - \nu_1$. Similarly, degenerate FWM can also occur (where $\nu_1 = \nu_2$) with new frequency components appearing both to the blue and red side of ν_1. The general effect of these processes is to broaden and modify the optical spectrum, and depending on the application this may be undesirable in high peak power and multi-wavelength optical amplifier systems. Since the dependence of the refractive index is given by $n(I) = n_0 + n_2 I$ where n_0 is the linear refractive index, I is the intensity of the optical signal, and n_2 is the nonlinear refractive index, suppression of these parametric phenomena will require a minimization of n_2.

From the above descriptions, a materials solution to these performance-limiting parasitic phenomena rely on developing optical fibers using materials with reduced (if possible, zero) photoelasticity (p_{12}), peak Raman gain, nonlinear refractive index, and thermo-optic coefficients. Though one composition that exhibits all of these effects would be ideal, it generally is not necessary since each phenomena has a different threshold condition and so often is application specific. The authors therefore suggest a more thoughtful consideration of the fiber's compositions based on the specific application and anticipated optical power levels.

2.2. Materials Solutions

As noted above, stimulated Brillouin scattering (SBS) and stimulated Raman scattering (SRS) are two of the principal nonlinearities that plague continued power scaling-up in fiber-based laser systems. The sections below provide an overview of selected optical fiber material solutions to mitigate and, in the case of SBS potentially negate entirely, these parasitic effects.

2.2.1. Intrinsically Low Brillouin Gain Glasses and Optical Fibers

The fundamental question at play is less so "what materials can be used to lower the Brillouin gain" but, rather, "do the materials that reduce Brillouin gain (when added into silica) form a sufficiently stable glass to draw into optical fiber?" While this question may seem straight-forward to answer, it is not always the case. Firstly, glass stability is a kinetic consideration and so the method of fabrication needs to be determined. For the purposes of this work, the conventional method of manufacturing will be one of the chemical vapor deposition methods employed in the commercial production of optical fiber; *i.e.*, outside vapor deposition (OVD), modified chemical vapor deposition (MCVD), vapor axial deposition (VAD), or plasma-assisted chemical vapor deposition (PCVD). In each case, the combination of temperature cycling between soot deposition and translation of the torch along the forming preform followed by subsequent densification and consolidation (and eventual fiber drawing) leads to a fairly extreme set of time and temperature conditions. Second, for practical reasons, the glass from which the fibers are made must be intrinsically low loss, mechanically robust, and compatible with the other fiber components used to make the fiber-based laser system. These conditions conspire to yield silica and high silica-content glasses as the standard family from which practical optical fibers are made.

This, of course, is not an impediment as silica is, and has been for millennia, a marvel when it comes to glass stability and properties. However, in returning to the more fundamental question of which glass systems could be used as the core phase in high performance optical fibers, the answer turns out to be a fairly small range of compositions. From a materials perspective, the principal additions to silica in commercial optical fibers are GeO_2, which raises the refractive index and can enhance photosensitivity, Al_2O_3, raises the refractive index and can enhance the solubility of active rare-earth dopants, P_2O_5, which raises the refractive index and reduces glass viscosity, B_2O_3, which reduces the refractive index and increases the expansion coefficient, and F, which reduces the refractive index and glass viscosity. Importantly, there are limits: for GeO_2 [25] and P_2O_5 [26], there is a complete solid solution with the base SiO_2 such that liquid-liquid or solid-solid immiscibility is not problematic and the optimum dopant concentration is mainly determined by its impact on refractive index, thermal expansion, and viscosity. For a surprisingly wide range of other additives into silica, such immiscibilities exist and greatly restrict the compositions from which optical fibers can be made. This material limitation has the added detriment of restricting the range of useful properties that the glass can exhibit. Accordingly, 'mixing the unmixable' can only be accomplished through novel manufacturing methods that obviate the combinations of time and temperature that conspire to facilitate the phase instabilities [27]. In the sections that follow, the fibers discussed are made using the "molten core" approach, which is described briefly in Section 3 below and references therein. For the purposes of this Review, suffice it to say that the molten core approach is a super-liquidus (or super-consulate point) process where the fiber is drawn at a temperature that exceeds the melting point (or immiscibility limit) of the core phase so that it is a homogeneous liquid. The cladding glass is selected such that it possesses the appropriate viscosity to draw into high quality fiber at this same temperature. Since the core is molten, and generally reasonably fluid, there is expected to be some dissolution of the cladding glass by the core melt during the fiber drawing process. As such, the core of the resultant fiber, which quenches into a glass as the fiber cools, is an 'interaction product' of the initial precursor composition and that of the cladding glass. Given that pure SiO_2 arguably is the most refractory of the conventional glasses, not to mention naturally compatible with commercial optical fiber components, it tends to be the cladding material of choice. Accordingly, the dissolution of some silica into the core melt during molten core processing has the (generally) added benefit of enhancing glass-forming stability of the resultant core.

In specific regard to suppressing stimulated Brillouin scattering (SBS), the natural consequence of the discussion above is whether or not a material solution exists rather than address the issue by spatially distributing the propagating mode over a larger cross-section of the core, *i.e.*, a large mode area approach. As noted above, the Brillouin gain at a given wavelength is dependent upon the refractive index, n, Pockels photoelastic constant, p_{12}, mass density, ρ, acoustic velocity, V_a, and Brillouin spectral width, $\Delta\nu_B$, of the material through which the light it transmitted. In terms of glass modifiers to silica, one desires compounds that increase the refractive index (for optical waveguide formation), density, acoustic velocity, and Brillouin line-width while reducing the photoelastic coefficient. Divalent metal oxides, such as the alkaline earths (MgO, CaO, SrO, and BaO) and a range of trivalent metal oxides, such as Al_2O_3, Y_2O_3, and the lanthanide oxides (Ln_2O_3 where Ln are the rare-earth elements) are known to impart the aforementioned features into silica glasses. However, in each case, there are quite limited compositional ranges over which miscibility exists in the silicate melt. A classic example of this, relevant to optical fiber systems, is the Al_2O_3-SiO_2 system, which exhibits a well-known two-phase, liquid-liquid immiscibility region for binary aluminosilicates [28]. Accordingly, the concentration of Al_2O_3 that can be homogeneously mixed into SiO_2 given the processing conditions of conventional optical fiber preforms is about 12 weight percent. Unfortunately, the binary systems Y_2O_3-SiO_2 [29], MgO-SiO_2 [30], CaO-SiO_2, SrO-SiO_2 and BaO-SiO_2 [31], and ternary systems Y_2O_3-Al_2O_3-SiO_2 [32] and MgO-Al_2O_3-SiO_2 [30] also suffer from similar limitations though the exact immiscibility range differs for each system.

As introduced briefly above, the molten core method offers a powerful route to the direct fabrication of optical fibers using materials that otherwise would be very difficult, if not impossible to form into conventional glass preforms and draw. Given the inevitable dissolution of some cladding glass into the core melt, it is often beneficial to begin with a core precursor composition that possesses more than the amount of a given compound desired in the final fiber. Further, and for completeness, it is worth noting that said core precursor phase can be amorphous or crystalline (single or polycrystalline) since it melts during the fiber draw process and therefore has no "memory" of its starting structure. Accordingly, for fiber possessing high concentrations of the aforementioned alkaline earth oxides and sesquioxides the pure end member was employed. As an example, for high Al_2O_3-content fibers, pure sapphire was employed [33]. Similarly for high Y_2O_3-Al_2O_3, BaO- and MgO-Al_2O_3-content fibers, YAG ($Y_3Al_5O_{12}$), BaO and spinel ($MgAl_2O_4$) were employed [34,35,36]. Since such compounds, at the concentrations realized, typically phase-separate when mixed with silica, these fibers exhibited a range of useful properties that had not previously been observed including a record low Brillouin scattering [33], an athermal Brillouin frequency condition [33], and the identification of an atensic (zero stress-dependent) Brillouin frequency composition [35].

In specific regard to Brillouin scattering, one particularly interesting opportunity that presents itself by using these novel glass compositions enabled by the molten core method is the ability to realize optical fibers possessing zero, or near-zero photoelasticity. As noted above, the Brillouin gain is dependent on several factors, including a square-dependence on the p_{12}photoelastic coefficient. While silica glass possesses a p_{12} value that is positive, many of the materials noted above possess also about negative p_{12} values. The result is a glass that can balance the positive and negative photoelastic contributions of its constituent compounds to yield a fiber with very low Brillouin scattering. As a case in point, sapphire (Al_2O_3) possesses a slightly negative p_{12} value (−0.03). When the sapphire melts during the molten core process and some of the silica (p_{12} = 0.271) from the cladding dissolves into it, the resulting aluminosilicate core exhibited an SBS gain that was reduced by nearly 20 dB in comparison to conventional glasses [33]. So 'mixing the unmixable' via the molten core approach can be very useful for obtaining glasses not previously fabricated yet ones with quite extraordinary properties; particularly as they relate to Brillouin scattering. Table 1 provides the derived values for the BGC parameters from a series of novel material fibers made recently. As can also be seen from Table 1, there is a wide range of tailorability in the parameters influencing the Brillouin gain that judicious choice of materials can affect.

Table 1. Brillouin-Gain-Related Material Parameters from Selected Crystal-Derived All-Glass Optical Fibers.

Parameter	Unit	Silica (SiO_2)	YAG-Derived Fiber	Al_2O_3-Derived Fiber	MgO-Derived Fiber	BaO-Derived Fiber	Yb_2O_3-Derived Fiber	La_2O_3-Derived Fiber
V_a	m/s	5970	7649	9790	8731	3131	4110	3979
ρ	kg/m^3	2200	3848	3350	3322	4688	8102	5676
$\Delta\nu_B$	MHz	17	253	274	†	178	1375	181
n	–	1.444	1.868	1.653	1.810	1.792	1.881	1.877
p_{12}	–	0.271	0.022	−0.03	†	−0.33	−0.123	−0.027
Reference	–	22	34	22,33	36	35	37	49

† Not able to be measured from fabricated fibers.

2.2.2. Intrinsically Low Raman Gain Glasses and Optical Fibers

Unlike Brillouin gain, Raman gain is not linked to material properties that could be potentially driven to zero by choice of composition. However, by analogy to LMA fibers where the propagating optical mode is spread out over the fiber cross-section to reduce its intensity at a given spatial location, the distribution of the glasses' bond energies can be broadened thereby reducing the effective Raman gain at a given wavelength. As a case in point, all-glass optical fibers derived from yttrium aluminum garnet (YAG) were fabricated [24]. It was found that the Raman peak at 440 cm^{-1}, which is associated with the Si–O–Si stretching mode, did not change position with core glass composition but was much broader spectrally likely due to a larger distribution of Si–O–Si bond angles in these more highly modified glasses. The defect line peaks at 490 cm^{-1} (D_1) and 600 cm^{-1} (D_2), which are attributable to the 4- and 3-member ring 'breathing' modes respectively, diminished with the combined (Y_2O_3 + Al_2O_3) content in the resultant glass though their spectral position did not change suggesting that yttria and alumina may lessen the number of ring structures in the glass. These same trends were also later observed for the sapphire-derived (high alumina content) fiber. Raman spectra for the YAG-derived fibers can be found in [24] whileFigure 1 shows the set of Raman spectra obtained for the sapphire-derived fiber described above. As mentioned, the features of the spectra trend in ways similar to the YAG-derived fiber. In fact, the Raman gain spectra for the YAG- and sapphire-derived fibers are nearly identical, except that the sapphire-derived fiber appears to be missing the line observed at 950 cm^{-1}(and attributed to the oxygen hole center [24]) from the YAG-derived fibers [37].

The spectra obtained suggest that the core material possesses much more disorder than the pure silica cladding glass. As a consequence of these modifications to the glass structure, the spontaneous Raman scattering intensity from the yttrium-aluminosilicate fibers was found to increase linearly with increasing silica content [24] with a reduction of about 3 dB measured for a silica content of about 67.5 mol%. This trend is similar for the sapphire-derived fiber, with the concentrations of alumina being 26.9, 30.8, and 41.2 mol% for the three aluminosilicate fibers and a maximum reduction in Raman gain of about 2.5 dB for the highest alumina content (see Figure 1). While the reduction in Raman gain can partly be attributable to the replacement of silica with materials of relatively lower Raman gain, it is interesting to note that other common fiber dopant materials (e.g., GeO_2, P_2O_5, and B_2O_3) have Raman gain coefficients larger than that of silica. Thus, the YAG-derived fiber results highlighted here suggest that alumina and yttria impart intrinsically lower Raman gain into the glass. An additional benefit, therefore, of these low-silica content yttrium aluminosilicate glasses for optical fibers is a reduced Raman (and Brillouin) gain and therefore increased SRS (and SBS) threshold.

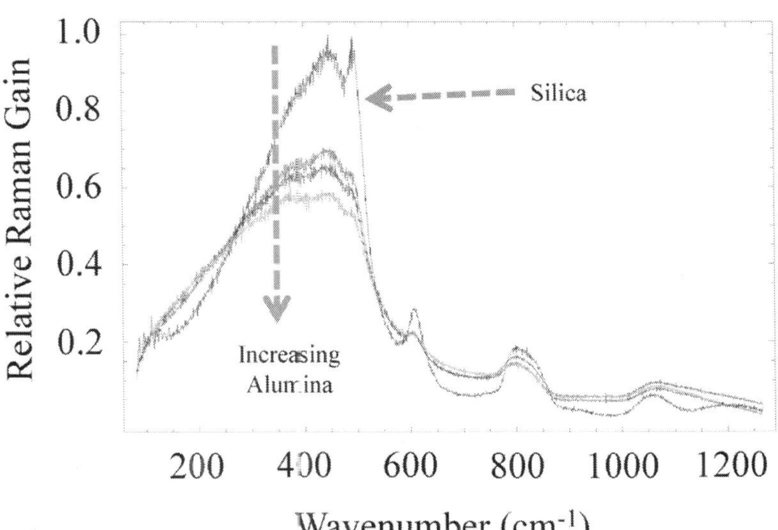

Figure 1. Raman gain spectra (normalized to silica) measured from the sapphire derived fiber for three different alumina concentrations. The spectrum appears to broaden and get weaker relative to the cladding (silica, black line).

It is important to note that even with the addition of 'high-Raman-gain' materials to silica it is still possible to achieve some level of reduction to the Raman gain. As an example, the Raman spectra for the baria-derived fiber [35] are shown inFigure 2. The baria concentrations in the three fibers are

10.4, 12.3, and 18.4 mol%. Relative to the cladding spectrum (pure silica), many of the conclusions about the shape of the spectra obtained from the YAG- and sapphire-derived fiber also seem to hold for the bariosilicate fibers. However, it is also clear that several new lines appear, including a prominent one near 1071 cm^{-1} likely attributable to the Ba-O bond. From the spectra and given the concentrations of baria in the fibers, it can be deduced that the strength of the aforementioned Raman line is relatively large in the glass. In fact, one can deduce by extrapolation that at a baria concentration in the vicinity of 25–30 mol%, the Raman gain will be larger in the baria-derived fiber than that in pure silica. However, at lower BaO concentrations, the contribution to the Raman gain spectrum by baria is at a different phonon frequency, not overlapping with a strong silica phonon line. Thus, as long as the concentration of baria is kept low enough such that its contribution to the Raman spectrum is weaker than silica, the absolute maximum Raman gain coefficient can be reduced. The bottom line is that, in the case of the bario-silicate glasses, the net Raman gain can still be reduced by up to 30% relative to silica.

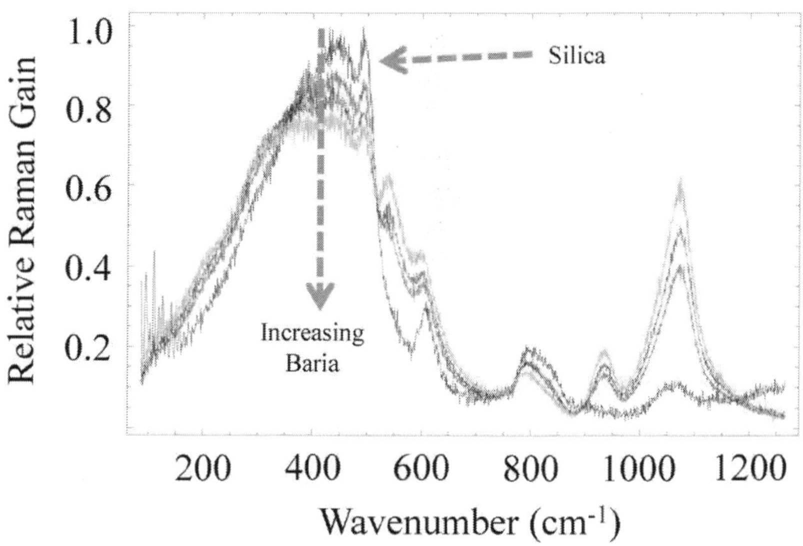

Figure 2. Raman gain spectra (normalized to silica) measured from the barium oxide (BaO) derived fiber for three different baria concentrations. Several new lines appear, likely attributable to the Ba-O bond, including a strong one near 1071 cm^{-1}. The silica spectrum (black line) was obtained from the fiber cladding of one of the fibers.

2.2.3. Optical Fibers with Enhanced Thresholds for Higher Order Mode Instabilities (HOMI)

As noted in the Introduction, there exists in active LMA fibers a power threshold where dynamic randomization of the mode distribution at the laser output is observed, known as "higher order mode instability" (HOMI), and is believed to result from a thermally-induced refractive index grating. Since the process is believed to be driven in part by the thermo-optic coefficient (dn/dT) the natural question is whether marked increases in the HOMI threshold can be obtained through judicious tailoring of the core material's thermo-optic coefficient?

While it has already been shown that YAG-derived fibers possess larger thermal conductivities than their silica counterparts [38] (as another potential way to reduce HOMI), the material dn/dT can also be tailored. Simply stated, if $dn/dT= 0$, such modal instabilities should be improved, if not completely removed. Much like the aforementioned work on intrinsically-low-Brillouin-gain glasses, combining materials with thermo-optic coefficients of opposite sign can give rise to a significant reduction in dn/dT, and possibly even its taking on a value of zero. Materials such as SiO_2, GeO_2 (dn/dT larger than silica), and Al_2O_3 (dn/dT similar to silica) have positive dn/dT, but this value can also be negative such as in P_2O_5 and B_2O_3 [39]. As an illustrative example, and based on experimental measurements on B_2O_3-doped optical fibers, the thermo-optic coefficient for a borosilicate glass is shown in Figure 3 including a composition where $dn/dT = 0$ is identified, along with points where the dn/dT is 75% and 50% that of silica. The additive materials model was utilized to generate the curve in Figure 3. The data point where dn/dT reached 50% that of silica is the lowest that the authors have observed to date.

Indeed if the HOMI threshold is proportional to dn/dT, and if dn/dT can be zeroed through careful materials selection, then the HOMI problem can, in principle, be negated. However, for the example above, this requires the use of large quantities of boria, which may not be compatible with conventional commercial methods of fiber fabrication. Therefore, targeting more modest increases in the HOMI threshold (25% or 50% for example) enables the use of these conventional fabrication methods, but while still taking advantage of significant improvements in performance. It is worth noting that it is not anticipated nonconventional fiber fabrication methods such as the molten core approach will represent limitations to fiber commercializability in the future. In the particular case of the molten core method, conventional draw processes are employed lessening the obstacles to technology transfer.

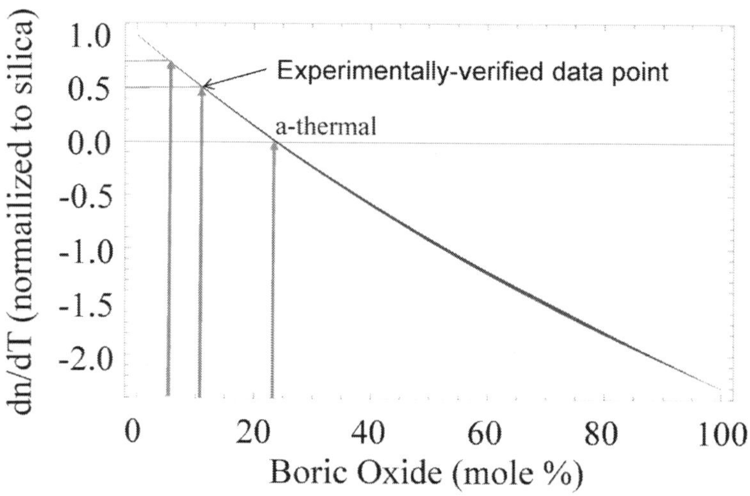

Figure 3. Thermo-optic coefficient of borosilicate glass with increasing boria content. A measured data point on a fabricated fiber is provided. The compositions where d*n*/d*T*reduces by 25% and becomes zero (athermal) are identified on the plot.

2.2.4. Nonlinear Refractive Index and Parasitics Depending Thereon

As noted in the Introduction, processes such as self-phase modulation (SPM) and four-wave mixing (FWM) result from the dependence of the refractive index on the optical intensity (the optical Kerr Effect). As with all other nonlinear optical processes, these may be utilized or be considered parastitics, depending on the requirements of an optical system. FWM and SPM may be useful, for example, for wavelength shifters or for generating of optical solitons, which could be useful for high bit-rate optical communications or short pulse lasers, respectively. However, their broadening and modifying of the optical spectrum may be undesirable in high peak power and multi-wavelength optical amplifier systems. The strength of these interactions is related to the nonlinear refractive index n_2, *i.e.*, to the material properties of the glass from which the fibers are made. To first order, the nonlinear index likely carries through the additive model via the refractive index (since, as mentioned above, $n(I) = n_0 + n_2 I$, where n_0 is the linear refractive index, I is the intensity of the optical signal, and n_2 is the nonlinear refractive index) much like the Pockels coefficients do through the strain and stress optic coefficients [22]. To date, however, this has not yet been verified.

There are numerous publications dedicated to the modeling of the nonlinear refractive index [40,41,42]. It is strongly dependent on the linear (nominal) refractive index of the material, and therefore lower refractive index materials tend to have lower n_2 values. An expression for n_2 in terms of widely used glass engineering coefficients, e.g., the Abbe number $\nu_d$and the refractive index at the Fraunhofer d-line, is [40]

$$n_2(esu) = \frac{K(n_\mathrm{d}-1)(n_\mathrm{d}^2+2)^2}{\nu_\mathrm{d}\sqrt{1.517+(n_\mathrm{d}^2+2)(n_\mathrm{d}+1)\frac{\nu_\mathrm{d}}{6n_\mathrm{d}}}} \tag{1}$$

where K is a material constant that is dependent upon the effective oscillator strength (or strength of the induced dipole moment) and shape of the potential well of the oscillator [40]. Upon inspection it is seen that materials with large Abbe numbers and low n_d are desirable for achieving low nonlinear refractive index values. As a result, classes of fluoride glasses, for example, have relatively low n_2 [42]. Unfortunately, the ranges of Abbe numbers and refractive indices available for most common glass-forming materials are limited (perhaps 50–100 for the former and 1.4 to 1.6 for the latter), and the effectiveness of reducing n_2 in this way is therefore also limited. Of perhaps greater interest is that heretofore a more 'chemical' approach may be taken in the design of materials with predetermined n_2. The nonlinear refractive index depends strongly on the polarizability of the material and, therefore, on the nature of the chemical bond (which strongly influence the parameter K). It has been found that in some covalent bonding systems, in contrast to ionic ones, the value of n_2 (via K[40]) might even take on negative values [43]. Therefore, if an abundance of covalent bonds can be engineered into the material, for example the Al-P bond in the $AlPO_4$ system [44], significant reductions in n_2 may be possible. This work has only just begun but offers exciting possibilities to control n_2-based nonlinearities.

2.3. Limitations

Knowledge of the appropriate phase behavior does not always guarantee success. As one example of this, consider the system: Yb_2O_3-SiO_2, which could be generalized to the lanthanide oxides. High rare-earth content glasses, particularly silicate glasses for their potential compatibility with silica-based optical fibers, are of interest for their magneto-optic properties. Specifically, high lanthanide oxide silicate glasses could be useful for Faraday isolators in high energy laser systems.

The Yb_2O_3-SiO_2 phase diagram exhibits a two-liquid immiscibility over a compositional range from about 75–98 mole percent SiO_2 for temperatures above about 1973K to the upper consolute temperature of 2473 K [45]. In this system a eutectic [L → Yb_2O_3 + $Yb_2Si_2O_7$] exists at a composition of about 38 mole percent SiO_2 and a temperature of 2125K (1852 °C). The liquidus temperature for this eutectic composition is below the conventional draw temperature for a silica-clad fiber using the molten core approach and so seemed an ideal way to obtain a high Yb-content silicate fiber. Accordingly, homogeneous powders of composition 38 mol% SiO_2/62 mol% Yb_2O_3 were synthesized, pressed into pellets, sleeved into a pure silica tube, and drawn at about 2000 °C.

Figure 4 provides a scanning electron microscope image of the resultant fiber; clearly observed is phase separation. Upon compositional analysis, the average composition as measured to be about 86 mol% SiO_2 (~50 weight percent), this SiO_2dissolving into the core melt from the pure silica cladding glass. Said composition is right in the center of the aforementioned liquid-liquid immiscibility and the subsequent phase separation was quenched into a solid form as the molten-core fiber cooled.

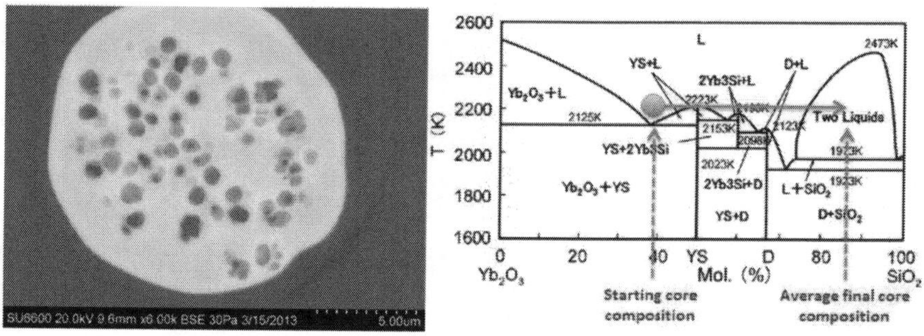

(a) (b)

Figure 4. Scanning electron micrograph of the cross-section of a pure-silica clad, Yb_2O_3-SiO_2 core fiber clearly showing (**a**) phase separation and (**b**) the phase diagram for the Yb_2O_3-SiO_2 system (reproduced with permission from [45], Copyright 2005 the Journal of the Ceramic Society of Japan). The phase diagram also depicts the starting core composition and the final average core composition following molten core fiber processing whereby additional SiO_2 dissolves into the core from the cladding and shifts the final core composition into the two liquid immiscibility region.

2.4. A Note on Costs

There is a natural tendency to assume that anything "specialty" is necessarily expensive due, in part, to the lack of an economy of scale associated with something that is not commodity. Here are two specific counter-examples to this assumption.

Described above was an optical fiber derived from sapphire. While sapphire is most commonly thought of in the form of a gem stone, industrial sapphire is quite common and inexpensive. Exemplifying this was the sapphire used in said work [33], which measured 4.2 mm in diameter by 100 mm in length (constituting a weight of 27.5 carats) and cost $175 (USD). The telecommunication-grade silica cladding tube costs about $2,000 (USD), bringing the total materials cost to less than $3,000 (USD; only in optics could silica cost 1100% more than alumina). For comparison, an equivalently-sized (27 carat) sapphire gemstone sold at auction for well over $15,000 (USD); the difference between sapphire as gemstone and sapphire as transparent industrial alumina is clear (pun intended). Obviously one needs a fiber draw tower (but not a lathe) to make the fiber but it was, in the grand scheme of achieving a record 19 dB suppression in SBS, very inexpensive. Photonic crystal fibers, which have not achieved this level of SBS suppression and are far more labor-intensive to make, and cost significantly more in material and labor costs.

A second example was the recent report of the first glass optical fiber to propagate light via transverse Anderson localization [46,47] as opposed to the conventional total internal reflection. The use of Anderson localization as a propagation modality opens the door to the use of optical fibers for spatially multiplexed high data capacity information links. The glass used in that work was drawn from a porous artisan glass known as "satin quartz", which costs $75 (USD) for a meter-length rod. As with the sapphire, one still needs a draw tower (but no lathe) and the appropriate know-how and expertise but those are constant no matter what is being drawn.

From the perspective of material and labor costs, these specialty fibers can be remarkably inexpensive. The difference is not cost but understanding of the enabling materials and glass science and chutzpah to try something new.

3. EXPERIMENTAL SECTION

The specific optical fibers whose properties are described herein have been reported on individually. For detailed information on the material and fiber fabrication and characterization, the reader is referred to the following References: YAG-derived fibers [24,34], sapphire-derived fiber [33], spinel-derived fiber [36], baria-derived fiber [35], ytterbia-derived fiber [48], and lanthana-derived fiber [49].

More generally, though, optical fibers conventionally are prepared via one of a series of chemical vapor deposition methods [50,51,52]. In such cases, the core and clad glasses possess reasonably similar thermal (particularly glass transition temperatures, T_g) and thermomechanical properties such that they can be co-drawn into a waveguiding fiber that possesses high strength and appropriate optical properties. This thermal and thermomechanical compatibility tends to imply that the compositions of said core and clad glasses are not overly different.

However, as noted above, most of the fibers treated in this work possess compositions that suffer from phase instabilities, such as liquid-liquid immiscibility, and so cannot be processed using the aforementioned conventional measures. As such, a new method had to be developed that yielded high quality optical fibers while still being industrially relevant. Based on the original work of Reference [53], the molten core approach was employed. References [4,54] provides a more detailed review of the molten core method and material solutions primarily for SBS and SRS mitigation but does not address issues associated with SPM, FWM, and HOMI (which is discussed in this work). Briefly, instead of relying on core/clad material compatibility, the molten core approach requires that the core phase melt at a temperature below where the cladding phase draws into fiber; in other words, during the fiber fabrication, the core is above its liquidus (or upper consulate point if an immiscibility exists). Silica is a preferred cladding glass for the molten core approach not only because it facilitates compatibility of the molten core-derived fibers with conventional silica-based fibers, but also it is a highly refractory glass such that many core phases exist that melt at temperatures below the draw temperature for pure silica (ranging from about 1925 to 2050 °C, depending on size of silica preform).

While advantageous in many cases, the cladding glass is attacked by the core melt and is somewhat dissolved. This makes the final post-drawn fiber an interaction product of the initial core phase and the cladding glass, depending upon the draw time and temperature as well as preform and fiber dimensions. In the specific case of SBS suppression, where most of the core phases chosen possess negative photoelastic constants, the dissolution of the silica (possessing a positive photoelastic constant) can give rise to a resultant multi-component glass with near-zero photoelasticity.

4. CONCLUSIONS

Provided herein has been a review of the present state of next generation optical fibers that meet the needs of present and future high energy laser systems from the specific perspective of SBS and SRS mitigation through a purely materials approach. Simple materials systems, including those in the Al_2O_3-SiO_2, BaO-SiO_2, Al_2O_3-Y_2O_3-SiO_2, and Al_2O_3-MgO-SiO_2families, have been shown to yield optical fibers of classic core/clad geometries that exhibit reductions in SBS by nearly 20 dB and SRS by about 3 dB, though much work remains. Also described were fibers with marked increases in the threshold for higher order mode instabilities (HOMI) though judicious tailoring of the core materials' thermo-optic coefficient, dn/dT.

Though most of the focus here has been placed on crystal-derived all-glass optical fibers fabricated using the cost-effective and industrially compatible molten core approach, the more general message is that materials clearly offer solutions to present and future problems plaguing the high energy fiber-based laser sector. Accordingly, significant opportunities exist for the optical materials community in terms of scholarly topics for study, funding to do so, and job prospects.

ACKNOWLEDGMENTS

Many have contributed to the work described herein—both experimentally as well as conceptually. The authors are especially indebted to Bob Rice (Dreamcatchers Consulting), Wade Hawkins (Clemson University), Courtney Kucera (Clemson University), Roger Stolen (Clemson University), Anna Peacock (University of Southampton), Mark Dubinskiy (US Army Research Lab), Josh Rothenberg (Northrop Grumman Corporation), and thought-leaders as relates to this opportunity for materials-related optical technology enablers, particularly John Mauro, Claudio Mazzali, and Mike Pambianchi, all of Corning Incorporated. Numerous grants and contracts financially supported various parts of this work including the US Department of Defense Joint Technology Office (JTO) through their High Energy Laser Multidisciplinary Research Initiative (HEL-MRI) program contracts W911NF-05-1-0517, FA9550-07-1-0566, and W911NF-12-1-0602, US Army Defense University Research Instrumentation Program (DURIP) award W911NF-07-1-0325, and the Northrop Grumman Corporation.

AUTHOR CONTRIBUTION

John Ballato analyzed data and wrote some of the manuscript. Peter Dragic conducted experimental work, analyzed data and wrote some of the manuscript. John Ballato and Peter Dragic jointly conceived of the work.

CONFLICTS OF INTEREST

The authors declare no conflict of interest.

REFERENCES

1. National Research Council. Optics and Photonics: Essential Technologies for Our Nation; The National Academies Press: Washington, DC, USA, 2013.

2. Baer, T.; Schlachter, F. Report to the OSTP—Lasers in Science and Technology 2010; Stanford University: Stanford, CA, USA, 2013.

3. Mazzali, C. The Timing for Specialty Optical Fibers. In Proceedings of the Workshop on Specialty Optical Fibers and their Applications, Optical Society of America Topical Digest, paper T3.3, Sigtuna, Sweden, 28–30 August 2013.

4. Ballato, J.; Dragic, P. Rethinking Optical Fiber: New Demands, Old Glasses. J. Am. Ceram. Soc. 2013, 96, 2675–2692.

5. Mauro, J.; Philip, C.; Vaughn, D.; Pambianchi, M. Glass Science in the United States: Current Status and Future Directions. Int. J. Appl. Glass Sci. 2014, 4, 1–14.

6. Stolen, R. The early years of fiber nonlinear optics. J. Light. Technol. 2008, 26, 1021–1031.

7. Fermann, M. Single-mode excitation of multimode fibers with ultrashort pulses. Opt. Lett. 1998, 23, 52–54.

8. Koplow, J.; Kliner, D.; Goldberg, L. Single-mode operation of a coiled multimode fiber amplifier. Opt. Lett. 2000, 25, 442–444.

9. Limpert, J.; Schmidt, O.; Rothhardt, J.; Röser, F.; Schreiber, T.; Tünnermann, A.; Ermeneux, S.; Yvernault, P.; Salin, F. Extended single-mode photonic crystal fiber lasers. Opt. Express 2006, 14, 2715–2720.

10. Ramachandran, S.; Nicholson, J.; Ghalmi, S.; Yan, M.; Wisk, P.; Monberg, E.; Dimarcello, F. Light propagation with ultralarge modal areas in optical fibers. Opt. Lett. 2006, 31, 1797–1799.

11. Sosnowski, T.; Kuznetsov, A.; Maynard, R.; Ma, X.; Zhu, C.; Hu, I.; Galvanauskas, A.; Koponen, J.; McComb, T. 3C Yb-doped Fiber Based High Energy and Power Pulsed Fiber Lasers. In Proceedings of the SPIE 8601, Fiber Lasers X: Technology, Systems, and Applications, Photonics West, San Fancisco, CA, USA, 2 February 2013.

12. Gaponov, D.; Février, S.; Devautour, M.; Roy, P.; Likhachev, M.; Aleshkina, S.; Salganskii, M.; Yashkov, M.; Guryanov, A. Management of the high-order mode content in large (40 μm) core photonic bandgap Bragg fiber laser. Opt. Lett.2010, 35, 2233–2235.

13. Kashiwagi, M.; Saitoh, K.; Takenaga, K.; Tanigawa, S.; Matsuo, S.; Fujimaki, M. Effectively single-mode all-solid photonic bandgap fiber with large effective area and low bending loss for compact high-power all-fiber lasers. Opt. Express 2012, 20, 15061–15070.

14. Dong, L.; Wu, T.; McKay, H.; Fu, L.; Li, J.; Winful, H. All-Glass Large-Core Leakage Channel Fibers. IEEE J. Sel. Top. Quantum Electron. 2009, 15, 47–53.

15. Laperle, P.; Paré, C.; Zheng, H.; Croteau, A. Yb-Doped LMA Triple-Clad Fiber for Power Amplifiers. In Proceedings of the SPIE 6453, Fiber Lasers IV: Technology, Systems, and Applications, San Jose, CA, USA, 20 January 2007.

16. Otto, H.; Stutzki, F.; Jansen, F.; Eidam, T.; Jauregui, C.; Limpert, J.; Tünnermann, A. Temporal dynamics of mode instabilities in high-power fiber lasers and amplifiers. Opt. Express 2012, 20, 15710–15722.

17. Jauregui, C.; Eidam, T.; Otto, H.; Stutzki, F.; Jansen, F.; Limpert, J.; Tünnermann, A. Physical origin of mode instabilities in high-power fiber laser systems. Opt. Express 2012, 20, 12912–12925.

18. Agrawal, G.P. Nonlinear Fiber Optics, 2nd ed.; Academic Press: San Diego, CA, USA, 1995.

19. Winkelmann, A. Ueber die specifischen Wärmen verschieden zusammengesetzter Gläser [On the specific heats of different composite glasses]. Ann. Phys. Chem. 1893, 49, 401–420. (In German).

20. Winkelmann, A.; Schott, O. Über die Elastizität und über die Zugund Druckfestigkeit verschiedener neuer Gläser in ihrer Abhängigkeit von der chemischen Zusammensetzung [On the elasticity and the tensile and compressive strength of several new glasses in their dependence on the chemical composition]. Ann. Phys. Chem. 1894, 51, 697–730. (In German).

21. Law, P.; Dragic, P. Wavelength dependence of the Brillouin spectral width of boron doped germanosilicate optical fibers. Opt. Express 2010, 18, 18852–18865.

22. Dragic, P.; Ballato, J.; Morris, S.; Hawkins, T. Pockels' coefficients of alumina in aluminosilicate optical fiber. J. Opt. Soc. Am. B 2013, 30, 244–250.

23. Stolen, R.; Ippen, E. Raman gain in glass optical waveguides. Appl. Phys. Lett. 1973, 22, 276–278.

24. Dragic, P.; Ballato, J. Characterisation of Raman gain spectra in Yb:YAG-derived optical fibres. Electron. Lett. 2013, 49, 895–897.

25. Baret, G.; Madar, R.; Bernard, C. Silica-Based Oxide Systems I. Experimental and Calculated Phase Equilibria in Silicon, Boron, Phosphorus, Germanium, and Arsenic Oxide Mixtures. J. Electrochem. Soc. 1991, 138, 2830–2835.

26. Tien, T.; Hummel, F. The System SiO2–P2O5. J. Am. Ceram. Soc. 1965, 48, 75–80.

27. Dragic, P.; Ballato, J. 120 Years of Optical Glass Property Calculations: From the Law of Mixtures and the Birth of Glass Science to Mixing the Unmixable. Opt. Photonics News 2014. in press.

28. MacDowell, J.; Beall, G. Immiscibility and Crystallization in Al2O3-SiO2 Glasses. J. Am. Ceram. Soc. 1969, 52, 17–25.

29. Hyatt, M.; Day, D. Glass Properties in the Yttria-Alumina-Silica System. J. Am. Ceram. Soc. 1987, 70, C283–C287.

30. Jung, I.; Decterov, S.; Pelton, A. Critical Thermodynamic Evaluation and Optimization of the MgO-Al2O3, CaO-MgO-Al2O3, and MgO-Al2O3-SiO2 Systems. J. Phase Equilibria Diffus. 2004, 25, 329–345.

31. Frantz, J.; Mysen, B. Raman Spectra and Structure of BaO-SiO2, SrO-SiO2, and CaO-SiO2 melts at 1600 °C. Chem. Geol.1995, 121, 155–176.

32. Zhang, Y.; Navrotsky, A. Thermochemistry of Glasses in the Y2O3-Al2O3-SiO2 System. J. Am. Ceram. Soc. 2003, 86, 1727–1732.

33. Dragic, P.; Hawkins, T.; Morris, S.; Ballato, J. Sapphire-derived all-glass optical fibers. Nat. Photon. 2012, 6, 629–635.

34. Dragic, P.; Law, P.; Ballato, J.; Hawkins, T.; Foy, P. Brillouin Spectroscopy of YAG-Derived Optical Fibers. Opt. Express2010, 18, 10055–10067.

35. Dragic, P.; Kucera, C.; Furtick, J.; Guerrier, J.; Hawkins, T.; Ballato, J. Brillouin Spectroscopy of a Novel Baria-doped Silica Glass Optical Fiber. Opt. Express 2013, 21, 10924–10941.

36. Mangognia, A.; Kucera, C.; Guerrier, J.; Furtick, J.; Dragic, P.; Ballato, J. Spinel-Derived Single Mode Optical Fiber.Opt. Mater. Express 2013, 3, 511–518.

37. Liu, Y.; Galvin, T.; Hawkins, T.; Ballato, J.; Dong, L.; Foy, P.; Eden, J.; Dragic, P. Linkage of Oxygen Deficiency Defects and Rare Earth Concentrations in Glass Optical Fiber Probed by Ultraviolet Absorption and Laser Excitation Spectroscopy. Opt. Express 2012, 20, 14494–14507.

38. Ballato, J.; Hawkins, T.; Foy, P.; Kokuoz, B.; Stolen, R.; McMillen, C.; Daw, M.; Su, Z.; Tritt, T.; Dubinskii, M.; et al. On the Fabrication of All-Glass Optical Fibers from Crystals. J. Appl. Phys. 2009, 105.

39. Shima, K.; Himeno, K.; Sakai, T.; Okude, S.; Wada, A.; Yamauchi, R. A novel temperature-insensitive long-period fiber grating using a boron-doped-germanosilicated-core fiber. In Proceedings of the Conference on Optical Fiber Communication, Dallas, TX, USA, 16–21 February 1997.

40. Boling, N.; Glass, A.; Owyoung, A. Empirical Relationships for Predicting Nonlinear Refractive Index Changes in Optical Solids. IEEE J. Quantum Electron. 1978, QE-14, 601–608.

41. Fournier, J.; Snitzer, E. The nonlinear refractive index of glass. IEEE J. Quantum Electron 1974, QE-10, 473–475.

42. Töpfer, T.; Hein, J.; Philipps, J.; Ehrt, D.; Sauerbrey, E. Tailoring the nonlinear refractive index of fluoride-phosphate glasses for laser applications. Appl. Phys. B 2000, 71, 203–206.

43. Agrawal, G.P.; Flytzanis, C. Delocalization and superalternation effects in the nonlinear susceptibilities of one-dimensional systems. Chem. Phys. Lett. 1976, 44, 366–370.

44. DiGiovanni, D.J.; MacChesney, J.B.; Kometani, T.Y. Structure and properties of silica containing aluminum and phosphorus near the AlPO4 join. J. NonCryst. Solids 1989, 113, 58–64.

45. Yamamoto, H.; Akiyama, K.; Hirata, T.; Murakami, Y. Dependence of Yb2O3/SiO2 Molar Ratio on High temperature Characteristics of Gas Pressure Sintered Si3N4. J. Ceram. Soc. Jpn. 2005, 113, 325–329.

46. Karbasi, S.; Hawkins, T.; Ballato, J.; Koch, K.; Mafi, A. Transverse Anderson localization in a disordered glass optical fiber. Opt. Mater. Express 2012, 2, 1496–1503.

47. Karbasi, S.; Frazier, R.; Koch, K.; Hawkins, T.; Ballato, J.; Mafi, A. Image Transport through a Disordered Optical Fiber Mediated by Transverse Anderson Localization. Nat. Commun. 2014, 5.

48. Dragic, P.; Ballato, J.; Morris, S.; Hawkins, T. The Brillouin gain coefficient of Yb-doped aluminosilicate glass optical fibers. Opt. Mater. 2013, 35, 1627–1632.

49. Dragic, P.; Litzkendorf, D.; Kucera, C.; Ballato, J.; Schuster, K. Brillouin Scattering Properties of Lanthano-Aluminosilicate-Core Optical Fiber. Appl. Opt. 2014. submitted.

50. MacChesney, J.; DiGiovanni, D. Materials Development of Optical Fiber. J. Am. Ceram. Soc. 1990, 73, 3537–3556.

51. Nagel, S.; MacChesney, J.; Walker, K. An Overview of the Modified Chemical Vapor Deposition (MCVD) Process and Performance. IEEE Trans. Microw. Theory Tech. 1982, 30, 305–322.

52. Izawa, T. Early days of VAD process. IEEE J. Sel. Top. Quantum Electron. 2000, 6, 1220–1227.

53. Ballato, J.; Snitzer, E. Fabrication of Fibers with High Rare-Earth Concentrations for Faraday Isolator Applications.Appl. Opt. 1995, 34, 6848–6854.

54. Morris, S.; Ballato, J. Molten Core Fabrication of Novel Optical Fibers. Bull. Am. Ceram. Soc. 2013, 92, 24–29.

8

OPTICAL FIBER EMBEDDED IN EPOXY GLASS UNIDIRECTIONAL FIBER COMPOSITE SYSTEM

Irina Severin [1,2,*], Rochdi El Abdi [2], Guillaume Corvec [2] and Mihai Caramihai [1]

[1] Department Materials Technology/Computer Science, Politehnica University of Bucharest, 313 Splaiul Independentei, Bucharest 060042, Romania

[2] Larmaur, ERL CNRS 6274, University of Rennes 1, Beaulieu Campus, Rennes Cedex 35042, France

ABSTRACT

We aimed to embed silica optical fibers in composites (epoxy vinyl ester matrix reinforced with E-glass unidirectional fibers in mass fraction of 60%) in order to further monitor the robustness of civil engineering structures (such as bridges). A simple system was implemented using two different silica optical fibers (F1—double coating of 172 μm diameter and F2—single coating of 101.8 μm diameter respectively). The optical fibers were dynamically tensile tested and Weibull plots were traced. Interfacial adhesion stress was determined using pull-out test and stress values were correlated to fracture mechanisms based on SEM observations. In the case of the optical fiber (OF) (F1)/resin system and OF (F1)/composite system, poor adhesion was reported that may be correlated to interface fracture at silica core level. Relevant applicable results were determined for OF (F2)/composite system.

KEYWORDS

optical fiber; epoxy/E-glass composite; manufacturing; interface adhesion; pull-out test; Weibull plots; MEB

1. INTRODUCTION

Composite materials are now becoming accepted for use in many major structural applications, particularly in the space, aerospace, marine, civil engineering, and automotive industries. These materials with their high specific strength and stiffness are gradually replacing their metal counterparts due to reduced weight, improved wear and corrosion resistance, increased fatigue life, and the ability of the material and component to be formed at the same time. The wide ranges of structural composites are based on polymeric matrices with glass or carbon reinforcement [1].

Optical fiber sensors have gained much attention in recent years for a variety of physical and chemical measurements. Intense research has been performed to develop "special fibers" for integration in smart structures in order to replace conventional sensors [2–4] and to provide information on the continuous evolution of damaged structures under mechanical stress, thus monitoring structural robustness [5,6].

The efficiency of optical fiber sensors to continuously monitor structure deformation and reinforcement cracks depends mainly on strength transfer at the interface level between the embedded optical fiber and the composite structure [7,8]. In order to assess mechanical properties at interface level and composites performance, certain testing, such as the pull-out test, has been developed [9,10]. The test may be used to reveal the bonding quality between the optical fiber sensor and the composite structure. The interfacial adhesion strength τ_d appears as a critical factor for structural robustness monitoring being given by:

$$\tau_d = F_d / 2 \cdot \pi \cdot R \cdot L_e \tag{1}$$

where F_d is the maximum force corresponding to the linear zone of the force *versus* displacement curve; R is the core optical fiber radius and L_e the optical fiber embedded length.

In the context of recently developed optical fibers for distributed sensors and other smart structures, the reliability of optical fibers appears as a critical factor. Testing procedures and damage mechanisms identified on a series of experimental studies are given elsewhere [11].

The present study has the aim to manage a reliable fabrication process of embedding optical fiber in an epoxy/glass fiber composite and to obtain basic data on a simple system. In a further step, Bragg grating optical fiber is envisaged to be embedded in a similar composite structure in order to implement a smart structure sensitive to flexion stress.

2. EXPERIMENTAL PROCEDURE

2.1. Composite Fabrication

The polymer matrix used in this study, provided by the French company DFC-DCP Pultrusion (Creil, France), is a mixture of epoxy vinylester resin DERAKANE 470-36 and a catalytic system composed of Styrene, Perkadox 16 and Trigonox C. After homogenously stirring the epoxy resin, the powdered Perkadox 16 is firstly diluted in Styrene so as to form a homogeneous Styrene/Perkadox 16 system, and then the Trigonox C is added. For the experimental implementation, small matrix quantities were prepared, such as 30 g of epoxy resin and 0.66 g Styrene, 0.264 g Perkadox 16 and 0.108 g Trigonox C.

Composite samples were prepared, reinforcing the epoxy matrix with E-glass fibers Roving 4800 Tex having a density of 2.62 g/cm^3. A mass fraction of 63% glass fiber in epoxy matrix was envisaged, meaning an incorporation of 12 roving in the implemented composite sample. To calculate the roving number, the following relation is used:

$$N = 0.545833\, f \cdot V_{com} \tag{2}$$

where f is glass fiber volume fraction and V_{com} the composite volume. Roving 4800 Tex means 1 roving = 0.0048 g/mm. For the samples of 50 mm × 5 mm × 10 mm, 12 roving correspond to a volume fraction of 40% and a mass fraction of 63%, respectively. The composite samples were weighted after fabrication and effective mass fraction ranging between 55% and 65% were obtained.

In order to characterize the interfacial adhesion between the optical fiber and the epoxy vinylester/glass fiber composite material (OF/C), a simple system composed of a single optical fiber embedded in epoxy vinylester resin (OF/R) was preliminarily studied. The system is a unidirectional composite, where glass fibers have the sample length direction and the optical fiber is centrally embedded in the reinforcement direction.

Two different optical fibers elaborated by iX Fibers S.A.S. company (Lannion, France) were used: firstly, the fiber, denoted by F1, having a 80 µm silica core and two acrylate polymer layers of 172 µm secondary coating diameter and secondly, the fiber, denoted by F2, having a 80 µm silica core, but a single acrylate polymer layer of 101.8 µm fiber coating diameter.

As optical fiber surface has determined fracture to a large extent, external coating appears critical. This coating is polymeric in most cases, and modern optical fibers are coated by two different layers: a soft coating at glass surface and a hard coating at external surface. The coating makes a protection against scratches that occur in normal handling and it reduces water activity at glass surfaces. Polymeric coatings (e.g., epoxy-acrylate) are preferred in practice because they are more efficient at inhibiting surface defects. As one will notice later in our study, regarding MEB examinations (see Section 3.3), fiber F1 has the typical section with two layers, but fiber F2 was subjected to a thermal treatment and presents one layer surrounding the silica core.

The thermal treatment consisted of exposure at 200 °C curing temperature for 100–200 h. The fiber behavior appears similar to polyimide, those polymers containing chemical groups that promote solubility without affecting significantly mechanical properties. Moreover, curing at 200 °C for 100 h, leads to the lost of 1.5% of the mechanical strength as compared to 9% lost in the case of polyimide fiber. Fiber parameters are given in Table 1.

Table 1. iX Fiber references.

Parameter	F1	F2
Optical fiber reference	iXF-PMG-1550-80-V20	iXF-PMG-1550-80-HT
Attenuation@1550 nm, dB/km	1.15	1.68
Cutoff wavelength, nm	1300	1318
Bare fiber/coating diameter, µm	80/172	80/101.8

The sample preparation consisted of incorporating the glass fiber in the resin into a mold and embedding the optical fiber centrally positioned into the sample. As seen in Figure 1, the glass roving is placed in the mold and, superimposed layer by layer (6 roving), is carefully impregnated with the resin until half the mold is filled, and then a fine steel cylinder guide is placed on top in order to allow optical fiber insertion. The mold is filled in, as previously described (other 6 roving), and then the optical fiber is gently inserted in the fine guide. Before closing the mold with a controlled screw torque of 2 Nm, the steel guide is gently and carefully extracted. For safety reasons, all operations were performed into an extractor hood.

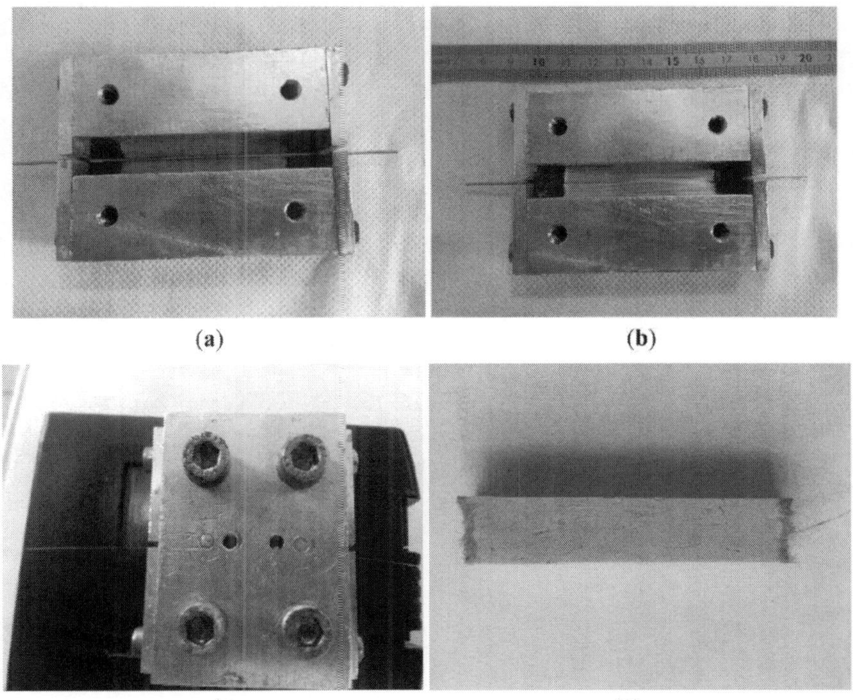

Figure 1. Composite manufacturing experimental procedure. (**a**) Pre-impregnated E-glass roving; and optical fiber steel guide placement in the mold (**b**) mold fill-in; (**c**) mold closure—see optical fiber ends and resin surplus evacuation system and (**d**) composite sample.

Then, curing in preheated oven at 80 °C temperature for 90 min was applied and samples were extracted after 60–90 min.

2.2. Pull-Out Testing

Following the procedure, series of F1 optical fiber embedded in resin and in composite, respectively and F2 optical fiber embedded in composite were prepared in order to perform pull-out tests. Using a LLOYD LR 50K tensile testing setup (Lloyd Instruments Ltd., Fareham, Hampshire, UK), a gripping force was applied, as seen in Figure 2a. The optical fiber end is rolled up on a pulley of 50 mm in diameter and covered with a powerful adhesive so as to prevent fiber slip during testing. When the optical fiber was pulled-out, the fiber breakage occurred at the bit. Thus, a notch was introduced into the specimen as seen in Figure 2b, at a distance of 8 mm from the sample end. The cross head speed was set at 1 mm·min^{-1}, which corresponds to a strain rate of about 0.02 min^{-1}. The de-bonding force F was considered as the maximum force preceding partial de-bonding. A scanning electron microscope (SEM) (Field Emission Scanning Electron Microscope-type JEOL JSM 6301F, Tokyo, Japan) was used to investigate the surface of optical fibers after pull-out tests.

(a) (b)

Figure 2. (**a**) Pull-out testing; and (**b**) sample prepared for pull-out tests.

2.3. Optical Fiber Dynamic Tensile Testing

A tensile bench Lloyds Instruments LR 50K (maximum: 100 N) was used. In order to be dynamically tensile tested, sample fibers were three tours rolled on the cylinder pulley having 50 mm in diameter. The pulleys were covered with a powerful double face adhesive; the mechanical properties of the adhesive layer appeared to be important controlling factors. Despite the standard conditions, for economy and testing time reasons, sample testing free length was chosen to 200 mm. A testing strain rate of 20, 100, 200, respectively 500 mm/min was chosen. These rates were selected in order to correspond to 10, 50, 100, respectively 250%/min as compared to sample main length. In the case of the reference fiber, at least 20 samples were tested and Weibull plots were traced, then n_d-stress (n_d, dynamic stress corrosion parameter) corrosion factor was calculated.

In order to briefly evidence several parameter influences (oven treatment for polymerization or catalytic system—Trigonox C) on optical fiber mechanical properties, series of F1 fiber were treated, then tensile tested and compared to as-received fiber.

Results treatment is then usually based on Weibull plots, despite some opponents concerning the adequacy of Weibull distribution use in case of tensile experiments [12–14]. Even if at least 50 samples are proposed for a reasonable estimation using Weibull, due to our mainly aim of comparing fiber mechanical strength subjected to different etching environments, the 20 series used for Weibull treatment appeared relevant.

3. EXPERIMENTAL RESULTS AND DISCUSSION

3.1. Optical Fiber Characterization

Based on our previous experimental studies, the optical fiber was dynamically tensile tested for four different strain rates of 0.1, 0.5, 1 and 2.5 min^{-1} corresponding respectively to the tensile testing cross speed of 20, 100, 200 and 500 mm/min. For each tested sample, we determined the stress to fracture, and then the results were treated through a statistical approach using the Weibull theory. The classical Weibull plots of the logarithm function of the cumulative failure probability (F_k) related to the logarithm of the stress to fracture (C) has allowed us to calculate and to compare the statistical parameters and the n_d-stress corrosion parameter.

It is assumed that fracture at the most critical flaw on a fiber leads to total failure. For brittle materials as silica optical fibers, strength results obtained from tensile tests present a significant scattering. Then, the statistical Weibull method is commonly used. This method leads us to obtain the mean stress value (strength at 50% fracture probability of the Weibull plot), the medium stress, the Weibull slope and the distribution of the critical flaw size in the sample. The statistical Weibull law gives a relationship between the probability F_K of fiber fracture and the applied stress (C). The evolution of $\ln[-\ln(1 - F_K)]$ according to $\ln C$ is called Weibull diagram.

The slope *p* of the curve ln*C* *versus* ln (where is the testing rate in µm/s) is related to the dynamic stress corrosion parameter n_d, a parameter characterizing the material capacity to resist to a stress. The accepted stress corrosion parameter is ~20 for high strength fiber.

Weibull plots corresponding to the fiber denoted by F1 (two polymer layer coating to protect the silica core) are given in Figure 3a.

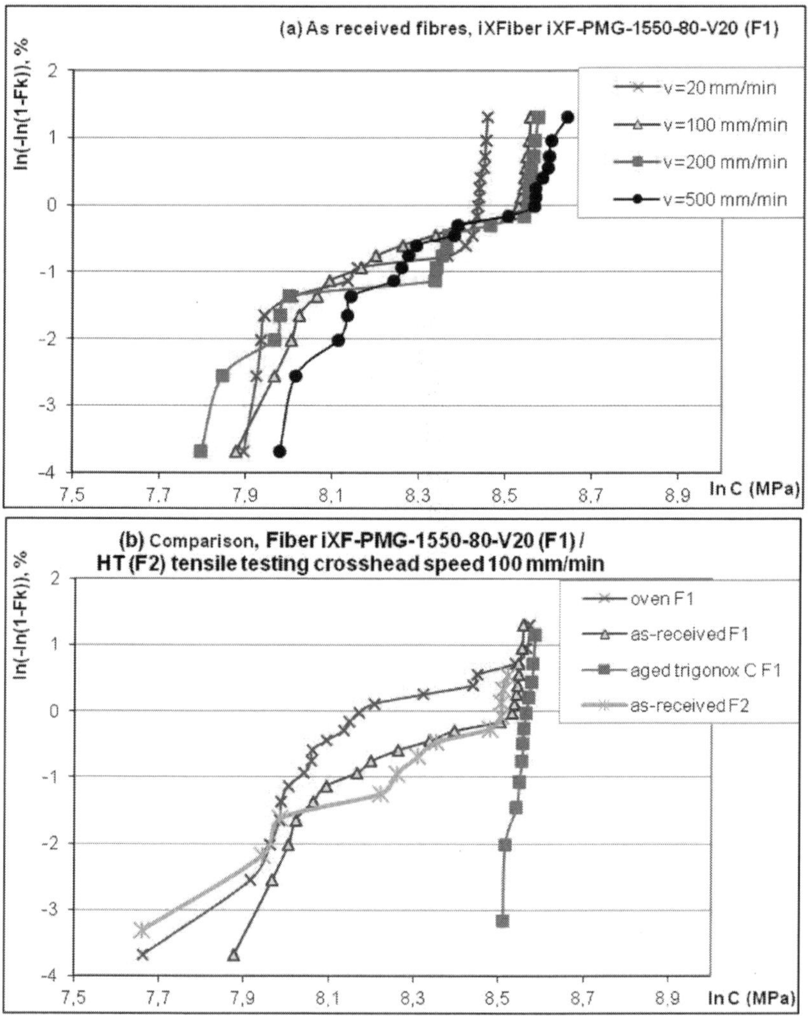

Figure 3. (**a**) Weibull plots for reference fiber F1 (double polymer coating) and (**b**) comparison of different etching factors tensile strength (F2—single polymer coating).

The medium values for the tensile stress correlated to the given strain rates were respectively 4034.3, 4183.7, 4328.5 and 4426.6 MPa. As compared to our previous dynamic tensile testing performed on Verrillon Inc. (North Grafton, MA, USA) and Alcatel optical fibers, one may conclude that F1 fiber presents quite broad dispersion and lower tensile stress. The stress corrosion factor $n_d = 33.8$ appears adequate with a linear interpolation of $R^2 = 0.98$.

Due to experimental reasons, the F2 fiber (single polymer layer coating) was tested for one speed (100 mm/min) and quite similar results are obtained, as seen in Figure 3b, with a medium stress of 4179.4 MPa, as compared to 4183.7 MPa for F1. The same broad dispersion and low tensile stress are reported.

The fabrication parameters influence the optical fiber tensile strength as seen in Figure 3b, thus heating in oven for 90 min at 80 °C led, as expected, to strength decrease, but the Trigonox C (resin compound) determined the strength increase with a more coherent and steeper distribution. Several testing performed with fiber F1 plunged for 7 days in water at room temperature (20 °C) confirmed our previous observations concerning slight increase of strength, but limited dispersion.

3.2. Pull-Out Testing—Interface Adhesion Characterization

The fabricated systems OF (F1)/Resin and OF (F1, F2)/Composite were prepared as described. The notch at 8 mm length from the sample end was cut to control the optical fiber embedding length L_e. The force (in N) *versus* displacement (in mm) curve was recorded and the maximum value was considered for comparison.

The debonding force F_d was taken as the maximum force preceding partial debonding.

In the case of optical fibers embedded in resin, noted OF (F1)/Resin, a poor interface adhesion is reported for all samples with a maximum of 0.98 N. Using the linear model Equation (1), an interfacial adhesion stress of 0.46 MPa is reported. The pull-out curves are similar to Figure 4.

The optical fiber F1 embedded in the composite presented poor adhesion interface, too, slightly higher than the resin, but still low. A maximum value of 1.32 N (see Figure 5a) was recorded that corresponds to an interfacial adhesion stress of 0.66 MPa. but in the most cases debonding forces of 0.8–

0.9 N were registered (see Figure 5b), leading to similar stresses as for non-reinforced matrix case.

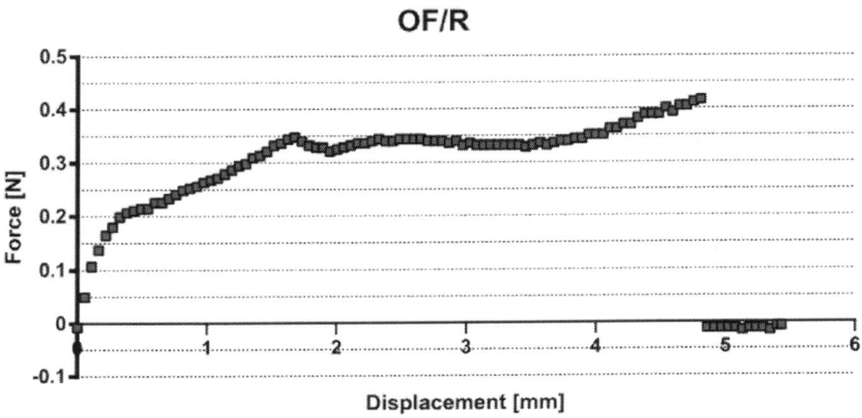

Figure 4. Pull-out record for optical fiber F1 embedded in epoxy-vinyl resin.

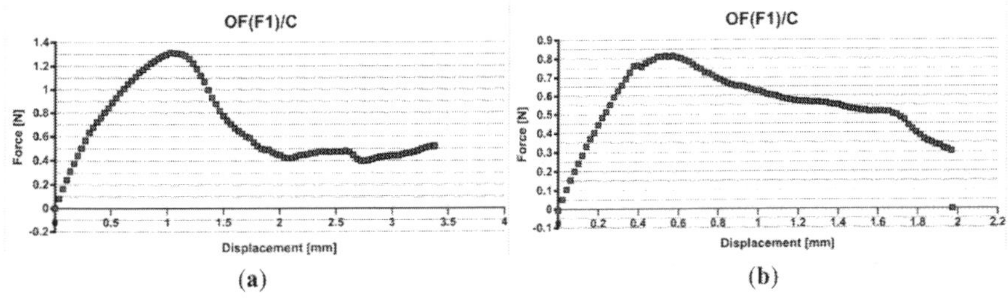

Figure 5. Pull-out record for optical fiber F1 embedded in epoxy composite/glass fiber. (**a**) Maximum 1.32 N and (**b**) maximum 0.8 N.

Finally, the optical fiber F2 embedded in the composite presented better adhesion at OF/composite interface (Figure 6).

A maximum value of 7.1 N (see Figure 6a) was recorded that corresponds to an interfacial adhesion stress of 3.53 MPa and several stick-slip phenomena were noticed (see Figure 6b), as previously reported [15], but for lower debonding forces.

The decrease of the force seems to be controlled by friction over the entire embedded length. The composite rigidity is higher than that of the non-reinforced resin. On the other hand, for the composite sample reinforced with glass fibers, one may expect less friction at the interface between the optical fiber and the composite than for non-reinforced resin in contact with the optical fiber coating. Thus, in the case of the composite sample with glass fibers, the interfacial debonding can propagate with a small instability and the obtained curve appears quasi-continual [15].

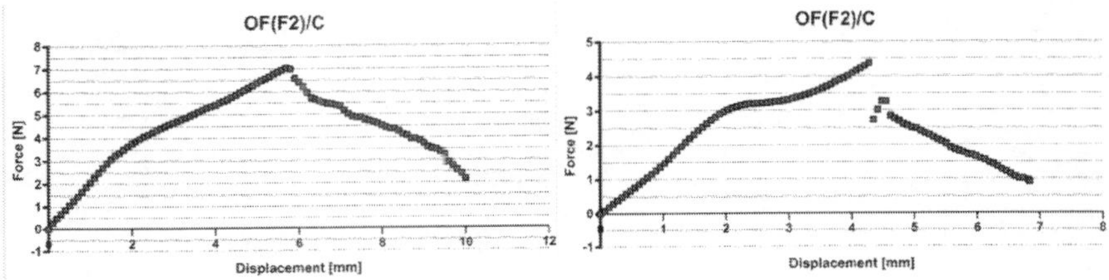

(a) (b)

Figure 6. Pull-out record for optical fiber F2 embedded in epoxy composite/glass fiber. (**a**) Maximum 7.1 N and (**b**) maximum 3.5 N.

3.3. SEM Observations—Interface Adhesion Characterization

An overall image of the composite end is given in Figure 7 with the detail of the optical fiber. The single polymer coating may be seen in this case of F2 optical fiber centrally embedded in the composite, as compared to the double polymer coating in the case of F1 optical fiber (Figure 8a).

In the case of the optical fiber (F1—double polymer coating) embedded in the resin, Figure 8a, examined after the pull-out test, the external polymer coating presented few resin matrix traces, but following the polymerization process the silica core interface appeared detached.

The physico-chemical nature of interfacial bonds depends on both elaboration temperature, and the time maintained at this elaboration temperature. The kinetics of interfacial bonding creation can be connected to a diffusive mechanism. The curing time enhances the formation of chemical bonds, and improves the diffusion mechanism at the interface optical fiber/resin. The adhesion at the interface results from the interdiffusion of macromolecular chains between the two polymer surfaces (resin and coating), thus forming an interphase. This interdiffusion could originate from a good compatibility between the acrylate coating and the vinyl ester resin, (both have hydroxyl groups, able to ensure a good wettability) or it could be the result of a macromolecular mobility, which leads to a tangle by reptation [15,16].

(a) (b)

Figure 7. (**a**) Micrograph of the optical fiber F2 embedded in the glass reinforced vinyl-ester composite; (**b**) detail of the optical fiber (see core and single polymer layer).

At the silica/polymer coating interface the coherence appears affected, as seen in Figure 8b, and this polymer detachment around the silica seems to be responsible for the low interfacial adhesion stress in case of resin and composite embedding F1 fiber. As noted, the resin filaments on the optical fiber external coating appeared more frequent in the composite case, Figure 8c.

Despite attentive precautions and rigorous fabrication among composite prepared samples, one may conclude that, following the pull-out test, half of the samples presented as-stripped optical fiber core (Figure 9a) or polymer coating with silica core. Losses because of detached coatings were due to core fracture and damaged adherence of inner soft polymer coating (Figure 9b).

For the F2 fiber embedded in the composite that has led to improved adhesion of the optical fiber, one may notice good cohesion at the silica core/polymer coating interface, see Figure 10a.

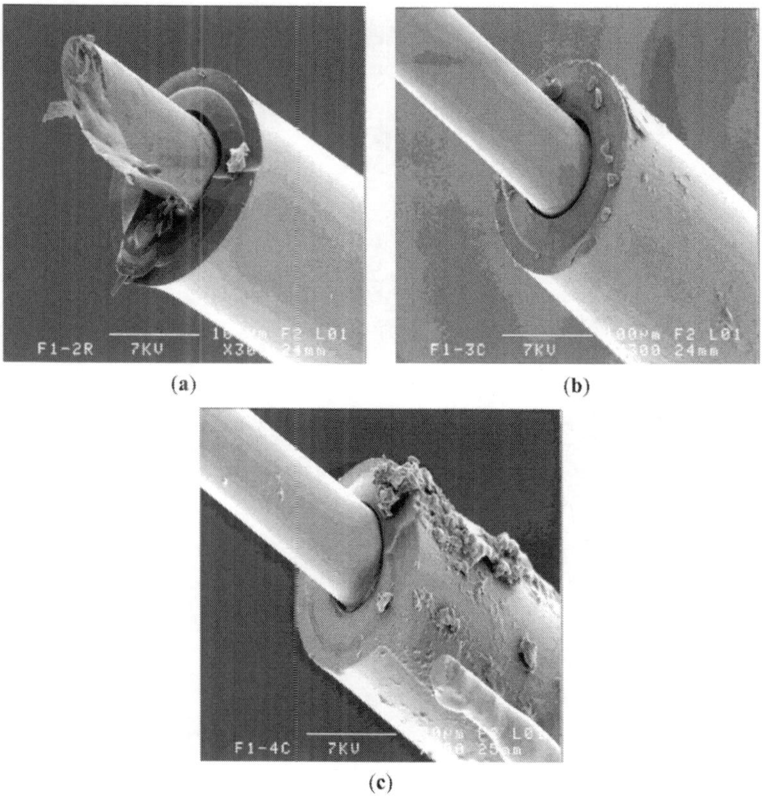

Figure 8. Interface fracture in F1 fiber embedded in (**a**) resin and in (**b**,**c**) composite.

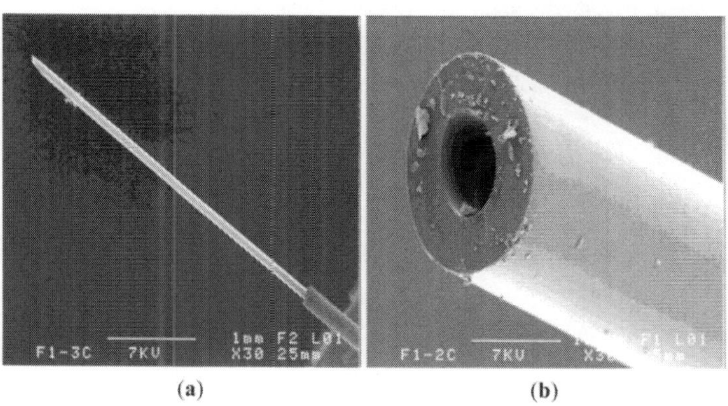

Figure 9. Micrograph of F1 fiber embedded in composite. (**a**) Core stripping; and (**b**) core damage.

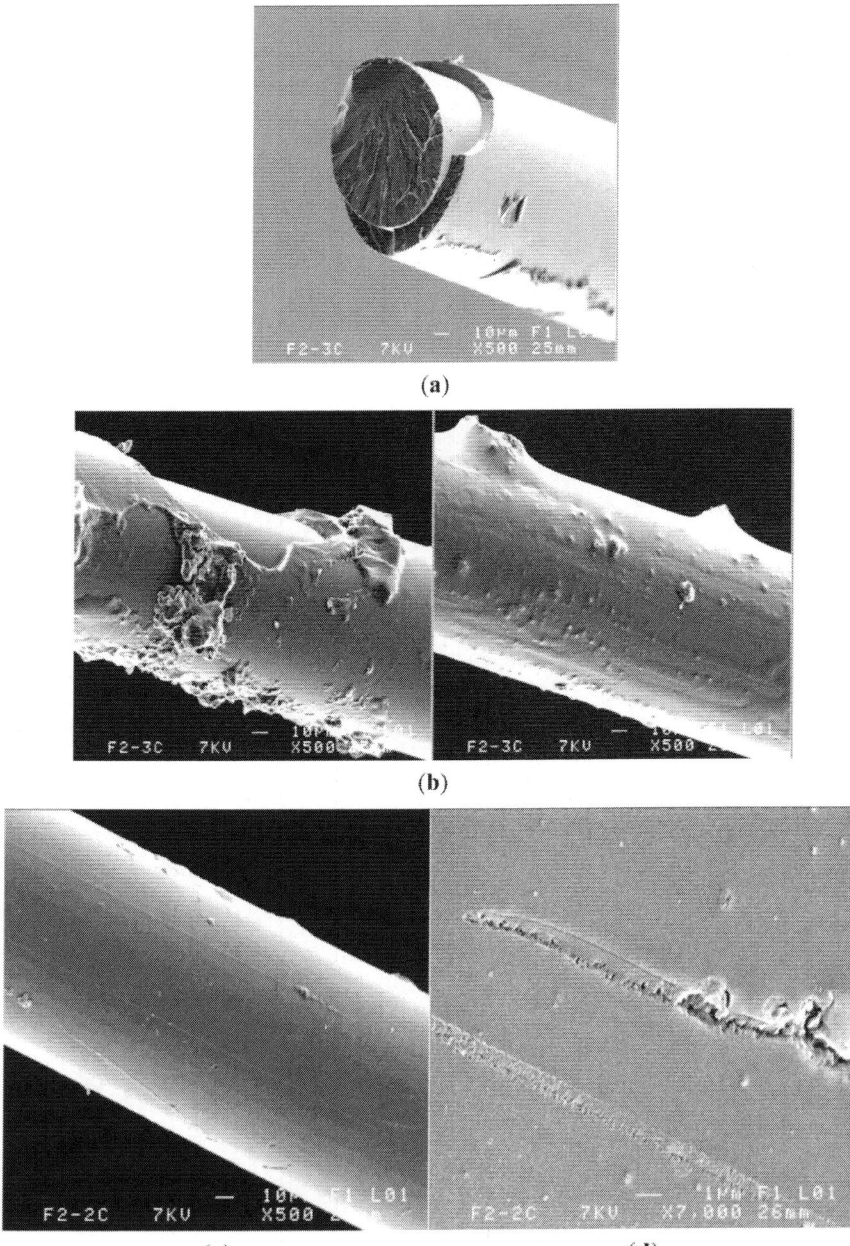

Figure 10. Micrograph of F2 fiber embedded in composite presents: (**a**) resin filaments; (**b**) polymer fracture, polymer detachment and glass reinforcements imprints; (**c**) scratches and (**d**) zoom of scratches.

In certain zones, maybe due to stress concentration, polymer fracture and detachment along the optical fiber (Figure 10b) was noticed, as were frequent resin traces (filaments and drops), sometimes oriented in the reinforcement direction.

Glass fiber reinforcements may scratch the polymer coating and certain imprints are also visible (Figure 10c), but this observation was not as frequent as previously reported [17].

SEM observations have confirmed that polymer detachment were more rare in the F2 fiber case than in the case of F1 fiber embedded similarly in the composite system.

4. CONCLUSIONS

As-received optical fibers F1 (double coating 172 µm diameter) and F2 (single coating 101.8 µm diameter) present similar mechanical strength with medium tensile strength between 4 and 4.4 GPa with broad dispersion and adequate stress corrosion factor (n_d = 33.8). Based on previous experimental studies, polymeric coatings are efficient to inhibit surface defects at the silica core interface.

The main difference between the investigated fibers is the thermal treatment applied in the case of F2 (single coating) that leads to a behavior close to polyimide.

As SEM observations exhibited, in the case of optical fiber with double layer polymer coating (F1), embedded in resin or composite, the soft polymer layer around the silica core retracted, probably due to the internal stresses following polymerization cycle of the resin matrix. Therefore, the interface has lost its coherence, leading to low interfacial adhesion stresses ranging from 0.4 to 0.66 MPa.

Optical fibers heat treatment at curing parameters revealed a strength decrease, but the resin catalytic system (Trigonox C) appeared favorable; the strength slightly increased and presented steeper distribution. Water at environment temperature has a similar effect after one week of exposure of the optical fiber, observations that appeared coherent to our previous results on other silica optical fibers

One may conclude that fabrication parameters acted synergistically on the optical fiber interfacial adhesion in the resin matrix, but certain precautions and a rigorous procedure should be followed to manage a coherent interface. Glass reinforcements of 55%–65%, fraction in mass, acted in the favor of the interfacial strength, the composite case exhibiting slightly better interfacial adhesion stress than the non-reinforced matrix resin, which may be explained through the harmful effect of internal stresses concentration in composite systems.

Finally, the optical fiber with a single layer coating (F2) subjected to a thermal treatment embedded in the composite system has allowed better quality at interface level. The polymer coated fiber appeared to be detached in certain zones following the pull-out test, scratched by reinforcements on its external surface and resin filaments. Drops were seen too, but the silica core/polymer coating revealed coherence and the interfacial adhesion stress is reported at 3.5 MPa.

Our objective to manage a reliable fabrication process of embedding optical fiber in an epoxy/glass fiber composite has been attainted and subsequent testing of Bragg gratings optical fiber (using the F2 single coating) should follow.

ACKNOWLEDGMENTS

The authors express their gratitude to DFC-DCP (Creil, France) and to iXFiber S.A.S (Lannion, France) for technical assistance and for material support. A special remark for Joseph le Lannic from C.M.E.B.A. for MEB observations and fruitful discussions.

CONFLICTS OF INTEREST

The authors declare no conflict of interest.

REFERENCES

1. Davidson, R.; Roberts, S. Optical Fiber Sensor Compatibility and Integration in Composite Materials. In *Comprehensive Composite Materials*; Elsevier: Amsterdam, the Netherland, 2000; Volume 5, pp. 591–606.

2. Berghmans, F.B. Reliability of optical fibers and components. *SPIE Proc* **2005**, *5855*, 20–23.

3. Wang, C.H.; Soufiane, A.; Majid, I.; Wei, K.; Drenzek, G. High reliability hermetic optical fiber for oil and gas application. *SPIE Proc* **2005**, *5855*.

4. Miao, P.; Kukureka, S.N.; Metje, N.; Chapman, D.N.; Rogers, C.D.F. Mechanical reliability of optical fibre for strain sensors. *SPIE Proc* **2005**, *5855*, 1044–1047.

5. Takeda, N. Structural health monitoring for smart composite structures systems in Japan. *Ann. Chim. Sci. Mater* **2000**,*25*, 545–549.

6. Takeda, N. Characterization of microscopic damage in composite laminates and real-time monitoring by embedded optical fiber sensors. *Int. J. Fatigue* **2002**, *24*, 281–289.

7. Zhou, G.; Sim, L.M. Evaluating damage in smart composite laminates using embedded EFPI strain sensors. *Opt. Lasers Eng* **2009**, *47*, 1063–1068.

8. Bettini, P.; di Landro, L.; Airoldi, A.; Baldi, A.; Sala, G. Characterization of the interface between composites and embedded fiber optic sensors or NiTiNOL wires. *Procedia Eng* **2011**, *10*, 3490–3496.

9. Gray, R.J. Analysis of the effect of embedded fiber length on fiber debonding and pull-out from an elastic matrix. *J. Mater. Sci* **1984**, *19*, 861–870.

10. DiFrancia, C.; Ward, T.C.; Claus, R.O. The single-fibre pull-out test, 1: Review and interpretation. *Compos. Part Appl. Sci. Manuf* **1996**, *27*, 597–612.

11. Poulain, M.; El Abdi, R.; Severin, I. Aging and Reliability of Single Mode Silica Optical Fibers. In *Optical Fibers Research Advances*; Schlesinger, J.C., Ed.; Nova Publishers: Hauppauge, NY, USA, 2007; pp. 355–368.

12. Kukureka, S.N.; Cairns, D.R. Statistical analysis for strength distributions in optical fibres. *SPIE Proc* **1999**, *3848*, 17–27.

13. Kukureka, S.N.; Cairns, D.R. Comparison of the distribution of estimated weibull parameters from optical fiber strength measurements and monte carlo simulations. *SPIE Proc* **2001**, *4215*, 90–97.

14. Mirer, T.; Ingman, D.; Zeifman, M. Non-equilibrum physical systems approach to modeling evolution of optical fiber reliability. *Phys. Lett* **2007**, *365*, 181–186.

15. Felcher, G.P.; Karim, A.; Russell, T.P. Interdiffusion at the interface of polymeric bilayers: Evidence for reptation. *J. Non-Cryst. Solids* **1991**, *131–133*, 703–708.

16. Bucknall, D.G.; Higgins, J.S.; Butler, S.A. Early stages of oligomer-polymer diffusion. *Chem. Eng. Sci* **2001**, *56*, 5473–5483.

17. Chean, V.; Matadi Boumbimba, R.; El Abdi, R.; Sangleboeuf, J.C.; Casari, P.; Drissi Habti, M. Experimental characterization of interfacial adhesion of an optical fiber embedded in a composite material. *Int. J. Adhes. Adhes* 2013,*41*, 144–151.

9

ANTHRACENE FIBERS GROWN IN A MICROSTRUCTURED OPTICAL FIBER FOR X-RAY DETECTION

Stanton DeHaven [1,*], Russell Wincheski [1] and Sacharia Albin [2]

[1] NASA Langley Research Center, Hampton, VA 23681, USA

[2] Department of Engineering, and Center for Biotechnology and Biomedical Sciences, Norfolk State University, Norfolk, VA 23504, USA

ABSTRACT

Anthracene fibers are grown inside a microstructured quartz matrix to form a multicore optical fiber for X-ray detection. A modified fiber growth method for single crystal anthracene from the melt via the Bridgman-Stockbarger technique is presented. The anthracene fiber is characterized by using spectrophotometry, Raman spectroscopy, and X-ray diffraction. These results show the anthracene grown in fiber has high purity and a crystal structure similar to anthracene grown from liquid, vapor, and melt techniques. As an X-ray detector, the output is 12%–16% efficient between the energy ranges of 40 and 10 keV. The effect of materials and fiber processing are discussed.

KEYWORDS

growth from melt; organic compounds; scintillating materials; optical fiber devices

1. INTRODUCTION

Anthracene is an organic semiconductor material that is used as a scintillator for ionizing radiation detection [1,2,3,4]. Its electroluminescent properties have been exploited as a base material for organic LEDs [5,6,7,8]. Anthracene has also been investigated as a sensor material for magnetic fields and pressure. Spectral fluorescence variations and quenching have been reported from magnetic field interactions with anthracene crystals [9,10,11]. Pressure applied to anthracene crystals affects the optical absorption and fluorescence spectra [12,13]. The light output of anthracene crystals is the highest of all organic scintillators with a decay time of approximately 32 ns [1,2,3].

The addition of quantum dots, carbon nanotubes, chemical coatings and liquids into microstructured optical fiber (MOF) provides various sensors for temperature, chemical, and X-ray detection [14,15,16,17,18,19,20]. These filled MOF devices operate based on various principles including fluorescence, absorption, and refraction index variations. Growing single crystal anthracene in quartz MOF allows utilization of the physical sensing properties of anthracene in an optical fiber.

The growth of single anthracene crystals has been extensively studied and is still a subject of research. Many experimental techniques have been explored to grow and characterize anthracene crystals [21,22,23,24]. Large anthracene crystals are typically grown from the melt in glass tubes using a modified vertical Bridgman-Stockbarger technique at very low growth rates [21]. The glass is subsequently etched away using HF acid to free the anthracene crystals.

These large single crystals are typically used in scintillation applications but are fragile. Smaller anthracene crystals grown from solution and vapor become part of electro-optic and non-linear optical devices. The deposition of anthracene crystals allows organic semiconductors for electronic applications such as field effect transistors [22,23,24].

The single crystal growth method for large diameter anthracene from the melt conditions was found to be unsatisfactory for fibers of micron-size diameters. Previous work on fiber growth focused on single core organic crystal fibers suitable for second-order harmonic generation [25,26]. This paper presents improvements in multicore organic single crystal anthracene fibers of a few microns in diameter for X-ray detection.

The anthracene fibers are characterized using spectrophotometry, Raman spectroscopy, and X-ray diffraction. The MOF presented can guide light in both the anthracene and quartz microstructure, allowing more scintillation photons to be guided by the fiber compared to when filling the MOF with anthracene. Improvements to the crystal growth and fiber configuration provide higher measured X-ray detection efficiency than the maximum theoretical output when the fiber is configured as a filled MOF [20].

We have successfully grown anthracene fibers of up to 12 cm long inside the MOF. The benefits of anthracene fibers integrated with the quartz MOF include increased photon collection, ease of handling due to the quartz support, enhanced functional reliability and flaw tolerance from multiple fibers in the quartz.

2. EXPERIMENTAL SECTION

Figure 1 shows the schematic diagram of the experimental arrangement for growing single crystal anthracene in the quartz MOF. The tube furnace is separated into two zones, an upper 406 mm section for melting and a lower 203 mm section for freezing. The upper zone uses a higher temperature for thermal stability. The typical zone temperatures are 230 °C for the upper (T_m) and 70 °C for the lower (T_f) measured using K-type thermocouples. The tube furnace and controller regulates temperature to ±1 °C. A quartz tube is placed between the furnace heating elements and the chamber to enhance thermal stability. Segmented insulation on the outside of the quartz tube between high and low temperature zones provides a transverse temperature gradient.

The process chamber is constructed of 316 stainless steel tubing. This chamber holds the vat with fiber and allows process atmosphere control. The vat is constructed from a capped brass fitting 3 mm in diameter and 8 mm deep and acts as a large reservoir holding hundreds of times the amount of anthracene used in the quartz MOF. The quartz MOF extends to the bottom where capillary action fills the MOF when the anthracene melts. The anthracene used for the melt is >99% pure scintillation grade with no additional purification.

The vat is initially filled with anthracene and sealed and the quartz MOF is inserted through a small hole in the cap. The MOF is sealed to the cap and the assembly placed in the process chamber, which is then placed in the furnace and evacuated between 2×10^{-2} and 5×10^{-2} Torr for 48 h. Before energizing the furnace, 1.14 atm of high purity argon is added to the chamber.

The chamber is initially placed in the upper (melt) zone of the furnace; its temperature rises with the furnace temperature and is held at the 230 °C for one hour to melt the anthracene and allow the capillary action to fill the MOF.

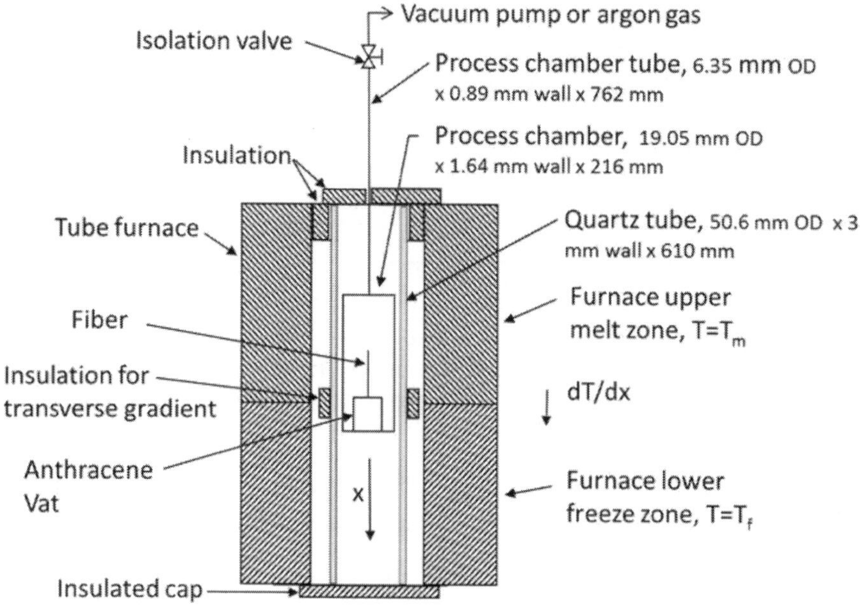

Figure 1. A schematic diagram of the furnace and chamber cross section of the vertical Bridgman-Stockbarger technique for growing single anthracene crystals inside microstructured quartz.

Subsequently, the chamber translation begins and proceeds until a fixed distance is completed. After processing, the chamber is allowed to cool before removing the MOF. Visual inspection is then performed with an optical microscope incorporating side illumination from a 385 nm UV light source. After growing anthracene in the MOF, aluminum or PTFE is applied on the exterior via sputtering to investigate photon collection with these coatings and the effect on fiber handling.

The aluminum coating on the MOF was applied using DC magnetron sputtering. The argon pressure was 20×10^{-3} Torr and the power was 350 watts. Approximately 0.1 micron thickness was applied of 99.99% pure aluminum.

The PTFE coating was applied using RF sputtering. An initial argon pressure between 5×10^{-3} and 20×10^{-3} Torr was applied at 250 watts power to initiate a glow. The argon flow was reduced and then shut-off after several minutes once the glow stabilized [27]. Approximately 0.1 micron thickness was applied of ASTM D1710 PTFE.

3. RESULTS AND DISCUSSION

An image of the quartz MOF used in our experiment is shown in Figure 2 and contains 138 air inclusions of 2.5 microns in diameter. After growth, the fiber has ~1 microgram/cm of anthracene inside the quartz MOF calculated from the fiber geometry and solid anthracene density of 1.25 g/cm^3. The growth experiments from the melt were conducted using a modified Bridgman-Stockbarger technique with a stainless steel chamber holding the quartz fiber, anthracene, and argon gas blanket during processing as in our previous work [28]. This chamber was positioned in a tube furnace that provided a temperature gradient similar to the method described in [25]. A sealed vat supports the MOF fiber and provides a molten pool of anthracene where capillary action fills the quartz from the pool; a stainless steel process chamber allows vacuum and back-filling with high purity argon; an outer quartz tube with insulation induces a radial temperature gradient between the furnace and process chamber.

The Bridgman-Stockbarger technique uses a controlled temperature gradient for crystal growth from the melt at a minimum free energy state. The temperature gradient is longitudinal along the crystal growth direction. In the case of large crystal growth, the heat transfer occurs primarily through the anthracene. However, the microstructured quartz containing anthracene has the primary heat transfer path occurring through the quartz due to the relative size of the anthracene to the quartz and the higher thermal conductivity of quartz.

The approach to determine the range of speeds and the temperature gradient were based on experimental observations for void-free anthracene fiber. The chamber translated at a uniform speed which varied from 40 down to 3 mm/h. The lower speeds were typical for large crystal growth having a smaller temperature gradient [21].

Initially, the growth conditions for large crystal anthracene were tried but failed to grow the anthracene fibers. One explanation may be that the extremely low mass of both the microstructured anthracene and quartz require highly accurate temperature regulation for low undercooling growth; or perhaps the small amount of surface area and material in the fiber requires an extremely slow (<<1 mm/h) growth rate.

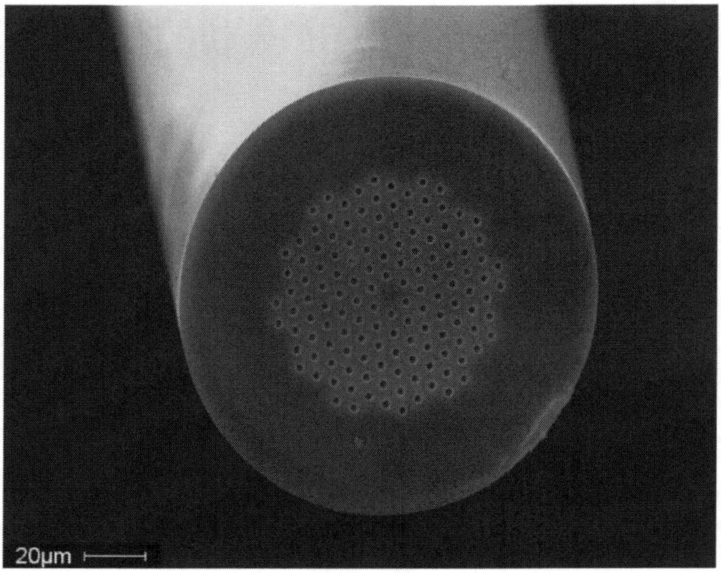

Figure 2. Microstructured quartz optical fiber cross-section SEM image taken at 30 keV. The fiber diameter is 125 microns with 138 air filled inclusions of 2.5 microns each. These inclusions become filled with anthracene.

Speeds from 20 to 40 mm/h with upper and lower zones of 230 and 70 °C, respectively, produced the best fiber with few voids and high optical clarity. Experimental results showed two speed variations produced no benefit, nor did speeds over 40 mm/h or very low translation speeds. The low speed ranges considered in previous work did not produce continuous lengths of anthracene fibers [25]. Additionally, no satisfactory results were found at translation speeds below 15 mm/h using high or low temperature gradients. Figure 3 shows an image of anthracene fiber using incandescent light with voids along the growth direction. A high quality anthracene fiber is translucent and UV light is required to assess if voids exist by looking for scattering bright spots from defects.

The chamber material has a direct effect on void formation, and when constructed from higher thermal conductivity, materials did not yield anthracene crystals of lengths longer than 1 mm. Using a lower conductivity stainless steel chamber reliably and repeatedly produced approximately 12 cm continuous fibers. High quality anthracene fiber grown in the MOF is shown using visible and 385 nm UV light in Figure 4. During anthracene fiber growth, an air gap is formed at one end of the MOF from contraction due to solidification of anthracene. This effect occurs due to the higher density of solid anthracene compared to liquid and plays a role in light guidance and the X-ray detection efficiency of the fiber. Anthracene filled MOF is formed by removing the air gap section.

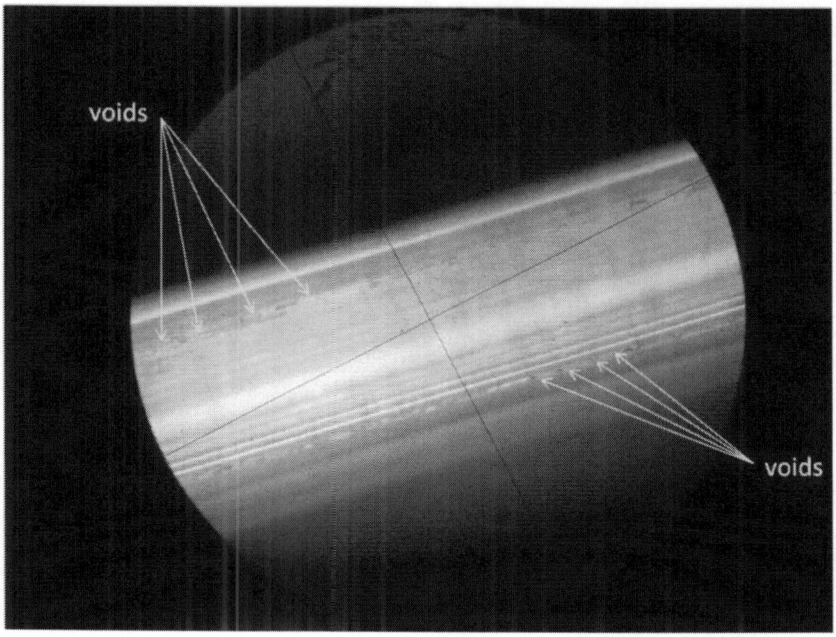

Figure 3. View of anthracene fibers in MOF at 500× showing voids along growth direction. Anthracene fibers grown without voids are translucent and not visible using incandescent light.

3.1. Material Characterization: Fluorescence Spectra

The anthracene was characterized using spectrophotometry, Raman spectroscopy, and X-ray diffraction. The results were compared with those for solid crystalline anthracene data. Anthracene MOF excitation and emission data taken using a Hitachi F4500 luminescence spectrophotometer is shown in Figure 5. A 3-cm long fiber was centered in a fluorescence quartz cuvette suitable for the spectrophotometer.

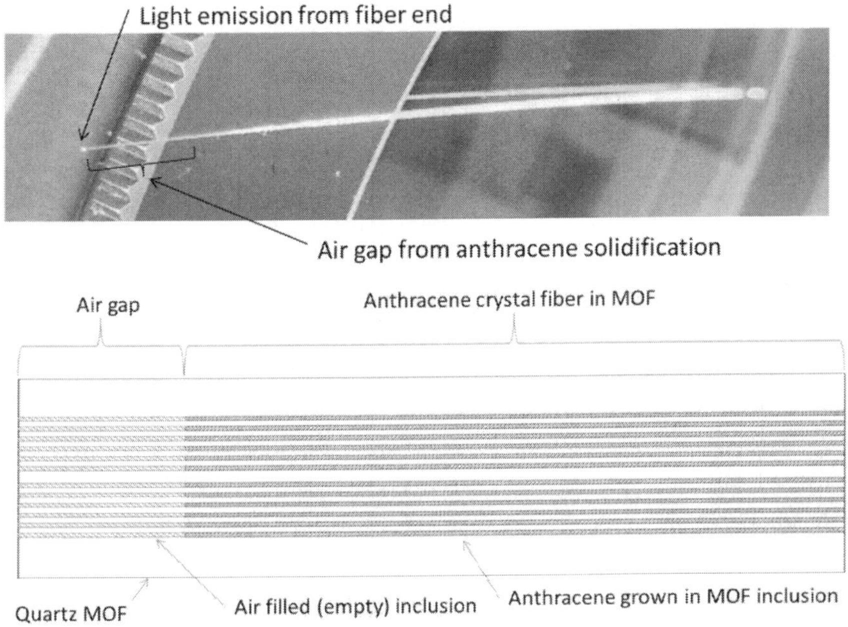

Figure 4. Anthracene in MOF: shown on a microscope stage illuminated with 385 nm UV light showing the air gap formed from solidification during the growth process and blue (420 nm) light from the fiber end; cross section diagram showing air gap at emission end.

The excitation spectrum was collected for a fixed emission wavelength of 420 nm and the emission spectrum was collected for a fixed excitation wavelength of 350 nm. The excitation spectrum shows very good agreement with previous crystal anthracene data [12,29]. The spectral peak and shape of the emission curve with a peak at 420 nm and a secondary peak at 445 nm compare well with the relative amplitude of microcrystalline anthracene published data [29].

3.1.1. Material Characterization: Raman Spectroscopy

A Kaiser optical Raman RxN microscope system was used to collect the Raman spectral data from the MOF end. An optical microscope incorporated into the Raman spectrometer allowed selecting the locations for excitation where data was collected using a 785-nm wavelength laser and compared with data for crystalline anthracene [30,31,32,33,34]. Two orthogonal polarization angles of the laser were used, both in the plane of the MOF end. However, the polarization angle had little effect on the Raman modes and resultant spectra. The lack of laser polarization sensitivity is attributed to the lack of preferred orientation of the anthracene in the MOF end.

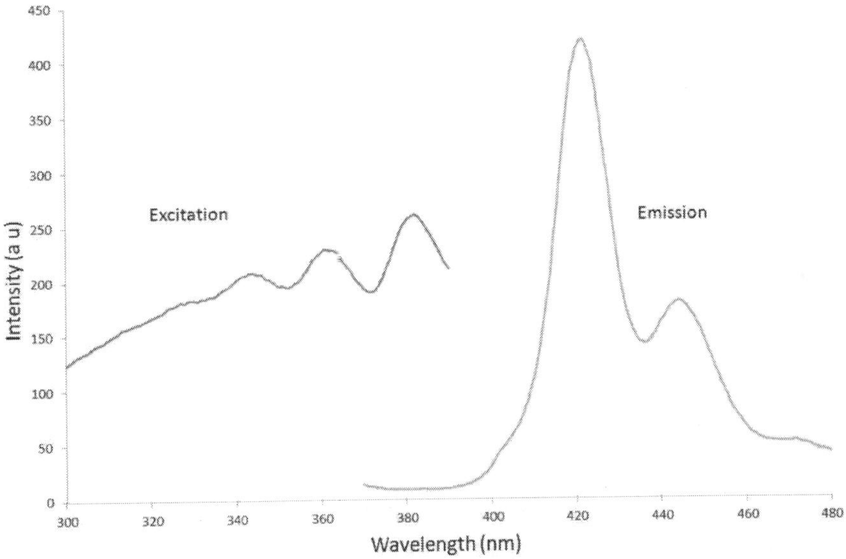

Figure 5. Excitation and emission spectra are shown from left to right for anthracene fibers in quartz. The emission spectral profile agrees with the data for microcrystalline anthracene from Ref [29].

Figure 6 shows the Raman spectra with the scaled anthracene crystal data from Table 1 of reference [30] for comparison. These peaks correspond to scattering associated with vibrational modes of the molecular solid [34]. The wavenumbers agree within ±1 cm^{-1}. The amplitude of the predominant peak at 1403 cm^{-1} was scaled to that of reference [30] with the result that 395 and 753 cm^{-1} peaks have attenuated amplitude as compared to data in [30]. The relative amplitude is the same; the peak at 395 cm^{-1} is the second highest with the third highest at 753 cm^{-1}. The 395, 753, 1007, 1403, 1482, and 1557 cm^{-1}wavenumber peaks are the highest amplitude corresponding to the a_g free molecular (D_{2h}) Raman active symmetry and agree with theoretical wavenumber calculations and previous experimental relative intensity measurements [30,33]. These data demonstrate that the anthracene fiber has high purity and the molecular structure is similar to that of anthracene crystal grown from solution.

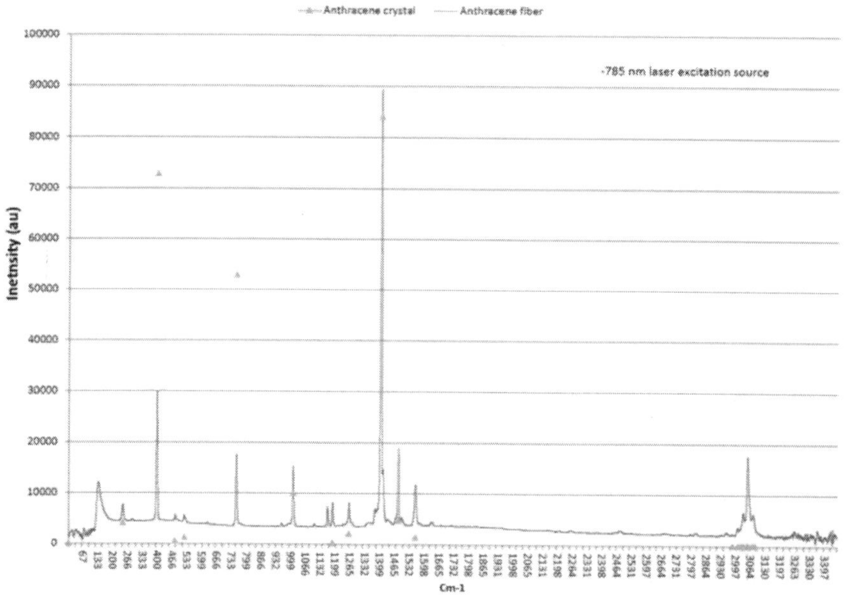

Figure 6. Raman spectra from anthracene fiber in the quartz matrix. Fiber data has ±1 cm^{-1} agreement with anthracene crystal data spectrum (indicated by triangular markers) [30].

3.1.2. Material Characterization: X-ray Diffraction

A Panalytical X-ray diffractometer was used to collect the X-ray diffraction (XRD) data. The anthracene fibers were released from the quartz MOF by etching in HF acid. Approximately 60 pieces of MOF fiber 1-cm in length were collected to make the XRD sample. The bare anthracene fibers were placed on a (100) silicon wafer in a rectangular pattern to simulate a thin film for surface analysis and a 10 mm × 3 mm sample was made for the diffractometer [24,35]. Figure 7 shows an image of the anthracene fibers at 50× magnification and illuminated with 385 nm UV light to produce blue (420 nm) florescence emission.

Figure 8 shows the peaks from the anthracene X-ray diffraction scan. The XRD data for anthracene has excellent spectral agreement with the powder XRD data file for crystalline anthracene within ±0.04 degrees. The peak amplitudes in Figure 8differ from the powder data due to the preferential orientation of the anthracene; however, the peaks at 19.295, 21.305, and 25.285 2θ angles with respective Miller indices of (100), (−102), (−202) agree with ICCD powder diffraction data [36]. Also shown are the peaks from NaF resulting from the HF acid neutralized by NaOH and the (100) peak from the silicon wafer [37].

Figure 7. Anthracene fibers extracted from the quartz cladding and arranged as a thin film for XRD surface analysis at 50× magnification under UV light. The 2.5 micron diameter fibers form a yarn of anthracene fibers.

The anthracene fiber has the same crystalline structure of the bulk powder as shown by Bragg diffraction angles. The main (100) geometric orientation peak is in the growth direction and is consistent with XRD data from high purity anthracene crystals grown from solution and the vapor phase [23,24]. It can be calculated using these Bragg angles that d_{100} = 9.18 Å, which is less than the single crystal powder diffraction length (d_{100} = 9.21 Å, P21/c) and solution growth crystal length (d_{001}= 9.23 Å, P21/a) but larger than the vapor phase length (d_{001} = 9.14 Å, P21/a) [23,24,36]. It is also noted that the Miller indices in Figure 8 for single crystal data use the P21/c space group orientation. However, most anthracene literature uses the P21/a space group notation [22,23,24]. The two space groups are equivalent for the anthracene monoclinic crystal [38].

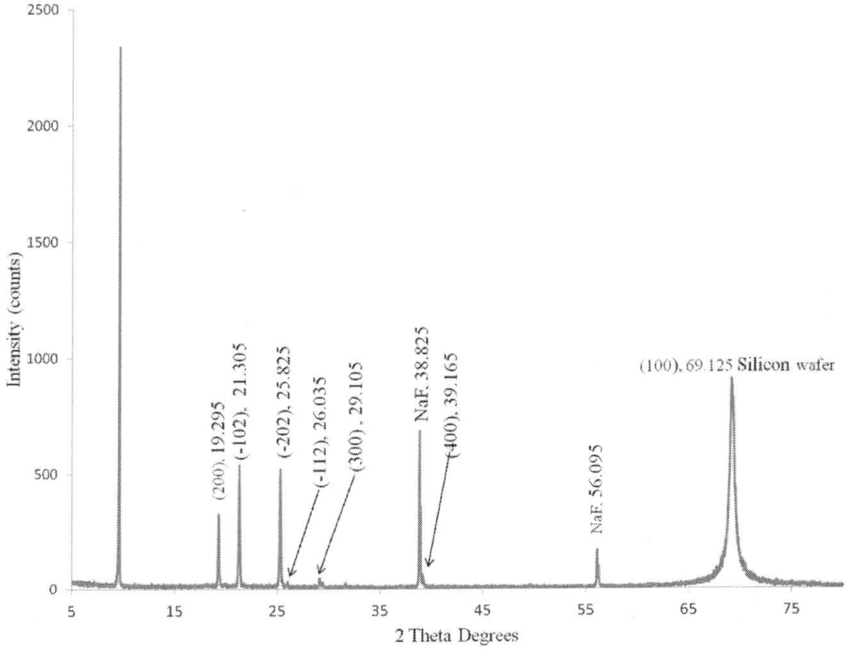

Figure 8. Anthracene X-ray diffraction (XRD) from bare anthracene fibers etched from quartz on a (100) silicon wafer. The (100) anthracene peak at 9.625 2 Theta degrees corresponds to the growth direction. Diffraction angles for peaks all correlate to the anthracene XRD card data [36], sodium fluoride [37], and silicon. The crystal orientation was taken from the ICCD data file [36].

3.2. X-ray Efficiency

The efficiency was measured using the setup and procedure previously described for liquid filled optical fiber and the results are compared with those from previous anthracene fiber efficiency [20,28]. Here, a 40 kV, 4 W tube was the source with a photomultiplier photon counting arrangement for the fiber output. The results in Figure 9 show improved fiber efficiencies of 16.6% at 10 kV to 12.0% at 40 kV for uncoated MOF with an air gap as shown in Figure 4. These values exceed the maximum theoretical light output of anthracene filled MOF.

The maximum theoretical output of anthracene filled MOF is based on work from liquid filled fiber where anthracene, using an effective refractive index of 1.62, is used instead of liquid [20]. A calculation of the photon capture probably in anthracene, with the quartz (fused silica) MOF acting as a cladding, is ~5% using Equations (2), (4), and (5) of reference [20]. This capture probability and representation of the anthracene in the MOF as bundle of optical fibers, each transmitting light through total internal reflection, provides the theoretical maximum output shown in Figure 9. Since the quartz MOF acts as a cladding, any coating has no effect on the theoretical maximum light output.

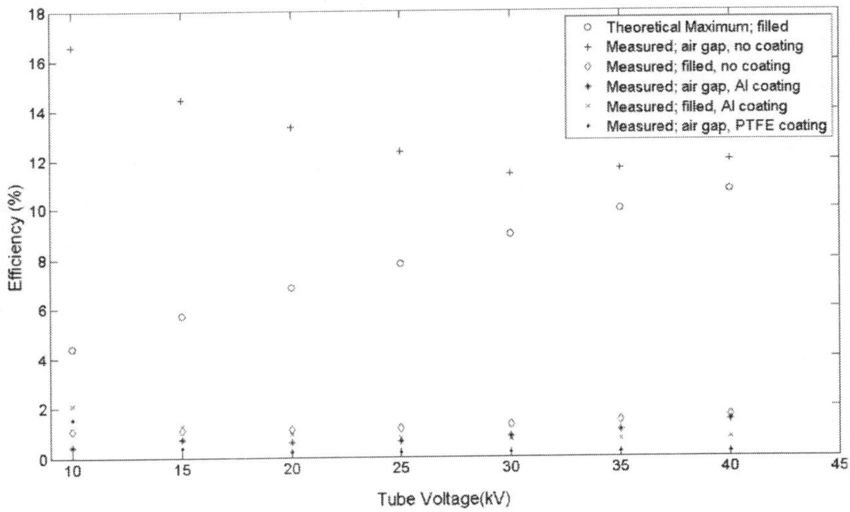

Figure 9. Anthracene increased efficiency from improved crystal growth and fiber configuration. The effect of coatings on fiber efficiency shows aluminum and PTFE thin films yield near the same efficiency with or without an air gap.

The improved X-ray efficiency of anthracene MOF occurs when 420 nm wavelength scintillation light travels through the air filled MOF quartz as shown in Figure 4. With ~5% of the scintillation light theoretically captured in anthracene, most (~95%) must travel through the quartz MOF. The capture of scintillation light through the quartz MOF allows for the improved efficiency. This improved efficiency, which involves complex three-dimensional optics, was observed for uncoated anthracene MOF. Coatings were applied and the resulting efficiency measured to aid in understanding the optics and the effect of coatings.

The use of aluminum and PFTE sputtered coatings was motivated to investigate the optical behavior of a highly reflective thin film from aluminum and a lower index of refraction thin film from PTFE, respectively.

The effect of coatings with both an air gap and filled fiber shows low efficiencies, near that of uncoated filled fiber in Figure 9. Both coatings improve handling of the fibers.

Considerations in the sputtering of aluminum and PTFE were the vacuum and UV light effects on the anthracene. It was observed that anthracene fibers removed from the quartz and exposed to a 30 keV SEM electron beam ceased fluorescing when exposed to UV light. However, anthracene continued to fluoresce after sputter coating with both aluminum and PTFE.

Based on the experimental data in Figure 9, the improved efficiency of anthracene fiber with the air gap and no coating is a combination of several factors: the anthracene crystal quality, the air gap formed during anthracene solidification, and uncoated (bare) quartz. Although minor imperfections in the anthracene are found in all fibers, inspection of the fiber using visible and UV light must show that the fibers are free from voids and flaws shown in Figure 3 and an air gap exists between the light emitting end of the fiber and the anthracene as shown in Figure 4. These conditions must exist for a significant increase in the light output from the fiber. For example, whenever anthracene is grown in the MOF the beginning segment always has voids and must be removed; the opposite end has an air gap. These conditions provide an increase in efficiency of approximately 1300% at 40 kV and over 1600% at 10 kV compared to previous work as shown in Figure 9 [28].

4. CONCLUSIONS

Anthracene fibers are successfully grown from the melt in quartz MOF using a modified Bridgman-Stockbarger method. Void free fibers approximately 12 cm in length are grown consistently. Microstructured anthracene crystal growth conditions vary significantly from those of bulk growth. At temperatures of 230 and 70 °C for the melt and freeze zones, respectively, a growth rate of 25 mm/h is achieved. The Raman spectra shows that the anthracene fiber is a high purity molecular crystal; the XRD data shows the single crystal anthracene fiber has a (100) crystal orientation along the growth direction. Higher photon collection efficiency fibers are obtained by including the air gap formed during growth solidification and having no coating on the MOF quartz.

ACKNOWLEDGMENTS

The authors wish to acknowledge Phillip Williams of NASA Langley for assistance in collecting the Raman spectrum data and Jonathan Skuza with Yeonjoon Park of the National Institute for Aeronautics for assistance in collecting the XRD data and the powdered XRD database.

AUTHOR CONTRIBUTIONS

Stanton DeHaven conceived, devised and carried out the growth experiments, X-ray measurements, helped carry out the sputter coating experiments and wrote the paper; Russell Wincheski carried out and refined the sputter coating experiments, edited the paper and helped interpret some of the experimental data; Sacharia Albin advised Stanton DeHaven on experimental results and edited the paper.

CONFLICTS OF INTEREST

The authors declare no conflict of interest.

REFERENCES

1. Knoll, G.F. Radiation Detection and Measurement; John Wiley & Sons: New York, NY, USA, 2000; pp. 219–261.

2. Birks, J.B. Theory and Practice of Scintillation Counting; Paragon Press: New York, NY, USA, 1964.

3. Birks, J.B. The fluorescence and scintillation decay times of crystalline anthracene. Proc. Phys. Soc. 1962, 79, 494–496.

4. Yokoi, K.; OHBa, Y. Recovery of X-ray radiation damage in anthracene single crystals. Chem. Phys. Lett. 1978, 56, 560–562.

5. Sano, M.; Pope, M.; Kallmann, H. Electroluminescence and band gap in anthracene. J. Chem. Phys. 1964, 43, 2920–2921.

6. Pope, M.; Kallmann, H.P.; Magnante, P. Electroluminescence in organic crystals. J. Chem. Phys. 1962, 38, 2042–2043.

7. Williams, D.F.; Schadt, M. DC and pulsed electroluminescence in anthracene and doped anthracene crystals. J. Chem. Phys. 1970, 53, 3840–3847.

8. Brutting, W. Physics of Organic Semiconductors; Wiley-VCH Verlag,Strauss GmBH: Morlenbach, Germany, 2005.

9. Ern, V.; Merrifield, R.E. Magnetic field effect on triplet exciton quenching in organic crystals. Phys. Rev. Lett. 1968, 21, 609–611.

10. Belaid, R.; Barhoumi, T.; Hachani, L.; Hassine, L.; Bouchriha, H. Magnetic field effect on recombination light in anthracene crystal. Synth. Met. 2002, 131, 23–30.

11. Soos, Z.G. Zeeman populations of optically produced triplet excitons in anthracene. J. Chem. Phys. 1969, 51, 2107–2112.

12. Wiederhorn, S.; Drikamer, H.G. The effect of pressure on the near-ultraviolet spectra of some fused-ring aromatic crystals. J. Phys. Chem. Solids 1959, 9, 330–334.

13. Hummer, K.; Puschnig, P.; Ambroschdraxl, C.; Oehzelt, M.; Heimel, G.; Resel, R. Calculated optical absorption of anthracene under high pressure. Synth. Met. 2003, 137, 935–936.

14. Sparks, J.R.; Esbenshade, J.L.; Rongrui, H.; Healy, N.; Day, T.D.; Keefer, D.W.; Sazio, P.J.A.; Peacock, A.C.; Badding, J.V. Selective semiconductor filling of microstructured optical fiber. J. Lightware Technol. 2011, 21, 2005–2008.

15. Jorge, P.; Martins, M.A.; Trindade, T.; Santos, J.L.; Farahi, F. Optical fiber sensing using quantum dots. Sensors 2007, 7, 3489–3534.

16. Larrion, B.; Hernáez, M.; Arregui, F.J.; Goicoechea, J.; Bravo, J.; Matías, I.R. Photonic crystal fiber temperature sensor based on quantum dot nanocoatings. J. Sens. 2009, 7, 797436–1.

17. Pisco, M.; Consales, M.; Cutolo, A.; Penza, M.; Aversa, P.; Campopiano, S.; Giordano, M.; Cusano, A. Hollow fibers integrated with single wall carbon nanotubes: Bandgap modification and chemical sensing. Sens. Actuators B 2008,129, 163–170.

18. Town, G.E.; Yuan, W.; McCosker, R.; Bang, O. Microstructured optical fiber refractive index sensor. Opt. Lett. 2010, 35, 856–858.

19. Galian, R.E.; Laferriere, M.; Scaiano, J.C. Doping of photonic crystal fibers with fluorescent probes: Possible functional materials for optrode sensors. J. Mater. Chem. 2006, 16, 1679–1701.

20. DeHaven, S.L.; Albin, S.; Kelliher, W. Liquid filled microstructured optical fiber for X-ray detection. Opt. Express 2010,18, 13754–13760.

21. Arulchakkaravarthi, A.; Lakshmanaperumal, C.K.; Santhanaraghavan, P.; Jayavel, P.; Selvan, R.; Sivaji, K.; Gopalakrishnan, R.; Ramasamy, P. Investigations on the growth of anthracene and trans-stilbene single crystals using Vertical Bridgman technique. Mater. Sci. Eng. 2002, B95, 236–241.

22. Madhurambal, G.; Srinivasan, P.A. Growth of high quality anthracene crystals by a simple solution technique. Cryst. Res. Technol. 2006, 41, 231–235.

23. Jo, S.; Yoshikawa, H.; Fujii, A.; Takegana, M. Surface morphologies of anthracene single crystals grown from vapor phase. Appl. Surf. Sci. 2006, 252, 3514–3519.

24. Li, H.; Zhang, D.; Duan, L.; Dong, G.; Wang, L.; Qiu, Y. Morphological structure and optical property of anthracene single crystal grown from solution. Jpn. J. Appl. Phys 2007, 46, 7789–7792.

25. Uenishi, N.; Uemiya, T.; Shimizu, Y.; Mizoguchi, A.; Oogaki, Y.; Hattori, Y.; Umegaki, S. Sumitomo Electric Industries,Method for Producing Organic Crystal and Crystal Growth Therefor. U.S. Patent 5363797, 15 November 1994.

26. Babai, F.H.; White, E.A.D. The growth of void-free crystal cored fibres of organic materials. J. Cryst. Growth 1980, 49, 245–252.

27. Biederman, H.; Ojha, S.M.; Holland, L. The properties of fluorocarbon films prepared by R.F. sputtering and plasma polymerization in inert and active gas. Thin Solid Films 1977, 41, 329–339.

28. DeHaven, S.L. Microstructured optical fiber for X-ray detection. In Proceedings of the Review of Progress Quantitative Nondestructive Evaluation 29 AIP Conference, Kingston, RI, USA, 26–31 July 2009; Volume 1121, pp. 541–548.

29. Birks, J.B.; Wright, G.T. Fluorescence spectra of Aromatic hydrocarbons. Proc. Phys. Soc. 1954, B67, 657–663.

30. Rasanen, J.; Stenman, F.; Penttinen, E. Raman scattering from molecular crystals—II Anthracene. Spectrochim. Acta1973, 29A, 395–403.

31. Sterin, K.E.; Aleksanyan, V.T.; Zhizhin, G.N. Raman Spectra of Hydrocarbons; Pergamon Press: New York, NY, USA, 1980; p. 326.

32. Abasbegovic, N.; Vukotic, N.; Colobo, L. Raman spectrum of anthracene. J. Chem. Phys. 1964, 41, 2575–2577.

33. Suzuki, M.; Yokayama, T.; Ito, M. Polarized Raman spectra of naphthalene and anthracene single crystals. Spectrochim. Acta 1968, 24, 1091–1107.

34. Ferraro, J.R.; Nakamoto, K. Introductory Raman Spectroscopy; Academic Press: New York, NY, USA, 1994; pp. 1–134.

35. Vickerman, J.C. Surface Analysis; John Wiley & Sons: New York, NY, USA, 1997; pp. 313–391.

36. International Center for Diffraction Data (ICDD). Anthracene PDF # 00–039–1848. ICDD, Newtown Square, PA, USA, 1988.

37. International Center for Diffraction Data (ICDD). Sodium Fluoride PDF # 00–036–1455. ICDD, Newtown Square, PA, USA, 1986.

38. Cullity, B.D. Elements of X-ray Diffraction; Addison-Wesley: Reading, MA, USA, 1978; pp. 32–80.

10

A SELF-REFERENCING INTENSITY BASED POLYMER OPTICAL FIBER SENSOR FOR LIQUID DETECTION

David Sánchez Montero [1,*], Carmen Vázquez [1], Ingo Möllers [2], Jon Arrúe [3] and Dieter Jäger [2]

[1] Departamento de Tecnología Electrónica, Universidad Carlos III de Madrid, Leganés, 28911, Madrid, Spain

[2] ZHO/Optoelektronik, Universität Dusiburg-Essen, 47057 Duisburg, Germany

[3] Departamento de Electrónica y Telecomunicaciones, ETSII y IT, Euskal Herriko Unibertsitatea, Bilbao, Spain

ABSTRACT

A novel self-referencing fiber optic intensity sensor based on bending losses of a partially polished polymer optical fiber (POF) coupler is presented. The coupling ratio (K) depends on the external liquid in which the sensor is immersed. It is possible to distinguish between different liquids and to detect their presence. Experimental results for the most usual liquids found in industry, like water and oil, are given. K value increases up to 10% from the nominal value depending on the liquid. Sensor temperature dependence has also been studied for a range from 25 °C (environmental condition) to 50 °C. Any sector requiring liquid level measurements in flammable atmospheres can benefit from this intrinsically safe technology.

KEYWORDS

polymer optical fiber; POF coupler; intensity-based optical sensor; bending losses; liquid level

1. INTRODUCTION

Recently, Polymer Optical Fibers (POFs) have been widely used in industry and automotive networks. New technologies and applications like Very high bit-rate Digital Subscriber Line (VDSL), Fiber-to-the Home (FTTH) and Internet Protocol Television (IPTV) require reliable high-speed but low cost networks with simple installation, high security and low space consumption [1]. POF technology could satisfy these requirements [2] and, hence, in-home POF based networks are currently being installed.

Furthermore, the increasing in-home networking feasibility can enhance the automation and integration of sensor applications and systems. Recently, it is shown that one seventh of all fabricated sensors are used for home appliances [3]. This rate will increase with a higher demand on home automation and security applications by supporting higher integration and networking of devices in the home.

For home sensing, there are three different categories of sensors that could be required:

- Environment sensors: air pressure, temperature, wind, humidity,
- Monitoring sensors: health care, movement, lighting/illumination,
- Building sensors: statics, burglary/hold-up alarm, fire, gas and liquid detection, etc.

In the optical sensing field, POFs are also experiencing a big growth because they present numerous advantages such as easier handling (more flexibility) and lower cost compared to glass optical fibers. These are some reasons why new POF sensors have appeared and are still appearing, most of them based on optical power intensity detection.

On the other hand, optical fibers have been used for measuring liquid levels in many forms, some are non-intrusive sensors based on light attenuation while passing tank walls as in [4], but being only useful in transparent tanks. Others used monomode fibers with reduced aperture [5] for short distances or POF fibers for longer distances [6,7], or intrusive large arrays of individual fluid-sensitive transducers [8], or fibers with clad and unclad zones [9]. Transducers for point measurements in control level devices are reported in [10,11]. For increasing the sensor sensitivity, bends [12] plus cladding removal and partially core polished [13,14] can be done on POF fibers. But none of them are self-referencing intensity based sensors; so they need additional elements to avoid influence of undesirable intensity perturbations on the measurements. There is an increasing demand on developing self-referencing intensity based sensor networks in remote operation using optical fibers [15].

In this paper a novel self-referencing intensity-based polymer optical fiber sensor for liquid detection is presented. This sensor itself is a coupler that changes its coupling ratio (K) depending on the different refraction indexes around it (provided by different alternative liquids surrounding the sensor). In addition, the sensitive parameter K has no dependence of transmission power fluctuations or undesirable power losses at the link or in the sensor network; that is why we named it as a self-referencing sensor.

This paper shows a representative example that demonstrates the capabilities of sensors based on polymer optical fibers in home networks for measuring liquid levels in harsh environments such as oil/petrol tanks or bio-mass boilers to be used in condominiums and buildings. The work includes novel device, theoretical study and experimental results.

2. PRINCIPLE OF OPERATION AND SIMULATIONS

A scheme of the sensor is shown in Figure 1, where P_1 is the input port and P_2 and P_4 are, the direct and coupled output ports of the sensor, respectively. The coupling ratio of the device is defined as $K = P_2/(P_2 + P_4)$. T, R and g define, respectively, the Fresnel transmission coefficient, the curvature radius applied at the sensing area and the gap between the cores of the two fibers at the coupling area. Parameter ε indicates the fiber polishing depth.

By introducing a bend on a multimode fiber, the higher order propagating modes in the fiber are refracted because the angle of incidence increases in the interface core-cladding. This refracted light produces an increase of the power losses in the receiver. If the core refractive index, n_{co}, is approximately equal to the cladding refractive index, n_{cl}, it can be considered that the Fresnel transmission coefficient T, for the beams refracted in an optical fiber, is given by [12]:

$$T = \frac{4\cos\theta(\cos^2\theta - \cos^2\theta_C)^{1/2}}{[\cos\theta + (\cos^2\theta - \cos^2\theta_C)^{1/2}]^2} \quad (1)$$

where θ is the angle of incidence for a certain beam with the normal to the core surface and θ_c is the critical angle defined as $\theta_c = sin^{-1}(n_{cl} / n_{co})$. For $\theta \leq \theta_c$ the beam will be refracted from the fiber core increasing the power losses. From (1) it can be deducted that for the greater value of θ_c the greater value of T is obtained.

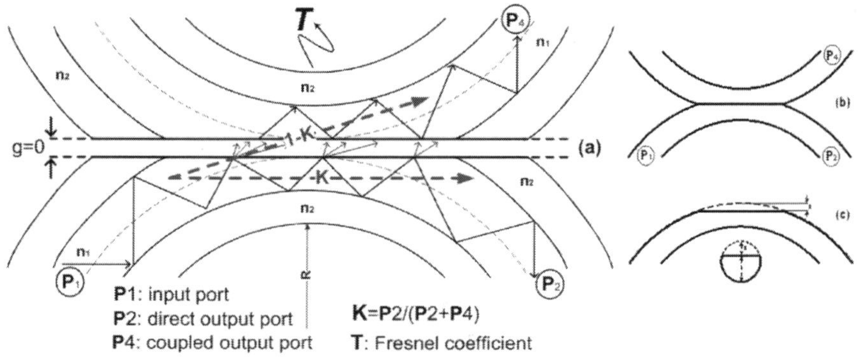

Figure 1. (a) Longitudinal section of the coupling (and sensing) area of the POF liquid sensor. (b) POF liquid sensor. (c) One arm sensor fiber after polishing.

Optical power is transferred and guided in the fiber cladding due not only to the curvature, but also to the polished core (see Figure 1). When the sensor is immersed into different liquids, these losses change because of the different refractive indexes surrounding the coupler which means a variation of the coupling ratio K of the sensor and, therefore, a variation of optical power at port P_4 at the reception stage.

If the surrounding liquid is oil, its refractive index (n_{oil} = 1.46) becomes very similar to the refractive index of the fiber cladding (n_{cl} = 1.417) and all the modes guide through the cladding are refracted outside the fiber. In that case, the value of T increases whereas the optical power at port P_4 decreases. Consequently, the coupling ratio K increases and the coupler becomes more directive. As the liquid refractive index decreases (n_{water} = 1.33 or $n_{air} \approx 1$) the ratio of the optical power guides again into the fiber core increases as well as the optical power received in port P_4, so K decreases. This change of K can be measured and, consequently, it is possible not only to distinguish the air-liquid change but also to detect the kind of liquid in which the sensor is immersed.

Use of a coupler as a sensor device intrinsically means a self-referencing sensor. The advantage of this technique is based on the fact that undesirable variations of the optical power at any point of the link between the emitter and the receiver do not change the value of *K*, because it equally affects both sensor outputs (P_2 and P_4) and consequently the ratio between them is unaltered. For the same reason it is neither necessary to modulate the source to eliminate the effects of the environmental light which brings considerably cost savings in sensor networks.

Ray tracing method can be used for calculating *T* coefficient in a polished fiber for different bending radius as well as for estimating *K* variations by treating light as rays [12]. In that case, light power can be considered to propagate along the fiber core within infinitesimal cross-section parallel rays and higher radiation losses are obtained due to meridional rays, being those rays contained in a plane defined by the core symmetry axis and the curvature centre of the bend. In order to reduce the computation time of simulations, the fiber cladding-liquid interface can be omitted to a first approach in optical fibers with cladding thickness of 10μm [13] leading to similar results as considering directly fiber core-liquid interface.

Figure 2 shows a simulation of *K* versus polishing depth for different bend radii (R = 8 mm, 7 mm and 6 mm) with water as external liquid surrounding the sensor. The aforementioned consideration has been taken into account in the simulation (directly fiber core-liquid interface). The point A corresponds with the fabricated prototype requiring less computational time requirements. Simulations and measurements for that prototype will be discussed later on.

3. SENSOR MEASUREMENTS

For manufacturing the sensor device two standard Step-Index PMMA (polymethylmethacrylate) POFs (n_{co} = 1.492 and n_{cl}= 1.417) were used, each one with a section where the cladding was removed and the core was polished (polishing depth was ε = 0.23 ± 0.01mm). Both cores were joined by gluing over the length of the sensing area (3 cm of polished fiber) with no-gap interface (g = 0) (see Figure 3).

The set-up scheme for the measurements is shown in Figure 4. A laser diode (Roithner, class III) operating at λ = 630 nm was used to illuminate the POF fiber and a two-input power meter was chosen at the reception stage. The POF sensor was placed in between two POF fibers and was immersed in a tank firstly with water and secondly with oil as the external liquids.

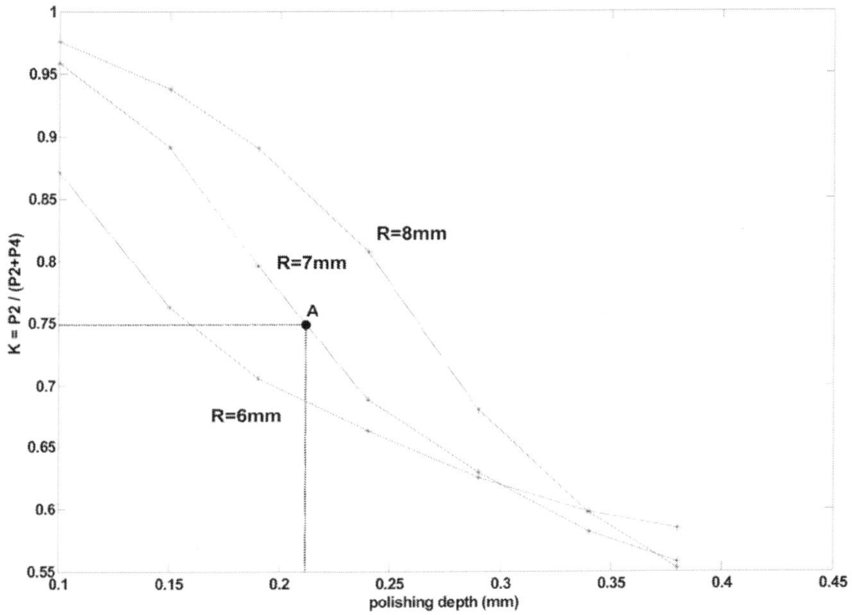

Figure 2. K versus polishing depth simulation for different bending radii (6, 7 and 8 mm) considering water as the external media.

Coupling ratio values for Carthamus tinctorius oil (n = 1.466) and water besides air surrounding the sensor have been experimentally measured having taken five measurements for each sensor configuration (each configuration is determined by the bending radius applied to the sensor and the type of liquid as external media). If we define repeatability as the percentage of the output parameter variation for a determined number of measurement cycles (under the same conditions), the worst case was found to be 1.02% (with R = 25 mm and air as external medium), which was a very good result. All these measurements were taken for an environmental temperature of T = 28.6 °C (average value) with a maximum deviation of ΔT = ±3 °C. Additionally, different bending radii (R = 60 mm, 25 mm, 12 mm and 7 mm) at the fiber sensing area were applied and investigated.

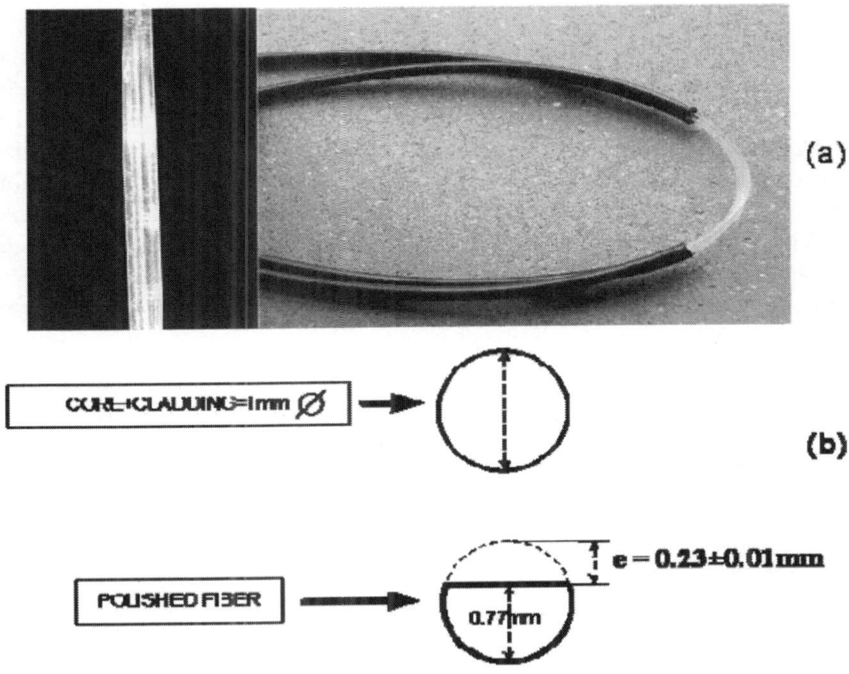

Figure 3. (a) Photograph of the manufactured sensor device and zoom of the sensing area (left side); (b) Schematic of cross section of the fiber core before and after polishing.

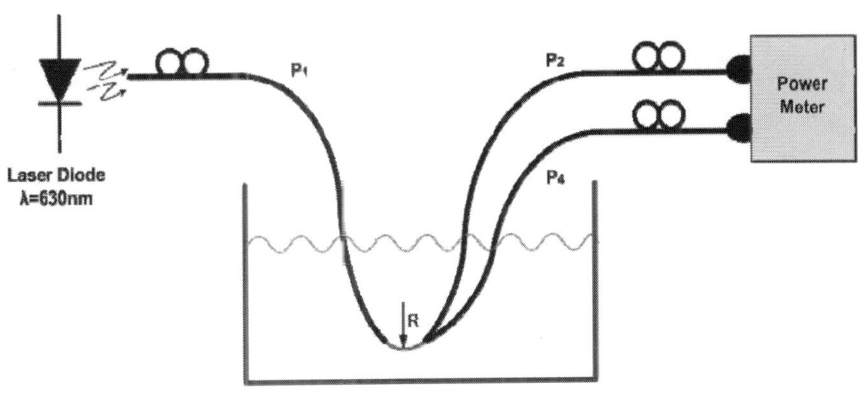

Figure 4. POF sensor set-up for determination of liquid level in tanks.

A self-referencing test was also done to prove the inherent self-referencing property of the proposed liquid level sensor.Figure 5 shows the average value and the measurement errors at each measurement point of the coupling ratio *K* versus different bending radius and different external media. Depending on the media surrounding the sensing area, the value of *K*changed for each fixed bending radius. With proper control electronics and detection schemes the liquid can be distinguished; highest sensitivity was obtained for a bending radius R = 25 mm. For this sensor configuration the following coupling ratios were obtained: K_a = 0.545 ± 0.0014 (air); K_w = 0.563 ± 0.0007 (water) and K_o = 0.605 ± 0.0017 (oil) with coupling ratio increments of ΔK = 0.018 (from air to water); ΔK = 0.060 (from air to oil) and ΔK = 0.042 (from water to oil). Measurement errors are given as standard deviation and are, in all cases, at least one order of magnitude below the average value.

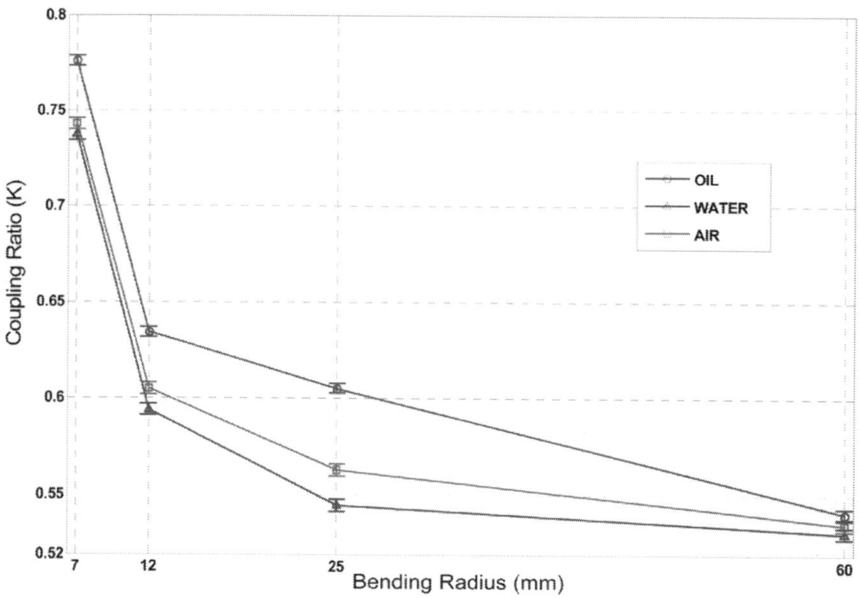

Figure 5. Measurements of the sensor coupling ratio *versus* bending radius, for different liquids. Errors in measurements, given as standard deviation, are also represented for each measurement point.

In order to make a comparison between measured and simulated results an arbitrary point was selected; obtaining a simulated value of K = 0.75, for ε = 0.23 mm, R = 7 mm and water as external medium (see point A in Figure 2). The measured value of the coupling ratio for a manufactured sensor with these parameters was found to have K = 0.744 (seeFigure 5). Both the simulated and the experimentally measured *K* values are congruent and show a good agreement between theory and measurements. The simulation technique used at the present paper can help to optimize the design of physic parameters in this type of sensors.

The independence of the sensor against hypothetical link losses due to its self-referencing characteristic was also tested as can be seen in Figure 6. From a nominal coupling ratio of K = 0.54 a maximum deviation of ΔK = ±0.002 (six measurements with a temporal space of half and hour between each other were taken per set-up) was obtained. The maximum link losses tested were 4dB and were made by bending the fiber into turns using Mode Scramblers.

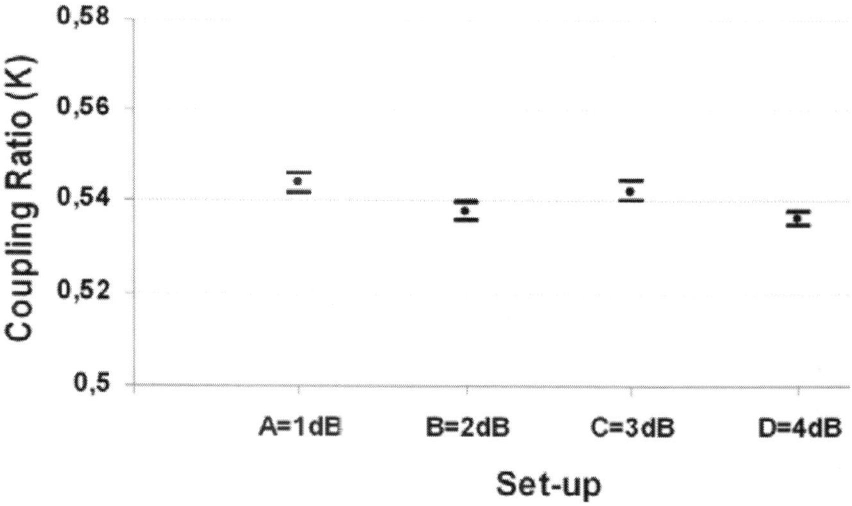

Figure 6. Measurements of coupling ratio of the sensor, for different link losses (up to 4dB from set-up A to set-up D).

4. SENSOR TEMPERATURE DEPENDENCE

The temperature dependence of the sensor device was tested inside a climatic chamber over a 25 °C temperature range, from T = 25 °C to T = 50 °C. Figure 7 shows the deviation of *K* at that temperature range. The sensor coupling ratio deviation was found to be around 1% (ΔK ≤ 0.006) from its nominal value. This deviation was mainly produced by the glue used for the sensor manufacturing; showing the low temperature dependence of the device as was derived from results shown inFigure 8.

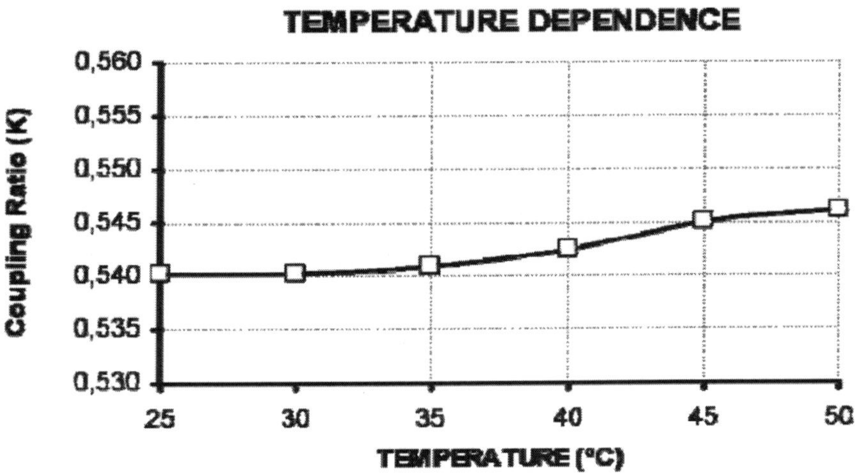

Figure 7. Temperature dependence of the coupling ratio of the sensor (from T = 25 °C to T = 50 °C).

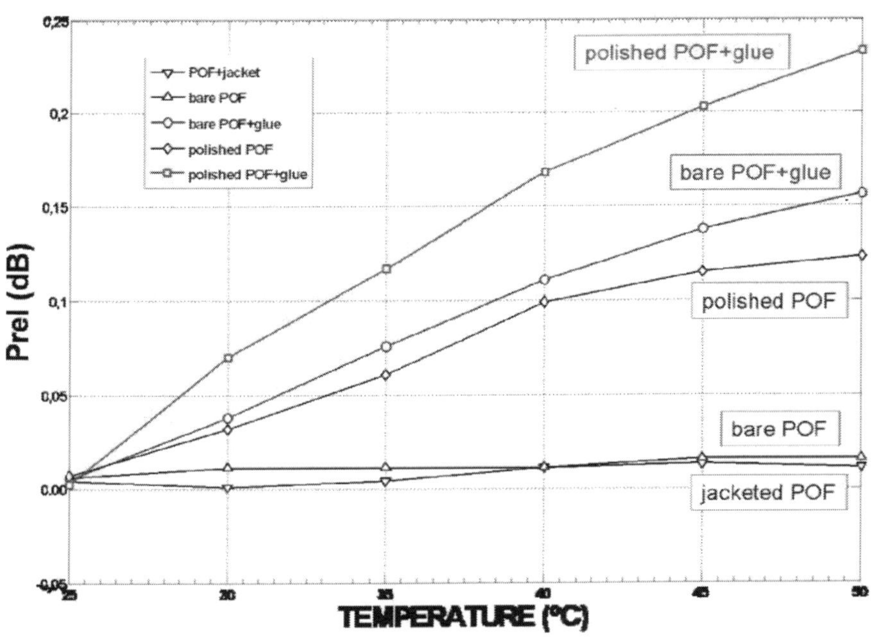

Figure 8. Temperature dependence for different fiber status from T = 25 °C to T = 50 °C.

Figure 8 shows the influence of temperature in a plastic optical fibre (POF) for different fibre status. Five tests were made: "jacketed POF" (fibre with no action taken on it), "bare POF" (stripped fibre), "polished POF" (fibre whose core was partially polished with a polishing depth of ε = 0.23 mm), "bare POF+glue" (stripped fibre with a glue layer over the cladding) and "polished POF+glue" (fibre partially polished with a glue layer on it). Although polishing the fibre core have significant effects when increasing the temperature (with regards to the stripped POF), it can be seen from Figure 8 that output optical power variations are mainly produced by the effect of gluing when considering the same fibre status, obtaining deviations of 0.15 dB between glued/not-glued bare fibre and 0.12 dB between glued/not-glued polished fibre.

5. CONCLUSIONS

A novel self-made POF self-referencing intensity-based sensor device for liquid detection has been manufactured and characterized. The sensing area has been obtained by a 3 step process: introducing bending losses, eliminating the fiber cladding and polishing a defined fraction of the fiber core. This technique increases the external media refractive index sensitivity of the sensor (highest sensitivity was obtained for a bending radius of R = 25 mm). Experimental results for the most usual liquids found in industry, like water and oil, were measured. The coupling ratio K of the sensor, with increments of ΔK = 0.018 (from air to water), ΔK = 0.060 (from air to oil) and ΔK = 0.042 (from water to oil), allows to distinguish between both the presence and type of the external liquid. Measurement variations were, in all cases, at least one order of magnitude below the average value of the coupling ratio. Measurements also showed a low temperature dependence of K, below 1% from its nominal value.

The self-referencing technique was also validated by showing the robustness of the sensor against undesirable losses at any point of the link, because the power ratio as well as the coupling ratio between both output ports of the sensor remains constant.

ACKNOWLEDGMENTS

This work was supported by BONE-project ("Building the Future Optical Network in Europe"), a Network of Excellence funded by the European

Commission through the 7th ICT-Framework Programme, and Spanish CICYT (TEC2006-13273-C03-03-MIC).

REFERENCES AND NOTES

1. Koonen, T. Fiber to the home/fiber to the premises: what, where and when? *Proc. IEEE* **2006**, *94*, 911–934.

2. Lee, S.C.; Breyer, F.; Randel, S.; Gaudino, R.; Bosco, G.; Bluschke, A.; Matthews, M.; Rietzsch, P.; Steglich, R.; van den Boom, H.P.; Koonen, A.M.J. Discrete multitone modulation for maximizing transmission rate in step-index plastic optical fibers. *J. Lightwave Technol* **2009**, *27*, 1503–1513.

3. Luo, R.C.; Yih, C.C.; Su, K.L. Multisensor fusion and integration: approaches, applications and future research directions. *IEEE Sens. J* **2002**, *2*, 107–119.

4. Buttriss, S. Liquid level monitor, US Patent 3,808,887,. July 5 1974.

5. Iwamoto, K.; Kawata, I. Liquid-level sensor with optical fiber. *App. Opt* **1992**, *31*, 51–54.

6. Vázquez, C.; Gonzalo, A.B.; Vargas, S.; Montalvo, J. Multi-sensor system using plastic optical fibers for intrinsically safe level measurements. *Sens. Actuat. A* **2004**, *116*, 22–32.

7. Pérez-Ocón, F.; Rubiño, M.; Abril, J.M.; Casanova, P.; Martínez, J.A. Fiber-optic liquid level continuous gauge. *Sens. Actuat. A* **2006**, *125*, 124–132.

8. Svirid, V.A.; León, V.; Khotiaintsev, S.N. A prototype fiber-optic discrete level-sensor for liquid propane-butane. *IEICE Trans. Electron* **2000**, *E83-C*, 303–308.

9. Betta, G.; Pietrosanto, A.; Scaglione, A. A Gray-Code based fiber optic liquid level sensor. *IEEE Trans. Instr. Meas* **1998**,*27*, 174–178.

10. Wang, A.; Gunther, M.; Murphy, K.; Claus, R. Fiber-optic liquid level sensors. *Sens. Actuat. A* **1992**, *35*, 161–164.

11. Raatikainen, P.; Kassamakov, I.; Kakanakov, R.; Luukkala, M. Fiber-optic liquid-level sensor. *Sens. Actuat. A* **1997**, *58*, 93–97.

12. Arrue, J.; Zubia, J.; Fuster, G.; Kalymnios, D. Light power behaviour when bending plastic optical fibres. *IEE Proc. Optoelect* **1998**, *145*, 313–318.

13. Lomer, M.; Arrue, J.; Jauregui, C.; Aiestaran, P.; Zubia, J.; López-Higuera, J.M. Lateral polishing of bends in plastic optical fibres applied to a multipoint liquid-level measurement sensor. *Sens. Actuat. A* **2000**, *137*, 68–73.

14. Vázquez, C. Sensor de fibra óptica autoreferenciado para la detección de líquido y/o medida de nivel de líquido.*Spanish Patent, P200,701,937,*. July 10 2007.

15. Montalvo, J.; Vázquez, C.; Montero, D.S. CWDM self-referencing sensor network based on ring resonators in reflective configuration. *Opt. Express* 2006, *14*, 4601–4610.

EFFECTS OF TEMPERATURE AND X-RAYS ON PLASTIC SCINTILLATING FIBER AND INFRARED OPTICAL FIBER

Bongsoo Lee, Sang Hun Shin, Kyoung Won Jang and Wook Jae Yoo *

School of Biomedical Engineering, College of Biomedical & Health Science, BK21 Plus Research Institute of Biomedical Engineering, Konkuk University, 268 Chungwon-daero, Chungju-si, Chungcheongbuk-do 380-701, Korea

ABSTRACT

In this study, we have studied the effects of temperature and X-ray energy variations on the light output signals from two different fiber-optic sensors, a fiber-optic dosimeter (FOD) based on a BCF-12 as a plastic scintillating fiber (PSF) and a fiber-optic thermometer (FOT) using a silver halide optical fiber as an infrared optical fiber (IR fiber). During X-ray beam irradiation, the scintillating light and IR signals were measured simultaneously using a dosimeter probe of the FOD and a thermometer probe of the FOT. The probes were placed in a beaker with water on the center of a hotplate, under variation of the tube potential of a digital radiography system or the temperature of the water in the beaker. From the experimental results, in the case of the PSF, the scintillator light output at the given tube potential decreased as the temperature increased in the temperature range from 25 to 60 °C. We demonstrated that commonly used BCF-12 has a significant temperature dependence of −0.263 ± 0.028%/°C in the clinical temperature range. Next, in the case of the IR fiber, the intensity of the IR signal was almost uniform at each temperature regardless of the tube potential range from 50 to 150 kVp. Therefore, we also demonstrated that the X-ray beam with an energy range used in diagnostic radiology does not affect the IR signals transmitted via a silver halide optical fiber.

KEYWORDS

fiber-optic sensor; plastic scintillating fiber; infrared optical fiber; X-ray; temperature

1. INTRODUCTION

Optical fiber-based sensors offer many advantages over conventional sensors, including small sensing volume, light weight, good flexibility, real-time and remote sensing, absence of electrical wires in the sensing probe, and immunity to electromagnetic interference (EMI) and radiofrequency interference (RFI). Accordingly, various fiber-optic sensors have been investigated and developed as promising candidates to overcome the shortcomings of existing physical, chemical, and biological sensors. In particular, for medical usage, fiber-optic dosimeters (FODs) and fiber-optic thermometers (FOTs) have been investigated to obtain radiation and temperature information, respectively.

FODs have been developed in conjunction with various scintillators to distinguish many kinds of radiation and to measure the related energies and dose distribution. Over the past 20 years, fiber-coupled organic scintillators including plastic scintillating fibers (PSFs) have been traditionally used for *in situ* dose measurement because of their desirable dosimetric qualities, such as a highly water equivalent characteristic, fast decay time, high spatial resolution, minimal perturbation of the radiation field, and the capability of providing real-time dose monitoring for *in vivo* dosimetry [1,2,3,4,5,6,7,8,9,10,11,12,13]. Recently, however, there has been considerable research that proves temperature affects the scintillator light output [14,15,16]. Buranurak *et al.* carried out an experimental study on temperature variation as a source of uncertainty in medical fiber-coupled organic plastic scintillator dosimetry [14]. For the BCF-12 and BCF-60, PSFs manufactured by Saint-Gobain Ceramic & Plastics, temperature coefficients of $-0.15 \pm 0.01\%/K$ and $-0.55 \pm 0.04\%/K$, respectively, were identified. The temperature dependence of BCF plastic scintillation detectors was also described by Wootton and Beddar [15]. In their experiment, BCF-12 exhibited a 0.09% decrease and BCF-60 a 0.50% decrease in measured dose per °C increase, relative to the dose measured at 22 °C.

Next, in the cases of FOTs, they normally use a silver halide optical fiber as an infrared optical fiber (IR fiber) to measure temperature by transmitting IR signals emitted from a heat source [17,18,19,20,21,22]. As a non-contact temperature sensor, those based on IR fibers have been fabricated to use in examination or treatment rooms where magnetic resonance imaging (MRI) or radiofrequency ablation (RFA) devices are installed [20,21]. Previous reports demonstrated that IR fiber-based FOTs have immunity to high EMI and RFI. However, it is also necessary to consider the effects of the radiation on the light output signals from FOTs based on IR fibers for application in hybrid medical devices, such as combined positron emission tomography (PET)-computed tomography (CT) systems and PET-MR imaging systems.

We have therefore studied the effects of temperature and X-ray energy variation on each light output signal generated from a FOD based on a PSF and a FOT using an IR fiber, respectively. In this study, two different types of sensing probes, which can produce scintillating light or transmit an IR signal, were fabricated to evaluate various non-ideal characteristics of the PSF and IR fiber. Using the fabricated two sensing probes, we measured the scintillating light generated from a PSF and the IR signal transmitted via an IR fiber during beam irradiation under different X-ray energy or temperature and compared the experimental results with the radiation dose and temperature obtained using a conventional dosimeter and a thermometer, respectively.

2. MATERIALS AND EXPERIMENTAL SETUP

2.1. Fabrication of a Fiber-Optic Dosimeter and a Fiber-Optic Thermometer

The main study concerns the impact of temperature or X-ray energy variation on the light output signals from a fiber-optic sensor based on a PSF or an IR fiber. Figure 1 and Figure 2 show the internal structures of the two different types of sensing probes, which can produce scintillating light and transmit an IR signal, respectively.

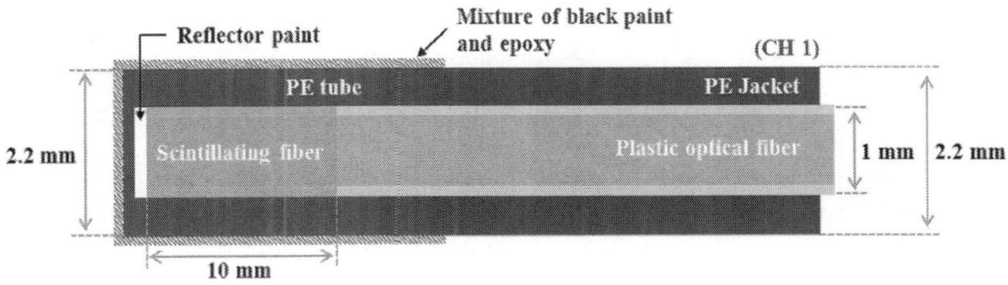

Figure 1. Internal structure of the dosimeter probe.

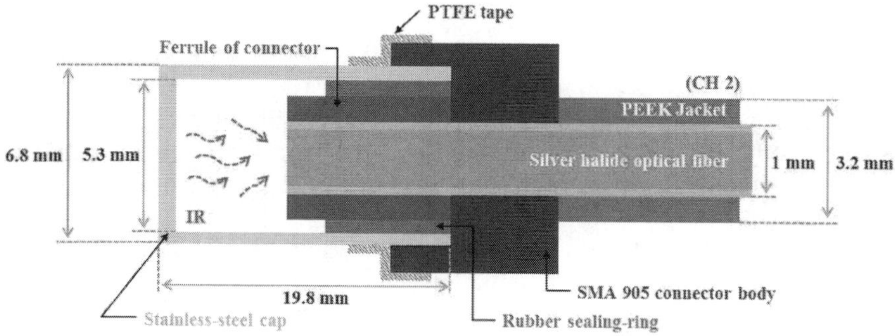

Figure 2. Internal structure of the thermometer probe.

First, we fabricated a typical dosimeter probe (CH 1) of a FOD to measure scintillating light signals induced by an X-ray beam. As a sensing element of a dosimeter probe, a PSF (BCF-12, Saint-Gobain Ceramic & Plastics, Hiram, OH, USA) was used to produce scintillating light with an emission peak in the blue range owing to X-rays. As an organic scintillator, BCF-12 is the most commonly used PSF in FOD [5,6,7,8,9,10,11,12,13,14,15]. This PSF has a core/single-clad structure and a cylindrical shape with an outer diameter of 1 mm. The thickness of the cladding is approximately 3% of the PSF diameter. The core of the PSF is synthesized with polystyrene (PS) and fluorescent dopants and the cladding is made of polymethylmethacrylate (PMMA). Accordingly, the refractive indices of the core and the cladding are 1.60 and 1.49, respectively, and the numerical aperture (NA) is about 0.58. The emission peak and the decay time of the PSF are 435 nm and 2.7 ns, respectively, and the operating temperature range is from −20 to 50 °C [23]. In order to transmit scintillating light, the PSF was coupled to a plastic optical fiber (POF: GH-4001, Mitsubishi Rayon, Tokyo, Japan). This optical fiber is multi-modal and has a step refractive index profile. The diameter of the core is 0.98 mm and with the cladding the diameter is 1 mm. The POF consists of a PMMA core with a refractive index of 1.49 and fluorinated polymer-based cladding with a refractive index of 1.402; thereby the NA of this POF is about 0.5. The maximum transmission loss of the POF is 170 dB/km when used with 650 nm collimated light and the operating temperature is from −55 to 85 °C [24]. The POF is covered by a black polyethylene (PE) jacket with an outer diameter of 2.2 mm to block external light noise and to protect the POF from ambient contamination. Figure 1 illustrates the internal structure of the dosimeter probe. In fabricating the typical dosimeter probe of the FOD, first, both ends of the PSF and POF, respectively, were polished with various types of lapping films (LFG series, Thorlabs, Newton, NJ, USA). Second, the proximal end of the 10 mm length PSF was mechanically connected to the distal end of the POF with a length of 6 m for transmitting scintillating light. Third, the distal end of the PSF was coated with a titanium dioxide (TiO_2)-based reflector paint (BC-620, Saint-Gobain Ceramic & Plastics) to increase the collection efficiency of the scintillating light. Fourth, the entire PSF was covered with a black PE tube and the outer surface of the dosimeter probe was additionally coated by a mixture of a black acrylic paint (Alpha Artist Acrylic Colors, Alpha Art Materials, Seoul, Korea) and an optical epoxy (DP-100 Plus, 3M, St. Paul, MN, USA) with a mass ratio of 3:7 to prevent the ingress of external light. Fifth, a subminiature type A (SMA) 905 connector was installed at the proximal end of the POF to enable the dosimeter probe to be connected with a photon-counting device. Lastly, the FOD system was completed by connecting the dosimeter probe with a photon-counting device, a multi-pixel photon counter (MPPC) module (C10751-03, Hamamatsu Photonics, Hamamatsu, Japan) with an internal MPPC device (S10362-11-100U, Hamamatsu Photonics) that has high sensitivity in a spectral response range from 320 to 900 nm and a peak sensitivity wavelength of 440 nm [7]. The output signals from the MPPC module with a threshold level of 0.5 photon equivalent (p.e.) and a gate time of 1 ms are transmitted via a universal serial bus (USB) line to a laptop running emulation software. During the beam irradiation, we obtained the total counts of the scintillating light generated from the PSF in the dosimeter probe.

Next, to measure IR signals induced by temperature, an embedded-type thermometer probe (CH 2) of a FOT was also constructed. As an IR fiber, we selected a silver halide optical fiber (PIR 900/1000, Art Photonics, Berlin, Germany) that can serve as a sensing element and an IR waveguide at the same time [17,18,19,20,21,22]. Diameters are 0.9 mm for the core only and 1 mm including the cladding. The refractive indices of the core and the cladding are 2.15 and 2.13, respectively, and the NA is about 0.25. This IR fiber is produced with silver bromide chloride (AgBrCl) polycrystalline, and the jacket is made of polyether ether ketone (PEEK) polymer. The silver halide optical fiber enables the transmission of mid-IR range from 4 to 18 μm. This IR fiber is also very flexible and durable over a temperature range from −200 to 250 °C and its melting point is 415 °C [25]. Figure 2 shows the internal structure of the thermometer probe, which is composed of a stainless-steel cap, a SMA 905 connector, a rubber sealing-ring, and a silver halide optical fiber with a length of 1 m. The stainless-steel cap was employed to protect the end surface of the silver halide optical fiber from harsh environments and to circumvent any emissivity effects of the measured heat source. The inner gap at the coupling interface between the stainless-steel cap and the ferrule of the SMA 905 connector was sealed by a rubber sealing-ring. In addition, the outer surface of the coupling interface between the stainless-steel cap and the SMA 905 connector body was also sealed with polytetrafluoroethylene (PTFE) tape (*i.e.*, Teflon). Unlike general non-contact types of FOTs [17,18,19,20,21], this type of thermometer probe can be directly embedded in a water environment and also can measure inner temperature in real time [22]. The completed thermometer probe was connected to a thermopile (A2TPMI334OAA060, Perkin Elmer, Wiesbaden, Germany) with a sensing range from 2 to 22 μm and an amplifier system [21,22]. The output voltage signals from the thermopile-amplifier system are transmitted to a data acquisition (DAQ) board (USB-6008, National Instruments, Seoul, Korea) and a laptop running the program LabVIEW (NI LabVIEW™ 2013, National Instruments).

2.2. Experimental Setup Using an X-ray System and a Hotplate

In order to measure the effects of the temperature or X-ray energy variation on the light output signals from the PSF or IR fiber, the laboratory tests were carried out using an experimental setup consisting of an X-ray tube (Rotanode™ E7252X, Toshiba Electron Tubes & Devices, Saitama, Japan) of a digital radiography (DR) system (Clear Vision DR 7000F, JPI Healthcare, Osong, Korea) [7] and a hotplate with a built-in magnetic stirrer (REXIM RS-6DR, AS-ONE, Osaka, Japan), as illustrated in Figure 3 and Figure 4.

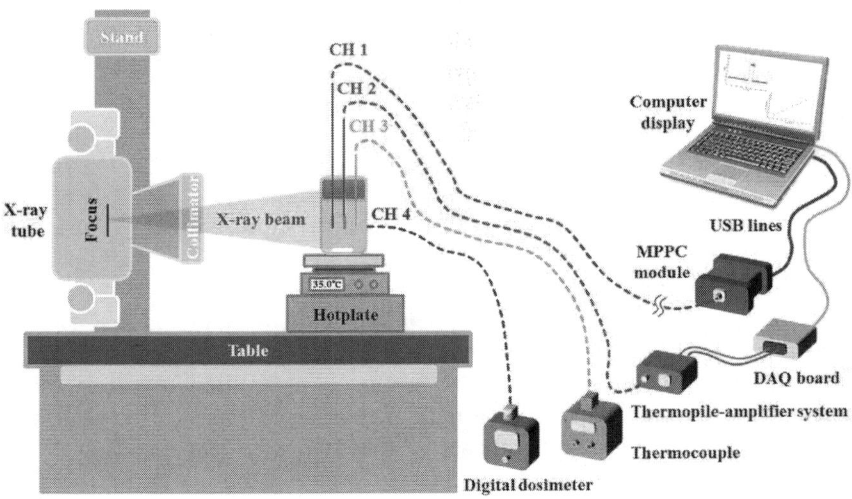

Figure 3. Whole experimental setup using a DR system and a hotplate.

In general, X-rays emitted from a DR system for diagnostic radiology are Bremsstrahlung radiation. Accordingly, to correct slightly different dose values of each X-ray beam with an identical tube potential, we used a semiconductor dosimeter (SCD), a finger-shaped dose sensor probe (CH 4), connected to a digital dosimeter (Pehamed, ALPHA plus, Sulzbach, Germany). The dose values measured by using the SCD are related to the light output signals of the FOT as well as the FOD. In the case of the IR output signals from the FOT, they were compared with the *in situ* temperature values directly obtained by using a K-type thermocouple probe (CH 3) of a thermometer (54II thermometer, Fluke, Everett, WA, USA) immersed in the water, as shown in Figure 3. Through random and repeated experiments, we measured the scintillating light and IR signals simultaneously according to the tube potential (kVp) of the X-ray tube and the temperature of the water.

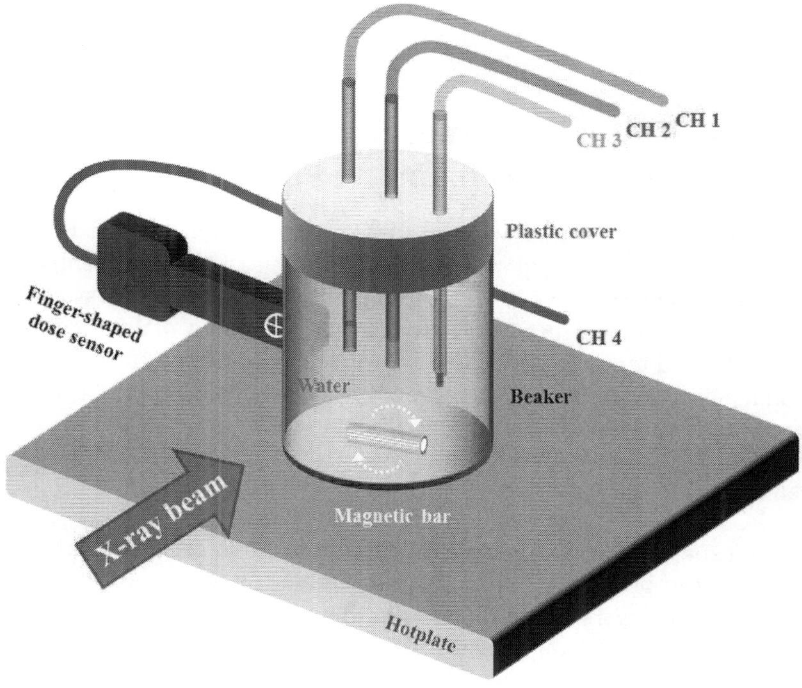

Figure 4. Detailed experimental setup using a hotplate with a built-in magnetic stirrer, a beaker of water and four different kinds of sensing probes.

A beaker of water was placed on the center of the hotplate and three different types of sensing probes (CHs 1~3) were held and inserted into the water through a plastic cover on the beaker. In the case of the SCD (CH 4), it was installed beside the beaker, as shown in Figure 4 [15]. For this experiment, the X-ray tube was rotated to irradiate the sensing probes in the water perpendicularly. As exposure parameters, the values of the current-time product (tube current × irradiation time), focus-to-surface distance (FSD) and irradiation field size were fixed at 5 mA·s (50 mA × 100 ms), 70 cm, and 20 × 20 cm^2, respectively. When the temperature of the water was maintained at the specific desired temperature controlled by the hotplate with a built-in magnetic stirrer, four sensing probes were irradiated by X-ray beams with different tube potentials. During beam irradiation, we measured the scintillating light generated from the dosimeter probe and the IR signal generated from the thermometer probe, simultaneously.

3. RESULTS

3.1. Effects of the Temperature Variations on the Light Output Signals from the PSF

In this study, we first measured the total counts of the scintillating light signals at a constant temperature while varying the tube potential in a range from 50 to 150 kVp. Here, the temperature of the water was changed in the clinical temperature range from 25 to 60 °C and maintained at the specific desired temperature with an uncertainty of ± 0.5 °C until achieving thermal equilibrium. Every experiment was carried out at a controlled room temperature of 25 ± 1 °C and all data (*i.e.*, counts) of the FOD were corrected by using absolute calibration factors of each X-ray beam at the given tube potential.

Various experimental studies on the plastic scintillator response to low-energy photon beam have been carried out, and it was reported that the plastic scintillators have non-linearity in their scintillation response to low-energy X-rays in the range used for diagnostic radiology [4,6,7,26,27,28]. From the present experimental results, the total counts of the FOD increased with a quadratic equation according to the increase of the tube potential related to the X-ray energy, as shown in Figure 5. This is consistent with results obtained by using FODs in previous studies. However, the scintillator light output at the given tube potential decreased as the temperature increased in the temperature range from 25 to 60 °C [14,15,16].

Figure 6 shows the relationship between the total counts of the scintillating light signals measured by the FOD and the X-ray doses measured by using the SCD as a function of the tube potential. Over the tube potential range of interest in diagnostic radiology, the total counts of the FOD increased with increasing X-ray dose due to the tube potential. In this test, it was found that the total counts curve has a gentle slope with increasing temperature and the difference between the total counts curve for 25 °C (*i.e.*, reference curve) and each curve for other temperature conditions gradually increases in accordance with the temperature rise. When the internal temperature of the beaker was higher than 45 °C, the outside temperature near the side of the beaker reached about 36 °C. In the case of the SCD, its operating temperature range is from 15 to 35 °C; consequently, the SCD had an over-response at a constant tube potential when its ambient temperature was higher than the maximum operating temperature (35 °C), as shown in Figure 6.

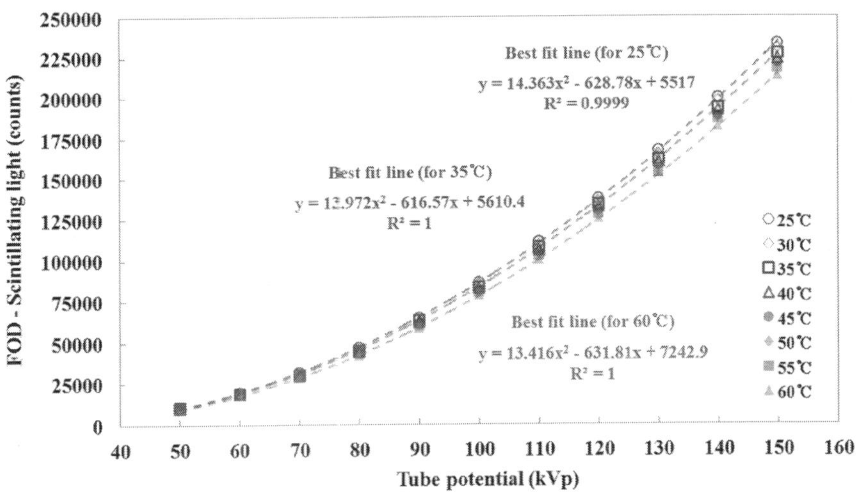

Figure 5. Total counts of the scintillating light signals measured by the FOD according to the tube potential under changing temperature of water.

Figure 7 plots the sensitivity of the FOD, in units of counts/mGy, as a function of the tube potential and it was calculated using the X-ray dose values obtained by the SCD [3,9]. The sensitivity of the FOD increased linearly with 0.678 ± 0.008%/kVp across the tube potential which is related to the X-ray dose. As can be seen in Figure 7, the sensitivity of the FOD at the given tube potential also decreased according to the increase of the temperature. The mathematical forms of each best fit line for 25, 35, and 60 °C are presented in Figure 5, Figure 6 and Figure 7. The square of the correlation coefficient (R^2), which refers to the level of agreement between the measured data and the fitting line, are also provided.

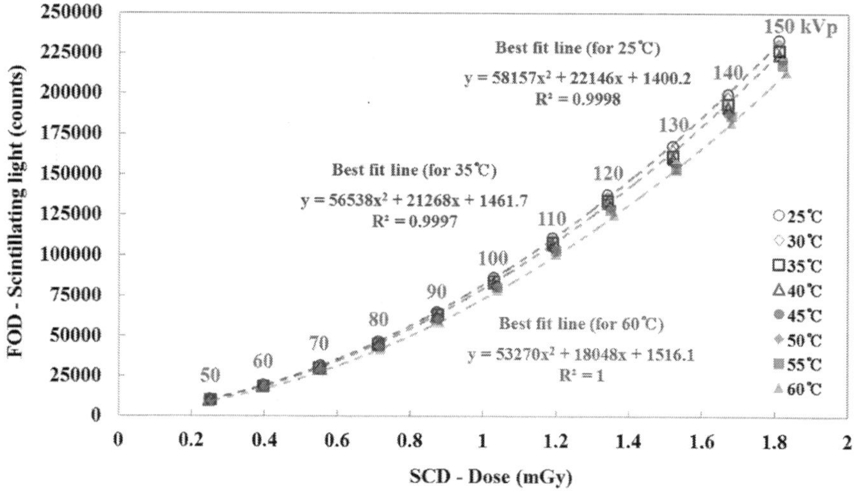

Figure 6. Relationship between the total counts of the FOD and the X-ray dose of the SCD as a function of the tube potential.

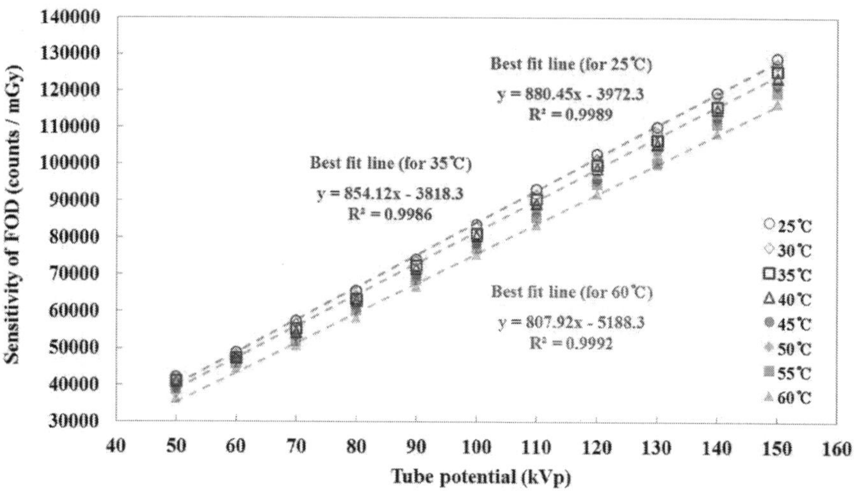

Figure 7. Sensitivity of the FOD as a function of the tube potential.

Figure 8 shows the total counts of the scintillating light signals measured by using the FOD according to the temperature variation of the water. Because the amount of scintillator light depends on X-ray energy, the total counts of the scintillating light signals at the given temperature increase as the tube potential increases, as shown in Figure 8. According to the temperature variation, the total counts of the scintillating light signals are also changed with an almost identical temperature dependence (T_d) coefficient (*i.e.*, the rate of change of the total counts per temperature variation) of −0.263 ± 0.028%/°C in the tube potential range of interest in diagnostic radiology, as shown in Figure 9. In this evaluation, it is calculated that the minimum and maximum values of T_d are −0.207 ± 0.006%/°C at 60 kVp and −0.310 ± 0.01%/°C at 50 kVp, respectively. In other words, although the PSF (*i.e.*, BCF-12) in the dosimeter probe has an almost identical T_d coefficient as a function of X-ray dose, the total counts of the scintillating light signals change more at higher X-ray dose according to the temperature variation. In conclusion, T_d is independent of energy variation although the scintillator light output is affected by both the T_d and the energy dependence (E_d) of the PSF. From these experimental results, we demonstrated that the T_d of the PSF is significant in the clinical temperature range [14,15,16]. Fortunately, the ambient temperature of most X-ray systems is normally controlled and maintained by a thermo-hygrostat in examination or radiotherapy rooms.

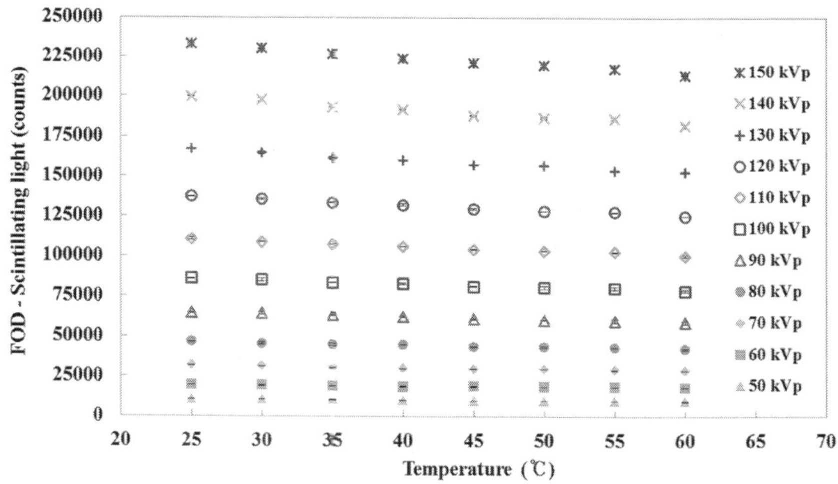

Figure 8. Total counts of the scintillating light signals measured by using the FOD according to the temperature variation of the water at each tube potential.

3.2. Effects of X-ray Energy Variation on the Light Output Signals from the IR Fiber

Next, we evaluated the effects of X-ray energy variation on the light output signals from an IR fiber. In this test, the exposure parameters were set to the same values as those of Section 3.1. Figure 10 shows the temperature of water, which was measured using a thermocouple, *versus* the output voltage of FOT as a function of the IR signal. The output voltage of the FOT increased as the temperature of water increased because the intensity of the IR signal transmitted through the IR fiber is proportional to the temperature of the heat source [17,18,19,20,21,22]. The mathematical form of the linear fit line to the curve is also presented in Figure 10 and R^2 was found to be 0.9982. Fortunately, the IR signal at the given temperature is fairly constant even when the tube potential was changed from 50 to 150 kVp.

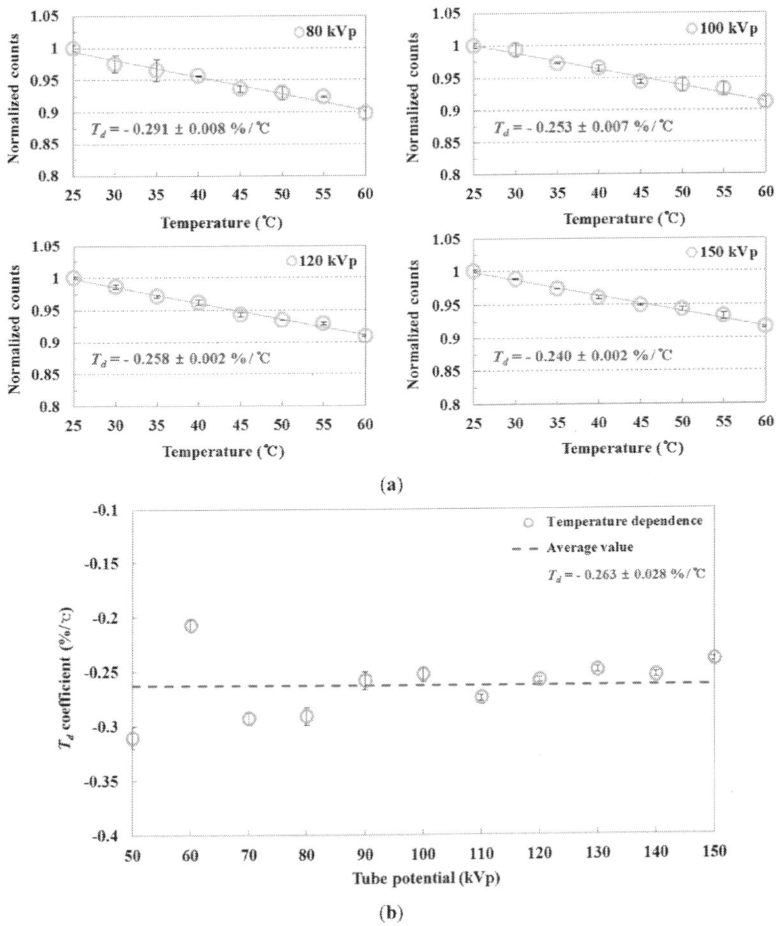

Figure 9. (**a**) Relationship between the temperature of the water and the normalized counts of the FOD when the tube potentials are 80, 100, 120, and 150 kVp, respectively; (**b**) T_d coefficients of the PSF at each tube potential.

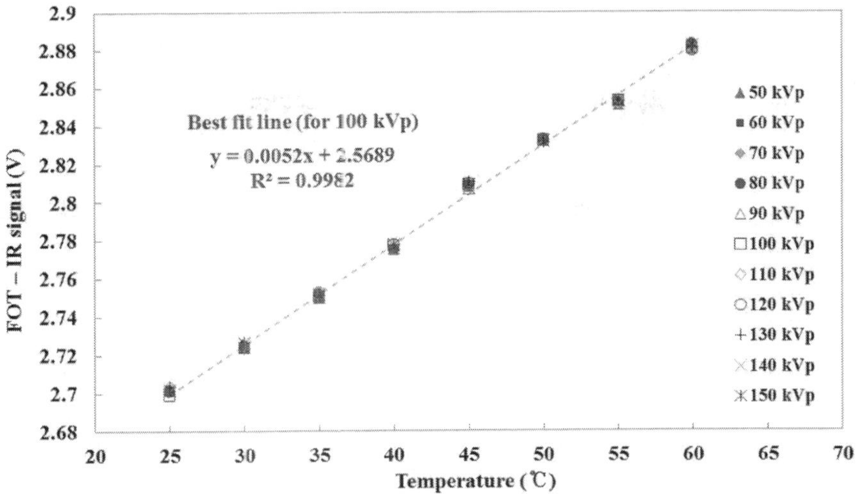

Figure 10. Relationship between the temperature of water and the IR signal measured by the FOT.

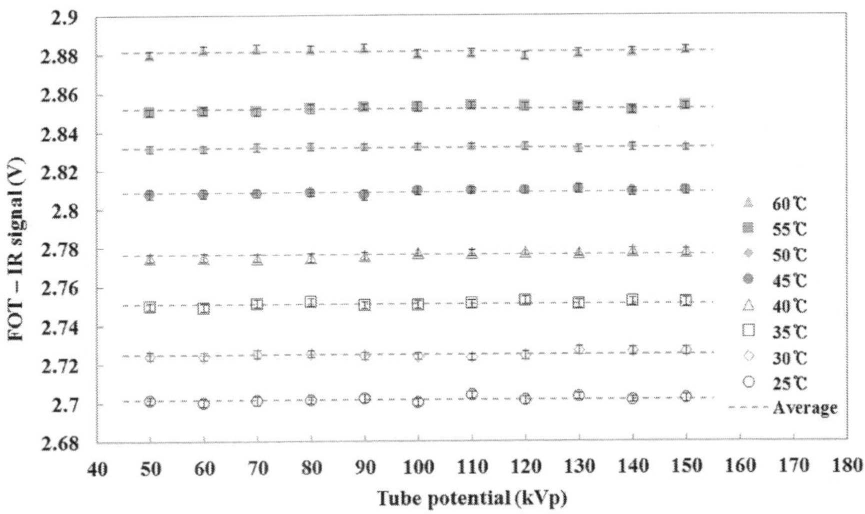

Figure 11. IR signal measurements of the FOT under changing tube potential conditions of the DR system.

Figure 11 shows the IR signals measured by using the FOT during X-ray beam irradiation under changing tube potential conditions. Although the tube potential was increased from 50 to 150 kVp, the intensity of the IR signal was almost uniform at each temperature and the FOT has a linear response with respect to the IR signal due to the temperature of the water. From the experimental results, we demonstrated that the intensity of the IR signal transmitted via the IR fiber (*i.e.*, silver halide optical fiber) is only affected by the temperature even when the tube potential is changed. In other words, the X-ray beam having an energy range used in diagnostic radiology does not affect the normal operation of the IR fiber. Because no effect due to the X-ray beam was observed, it was verified that reliable IR signal measurement is possible in medical and nuclear facilities with low-energy X-rays to monitor the temperature of a heat source.

4. DISCUSSION

From the two main sets of experimental results, first, the scintillating light signals decreased with an almost constant $T_d$coefficient of −0.263 ± 0.028%/°C in the tube potential range of interest in diagnostic radiology as the temperature increased in a clinical temperature range from 25 to 60 °C. Consequently, we demonstrated that a commonly used PSF, BCF-12, has a significant T_d response. According to the experimental results, the T_d is independent of energy variation although the scintillator light output is affected by both the T_d and the E_d of the PSF.

Prior to our study, Buranurak *et al.*, found that the BCF-12 and BCF-60 have T_d coefficients of −0.15 ± 0.01%/K and −0.55 ± 0.04%/K, respectively, in the temperature range from 15 to 40 °C with a 50 kVp X-ray beam [14]. From the experimental results, they suggested that the T_d coefficient reflects inherent property of a scintillator and it is not dependent on the radiation quality. Wootton and Beddar also studied on the T_d of BCF plastic scintillation detectors [15]. In their experiments, the measured doses using the BCF-12 and the BCF-60 decreased by approximately 0.09%/°C and 0.50%/°C, respectively, when the PSFs are irradiated with 100 monitor units (MU). According to their study, BCF-60 has greater T_d coefficient than BCF-12 and the generations of Cerenkov light in the PSF and optical fiber are temperature independent.

As mentioned earlier, the ambient temperature of most X-ray systems installed in examination or radiotherapy rooms is normally controlled and maintained for quality assurance (QA). Therefore, it is only necessary to consider the conversion factor of the scintillating light signals reflecting the body temperature of the patient.

Next, the intensity of the IR signal was almost uniform at each temperature regardless of the tube potential. From the experimental results, we also demonstrated that an X-ray beam having an energy range used in diagnostic radiology does not affect the IR signals transmitted through a silver halide optical fiber. It was thus verified that the FOT allows measurement of IR signals without any effect due to X-rays in medical and nuclear facilities that deal with low-energy X-rays.

5. CONCLUSION

To evaluate the influence of the temperature and X-ray energy variations on the light output signals from a PSF and an IR fiber, respectively, we fabricated two different types of sensing probes that can produce scintillating light or an IR signal. While varying the tube potential or the temperature of water, the scintillating light signals and the IR signals were measured simultaneously using the dosimeter probe of the FOD and the thermometer probe of the FOT, respectively. From the experimental results, we demonstrated that the BCF-12 has significant T_d response in the clinical temperature range and the X-ray beam with an energy range used in diagnostic radiology does not affect the IR signals transmitted via a silver halide optical fiber.

Based on the results of this study, it is anticipated that the typical FOD and FOT systems will be used to reliably measure the absorbed dose and temperature information, respectively, in hybrid medical devices including PET-CT and PET-MR imaging systems. Further studies will be carried out to evaluate the influence of the variation of temperature and radiation energy on the light output signals from PSF and IR fibers using a high-energy radiation source, such as a linear accelerator (LINAC) and a Co-60 therapy unit.

ACKNOWLEDGMENTS

This work was supported by Konkuk University in 2014.

AUTHOR CONTRIBUTIONS

Wook Jae Yoo made substantial contributions in design of the sensor, all experiments, data analysis, and manuscript preparation. Bongsoo Lee made significant contributions in experimental design, data analysis and manuscript preparation. Sang Hun Shin made contributions in fabrication of the sensor, experimental setup and data acquisition. Kyoung Won Jang made contributions in data analysis and sensor characterization.

CONFLICTS OF INTEREST

The authors declare no conflict of interest.

REFERENCES

1. Beddar, A.S.; Mackie, T.R.; Attix, F.H. Water-equivalent plastic scintillation detector for high-energy beam dosimetry: Part I, Physical characteristics and theoretical consideration. *Phys. Med. Biol.* **1992**, *37*, 1883–1900.

2. Beddar, A.S.; Mackie, T.R.; Attix, F.H. Water-equivalent plastic scintillation detector for high-energy beam dosimetry: Part II, Properties and measurements. *Phys. Med. Biol.* **1992**, *37*, 1901–1913.

3. Jones, A.K.; Hintenlang, D. Potential clinical utility of a fibre optic-coupled dosemeter for dose measurements in diagnostic radiology. *Rad. Prot. Dosim.* **2008**, *132*, 80–87.

4. Lessard, F.; Archambault, L.; Plamodon, M.; Despres, P.; Therriault-Proulx, F.; Beddar, S.; Beaulieu, L. Validating plastic scintillation detectors for photon dosimetry in the radiologic energy range. *Med. Phys.* **2012**, *39*, 5308–5316.

5. Hoerner, M.R.; Stepusin, E.J.; Hyer, D.E.; Hintenlang, D.E. Characterizing energy dependence and count rate performance of a dual scintillator fiber-optic detector for computed tomography. *Med. Phys.* **2015**, *42*, 1268–1277.

6. Yoo, W.J.; Shin, S.H.; Jeon, D.; Han, K.-T.; Hong, S.; Kim, S.G.; Cho, S.; Lee, B. Development of a fiber-optic dosimeter based on modified direct measurement for real-time dosimetry during radiation diagnosis. *Meas. Sci. Technol.* **2013**, *24*, 094022.

7. Yoo, W.J.; Shin, S.H.; Jeon, D.; Hong, S.; Sim, H.I.; Kim, S.G.; Jang, K.W.; Cho, S.; Youn, W.S.; Lee, B. Measurement of entrance surface dose on an anthropomorphic thorax phantom using a miniature fiber-optic dosimeter. *Sensors* **2014**,*14*, 6305–6316.

8. Archambault, L.; Beddar, A.S.; Gingras, L.; Lacroix, F.; Roy, R.; Beaulieu, L. Water-equivalent dosimeter array for small-field external beam radiotherapy. *Med. Phys.* **2007**, *34*, 1583–1592.

9. Hyer, D.E.; Fisher, R.F.; Hintenlang, D.E. Characterization of a water-equivalent fiber-optic coupled dosimeter for use in diagnostic radiology. *Med. Phys.* **2009**, *36*, 1711–1716.

10. Yoo, W.J.; Shin, S.H.; Jeon, D.; Hong, S.; Kim, S.G.; Sim, H.I.; Jang, K.W.; Cho, S.; Lee, B. Simultaneous measurements of pure scintillation and Cerenkov signals in an integrated fiber-optic dosimeter for electron beam therapy dosimetry.*Opt. Express* **2013**, *21*, 27770–27779.

11. Archambault, L.; Briere, T.M.; Ponisch, F.; Beaulieu, L.; Kuban, D.A.; Lee, A.; Beddar, A.S. Toward a real-time *in vivo*dosimetry system using plastic scintillation detectors. *Int. J. Radiat. Oncol. Biol. Phys.* **2010**, *78*, 280–287.

12. Lacroix, F.; Archambault, L.; Gingras, L.; Guillot, M.; Beddar, A.S.; Beaulieu, L. Clinical prototype of a plastic water-equivalent scintillating fiber dosimeter array for QA applications. *Med. Phys.* **2008**, *35*, 3682–3690.

13. Moutinho, L.M.; Castro, I.F.; Peralta, L.; Abreu, M.C.; Veloso, J.F.C.A. Development of a scintillating optical fiber dosimeter with silicon photomultipliers. *Nucl. Instrum. Methods Phys. Res. A* **2014**, *735*, 640–643.

14. Buranurak, S.; Andersen, C.E.; Beierholm, A.R.; Lindvold, L.R. Temperature variations as a source of uncertainty in medical fiber-coupled organic plastic scintillator dosimetry. *Radiat. Meas.* **2013**, *56*, 307–311.

15. Wootton, L.; Beddar, S. Temperature dependence of BCF plastic scintillation detectors. *Phys. Med. Biol.* **2013**, *58*, 2955–2967.

16. Beddar, A.S. On the possible temperature dependence of plastic scintillator response. *Med. Phys.* **2012**, *39*, 6522.

17. Eyal, O.; Scharf, V.; Katzir, A. Temperature measurements using pulsed photothermal radiometry and silver halide infrared optical fibers. *Appl. Phys. Lett.* **1997**, *70*, 1509–1511.

18. Sade, S.; Katzir, A. Spectral emissivity and temperature measurements of selective bodies using multiband fiber-optic radiometry. *J. Appl. Phys.* **2004**, *96*, 3507–3513.

19. Katzir, A.; Bowman, H.F.; Asfour, Y.; Zur, A.; Valeri, C.R. Infrared fibers for radiometer thermometry in hypothermia and hyperthermia treatment. *IEEE Trans. Biomed. Eng.* **1989**, *36*, 634–637.

20. Sade, S.; Katzir, A. Fiberoptic infrared radiometer for real time *in situ* thermometry inside an MRI system. *Magn. Reson. Imag.* **2001**, *19*, 287–290.

21. Yoo, W.J.; Cho, D.H.; Jang, K.W.; Shin, S.H.; Seo, J.K.; Chung, S.-C.; Lee, B.; Park, B.G.; Moon, J.H.; Cho, Y.-H.; *et al.* Infrared radiation thermometer using a silver halide optical fiber for thermal ablation. *Opt. Rev.* **2009**, *16*, 386–390.

22. Yoo, W.J.; Jang, K.W.; Seo, J.K.; Moon, J.; Han, K.-T.; Park, J.-Y.; Park, B.G.; Lee, B. Development of a 2-channel embedded infrared fiber-optic temperature sensor using silver halide optical fiber. *Sensors* **2011**, *11*, 9549–9559.

23. Saint-Gobain Ceramic & Plastics, Inc., Scintillating Optical Fibers (Hiram, OH, USA), 2014. Available online: http://www.crystals.saint-gobain.com/Scintillating_Fiber.aspx (accessed on 8 May 2015).

24. Mitsubishi Rayon Co., Specification Sheet—GH 4001 (Tokyo, Japan), 2001. Available online: http://i-fiberoptics.com/fiber-detail.php?id=47&sum=80 (accessed on 8 May 2015).

25. Art Photonics GmbH, Polycrystalline IR-Fibers & Cables (Berlin, Germany), 2014. Available online: http://www.artphotonics. de/products/special-fibers-and-fiber-cables/polycrystalline-ir-fiber-cables/(accessed on 8 May 2015).

26. Sporea, D.; Mihai, L.; Vata, I.; McCarthy, D.; O'keeffe, S.; Lewis, E. Characterization of scintillating X-ray optical fiber sensors. *Sensors* **2014**, *14*, 3445–3457.

27. McCarthy, D.; O'keeffe, S.; Lewis, E.; Sporea, D.G.; Sporea, A.; Tiseanu, I.; Woulfe, P.; Cronin, J. Radiation dosimeter using an extrinsic fiber optic sensor. *IEEE Sens. J.* **2014**, *14*, 673–685.

28. Williamson, J.F.; Dempsey, J.F.; Kirov, A.S.; Monroe, J.I.; Binns, W.R.; Hedtjarn, H. Plastic scintillator response to low-energy photons. *Phys. Med. Biol.* **1999**, *44*, 857–871.

12

OPTICAL FIBER SENSORS FOR AIRCRAFT STRUCTURAL HEALTH MONITORING

Iker García [1,*], Joseba Zubia [1], Gaizka Durana [1], Gotzon Aldabaldetreku [1], María Asunción Illarramendi [2] and Joel Villatoro [1,3]

[1] Department of Communications Engineering, E.T.S.I. of Bilbao, University of the Basque Country UPV/EHU, Alda. Urquijo s/n Bilbao 48013, Spain

[2] Department of Applied Physics I, E.T.S.I. of Bilbao, University of the Basque Country UPV/EHU, Alda. Urquijo s/n Bilbao 48013, Spain

[3] IKERBASQUE—Basque Foundation for Science, E-48011 Bilbao, Spain

ABSTRACT

Aircraft structures require periodic and scheduled inspection and maintenance operations due to their special operating conditions and the principles of design employed to develop them. Therefore, structural health monitoring has a great potential to reduce the costs related to these operations. Optical fiber sensors applied to the monitoring of aircraft structures provide some advantages over traditional sensors. Several practical applications for structures and engines we have been working on are reported in this article. Fiber Bragg gratings have been analyzed in detail, because they have proved to constitute the most promising technology in this field, and two different alternatives for strain measurements are also described. With regard to engine condition evaluation, we present some results obtained with a reflected intensity-modulated optical fiber sensor for tip clearance and tip timing measurements in a turbine assembled in a wind tunnel.

KEYWORDS

structural health monitoring; optical fiber sensors; fiber Bragg grating; long period grating; turbine condition monitoring

1. INTRODUCTION

All kinds of engineering infrastructures undergo aging, and damage appears as a consequence of the loads applied to them, so inspection and maintenance actions are required to predict and lengthen their lifetime thus avoiding catastrophic failures. Among these infrastructures, those related to aviation demand the highest levels of damage detection, since most of them are designed according to a damage-tolerant principle [1]. In order to get lighter structures, these are designed to withstand damages with certain characteristics. Nevertheless, damage-tolerant design involves a considerable amount of time and effort for structure inspection, because there are a wide variety of factors that can cause damage in aircraft structures, most of which are summarized in [2]. Due to the harsh conditions suffered by aircraft structures, periodic and scheduled inspection and maintenance tasks are essential for safe and efficient operation. The cost for the personnel needed to carry out these procedures is high, but the cost due to aircraft downtime during these time-consuming inspections is significantly greater [3]. Hence, automation of the inspection process is a point of capital importance to reduce inspection efforts. In this context, a structural health monitoring (SHM) system can be defined as a set of devices that provides information that allows us to locate, evaluate and predict the loading and damage conditions of a structure. SHM of aircraft structures can perform real time inspection, reducing costs and improving the reliability and performance of these structures [4]. The successful application of SHM to aircraft requires two conditions: it must provide reliable and accurate information about the condition of the structure, increasing the security of critical components and avoiding disasters, and it must be a profitable process for the airline operators by reducing economic losses caused by unproductive downtimes. Therefore, there is a great interest from industry and academia in developing SHM systems that meet both conditions. A wide range of potential SHM technologies is being developed to fulfill these conditions [1], and the most promising options are: electrical strain gauges and crack wires, acoustic emission methods, optical-based technologies, comparative vacuum monitoring [5] and MEMS.

Their intrinsic capabilities, such as insensitivity to electromagnetic radiation, light weight, small size, great sensitivity and resolution, and, above all, their suitability to be embedded into structures [6], make optical fiber sensors (OFSs) very appropriate to perform SHM [7]. There are three possible approaches in order to deploy a SHM system based on optical sensors [8]. The first one uses single-point sensors [9], which is the way in which most of the pressure or temperature sensors operate. The second one employs distributed sensing [10], where the measurand can be obtained at any point of an optical fiber. Finally, there are quasi-distributed systems that use a number of single-point sensors, allowing the sensing of large structures [11]. In this review article we summarize some of the results we have obtained in relation to aircraft SHM projects during the last few years. In Section 2.1, the basic principles of fiber Bragg gratings (FBG) and a quasi-distributed system for impact detection are presented. In Section 2.2, we describe the measurement of strain along a helicopter tail boom by using an elongation sensor, and also the determination of strain in a steel plate by means of a long period grating (LPG). Whereas the LPG is a single-point sensor, the elongation sensor is a distributed one. The performances of both sensors are compared with the results obtained using FBGs. Finally, in Section 3 the design and results of an optical sensor that has been used to carry out the SHM of an aircraft turbine are reported.

2. SHM IN AIRCRAFT STRUCTURES

OFSs for monitoring the strain in aircraft structures can be classified into the following categories [12]: intensity-based OFSs [13], interferometric OFSs [14], grating-based sensors and distributed OFSs [15]. Intensity-based OFSs were the first optical sensors employed to measure the strain, and they are the simplest and least expensive optical sensors for SHM. Interferometers work on the principle of the interference caused by a reflected monocromatic light with its original source. Several interferometer configurations have been proposed, being Fabry-Perot the most widely utilized. There are different techniques for implementing a distributed OFS; optical time domain reflectometry (OTDR) based on Rayleigh scattering, Raman optical time domain reflectometry (ROTDR) or Brillouin optical time domain reflectometry (BOTDR), based on Raman and Brillouin scattering, respectively. BOTDR resolution and acquisition time has been improved by Brillouin optical time domain analysis (BOTDA) using the phenomenon know as Stimulated Brillouin Scatter. Among grating-based sensors, FBGs are probably the most mature and widely employed optical sensors for SHM of engineering structures, due to their fast development achieved in recent years [16,17]. For the particular case of aircraft structures, even though FBGs have demonstrated to be a promising technology to monitor strain [18], strain gauges still remain being the most used method to perform strain measurements in operational aircraft structures. FBGs have important advantages over conventional strain sensors, namely:

a. Intrinsic benefits of optical fiber sensors such as light weight and small size, absence of electromagnetic interference, high sensitivity and resolution, *etc.* [19].

b. Suitability for being attached to a structure or embedded in composite materials [20].

c. Wavelength-encoded sensing in a way that is totally independent of the optical intensity [21], which confers them long-term stability without the need of recalibrating.

d. High multiplexing capability: since each FBG has a narrow-wavelength operating band, it is possible to multiplex several sensors in the same fiber, thus allowing for simultaneous multi-point measurements [22].

e. Different magnitudes can be measured using FBGs, such as strain, temperature, vibration or humidity.

Regarding strain measurement, FBGs show a similar behavior to strain gauges. However, even though strain gauges are robust, low cost and well known, they present important disadvantages such as their long-term signal drift or time-consuming and complex wiring [23], since two leads per sensor are required. The multiplexing capability of the FBGs overcomes the complex wiring drawback. This characteristic has been reported in several works [11,24], in which the multiplexing capability of the FBGs provides a size optimization for the practical implementation of a quasi-distributed SHM system. The additional possibility of being embedded in the structures allows the use of FBGs to identify damage in composite structures [20], as well as the monitoring of composite cure during its manufacture [25].

2.1. Fiber Bragg Gratings

An optical fiber grating can be defined as a periodic perturbation pattern in the refractive index of the fiber core, in such a way that certain wavelengths of the guided mode are diffracted either into other radiation modes or into cladding modes [26]. In the former case, the device is known as a short-period fiber grating or fiber Bragg grating. Such gratings have a sub-micron period and a total length ranging from 1 mm to 10 mm. A Bragg grating allows most of the power (P_o) to propagate forward. However, when the guided mode in a single-mode fiber reaches the grating, a certain portion of its incident power (P_i) is reflected at each grating plane. The electric fields are added up only if the Bragg condition given by Equation (1) is satisfied:

$$\lambda_B = 2 \cdot n_{eff} \cdot \Lambda \quad (1)$$

where λ_B is the resonant wavelength or Bragg wavelength, n_{eff} is the effective index of the mode and Λ is the grating period. The working principle of the FBGs is illustrated in Figure 1.

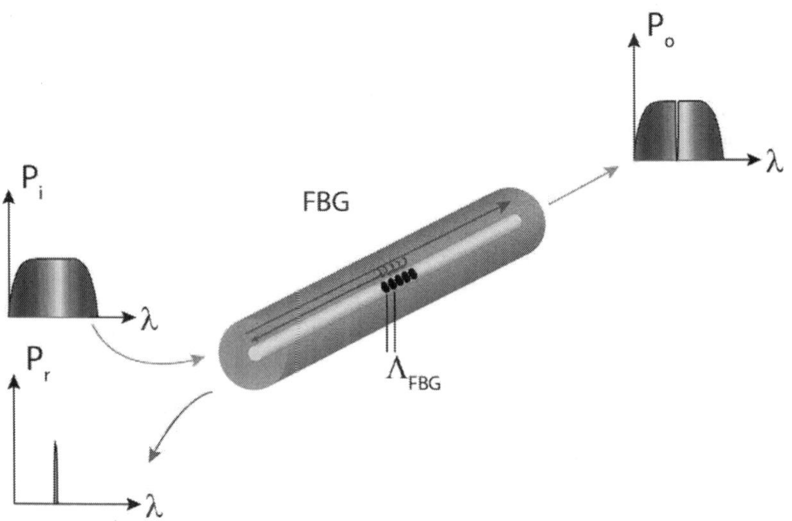

Figure 1. FBG concept and working principle.

There are two basic techniques to inscribe Bragg gratings in fibers. The first one uses ultraviolet light to change the refractive index of photosensitized fibers. The photosensitivity process typically consists in doping the fiber core with germanium, so that the fiber becomes receptive to UV-induced refractive index changes. Several methods have been employed to enhance the photosensitivity, such as high temperature treatment of hydrogen preforms, B-Ge co-doping, flame brushing, and high-pressure hydrogen loading of fibers [27]. The second fabrication technique makes use of a femtosecond laser to write a grating in non-photosensitive single-mode commercial fibers [28]. The main advantages of this technique are the shortening of the processing time, the reduced requirements for the fabrication process and the mechanical strength of the gratings. Currently, the most employed methods to inscribe the gratings are holographic techniques [29] and phased mask methods [30]. The point-by-point technique is usually employed when femtosecond lasers are used to inscribe the gratings [31].

As mentioned before, the measurand is encoded in the variations of the Bragg wavelength. Any change in the temperature or in the strain of the fiber causes a change in the grating period and/or in the effective index, shifting the central wavelength reflected by the grating. When the FBG is subjected to a tension, the change in the Bragg wavelength is given by Equation (2):

$$\frac{\Delta\lambda_B}{\lambda_B} = (1-P_e)\cdot\Delta\varepsilon \tag{2}$$

where ε is the longitudinal strain in the FBG and P_e is the effective photo-elastic constant of the fiber core material. In Equation (2), P_e can be calculated as:

$$P_e = \frac{n_{eff}}{2}\left[p_{12} - \upsilon\left(p_{11}+p_{12}\right)\right] \tag{3}$$

where p_{11} and p_{12} are the elasto-optic coefficients, which represent the change in the refractive index due to the applied strain, and υ is the Poison ratio. The temperature changes affect the Bragg wavelength in two ways. On the one hand, the expansion coefficient of the fiber material (α) characterizes the thermal expansion of the grating, and, on the other, the variations in the refractive index are determined by the thermo-optic coefficient (ζ). Therefore, the change in the Bragg wavelength with temperature can be expressed as:

$$\frac{\Delta\lambda_B}{\lambda_B} = (\alpha+\zeta)\cdot\Delta T \tag{4}$$

Combining Equations (2) and (4) we get that the total Bragg-wavelength variation can be calculated as:

$$\frac{\Delta\lambda_B}{\lambda_B} = (\alpha+\zeta)\cdot\Delta T + (1-P_e)\cdot\Delta\varepsilon \tag{5}$$

Thus, it is clearly seen that the Bragg wavelength presents a cross-sensitivity phenomenon (it is sensitive to both temperature and strain). For strain measurements, the effect of temperature has to be compensated. Different configurations have been reported for this purpose [32,33, 34,35,36,37,38,39,40,41,42,43,44]. However, when temperature gradients in the structure are not excessive, the use of a single FBG insulated from the effect of the strain is enough to compensate the effect of temperature in the rest of the FBGs, thus allowing us to discriminate the variations in the Bragg wavelength that appear as a consequence of strain. FBGs are excellent strain sensors when the load is applied in the axial direction of the sensor. In this situation, the FBG only undergoes contraction or elongation. If there is a transverse stress applied the fiber presents birefringence due to the variation experienced by the effective refraction index on each propagation axis [45]. As a consequence, the grating exhibits two different Bragg conditions, and the approximately Gaussian-shaped reflected spectrum of the FBG splits into two peaks, as represented in Figure 2 [46]. Another similar phenomenon appears when there is a non-uniform strain field, because, in this case, the spectrum is broadened and even split into several peaks [20], which makes it difficult to track the Bragg wavelength. Both problems can arise when the FBG is embedded in composite materials. In conclusion, FBGs constitute a mature technology and compared to strain gauges, provide multiple benefits and attractive properties, such as multiplexing capability, the possibility of being embedded in the structure, long term stability and a competitive cost per channel. Nevertheless, some issues, like non-uniform or multidirectional complex strain conditions and the lack of aircraft certification must be solved, so that FBGs can be used in SHM systems for aircraft structures.

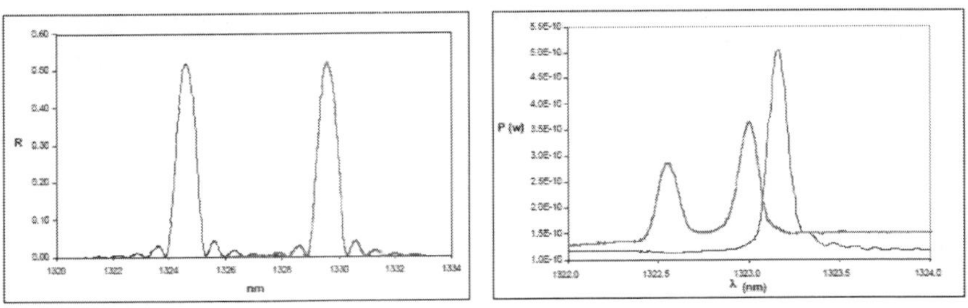

(a) (b)

Figure 2. (**a**) Ideal behavior of the reflected spectrum of an FBG subjected to a longitudinal strain; (**b**) Effect of a transverse stress on the reflected spectrum of the FBG. In all figures the blue line represents the non-perturbed spectrum of the FBG and the red one is the spectrum after the application of the load.

Another important concern is the effect of aging on the response of the FBGs under extreme conditions. In [47] an analysis of the influence of aging FBGs is presented. FBGs were bonded on three specimens using different adhesives to evaluate their performance against aging. In order to reproduce aging under real conditions, two of them were stored in a climate chamber at 70 °C and 90% relative humidity for one and two months, respectively. The damage produced in the specimen aged for two months can be observed in Figure 3.

Figure 3. Photograph of four FBGs bonded with different adhesives to the F114 steel specimen after being aged for two months in a climate chamber.

The results of the tests on the specimens previously aged for one and two months are very similar and, unexpectedly, they show that the readings of the FBGs are about 50% higher than those obtained before the aging. Since the maximum working temperature of the FBG was never exceeded, this discrepancy in the readings must be caused by the effect of humidity. Its influence can make the refractive index vary, shifting the Bragg wavelength and yielding much higher strain values than expected, so in order to avoid this problem, it is necessary to properly insulate the FBG from humidity. The results for two FBGs of the specimen aged for one month for 200 kN traction stress tests are displayed in Figure 4. The traction stress was applied in steps of 20 kN. In most cases, FBGs are employed in SHM applications to detect strain or temperature variations in a quasi-static way. However, if FBGs are used to detect impacts in aircraft structures, vibrations at higher frequencies must be analyzed [48]. We developed a proof-of-concept system for impact detection in composites using FBGs [49]. The objective was to obtain the arrival times of the surface mechanical waves produced by an impact in different points of the structure. Using these times, the impact location could be determined by triangulation. Two arrays of five FBGs were instrumented on a piece of Carbon Fiber Reinforced Polymer (CFRP). In order to detect the mechanical waves due to the impacts in as many directions as possible, nine of the FBGs were placed following the arrangement given in [50]. The remaining FBG was insulated from the vibrations and employed to compensate the effect of the temperature. The final layout can be observed in Figure 5. The test procedure included five different impact points; each impact was repeated three times with increasing energies from 0.245 J to 29.4 J.

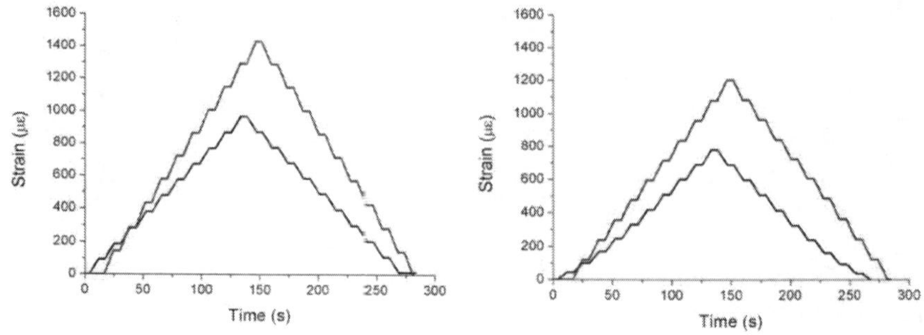

Figure 4. Spectra of two FBGs bonded on the specimen before (**black line**) and after (**blue line**) aging for one month in a climate chamber.

A 20 kHz interrogator was employed to obtain the response of the FBGs and Figure 6 shows the difference between the arrival times of the wavefronts generated by the impact for two FBGs.

2.2. Alternatives to Fiber Bragg Gratings

FBGs are probably the most well-known, developed and widely used OFSs for SHM in aeronautical structures. Their characteristics make them the most promising technique to evaluate the condition of these structures, but in order to have a broad vision of the possibilities that OFSs offer, it is necessary to consider different alternatives to FBGs that can provide suitable solutions to specific cases. In this section we present two different sensors as possible alternatives to FBGs. Their performances are analyzed and compared to those of the FBGs working in the same conditions.

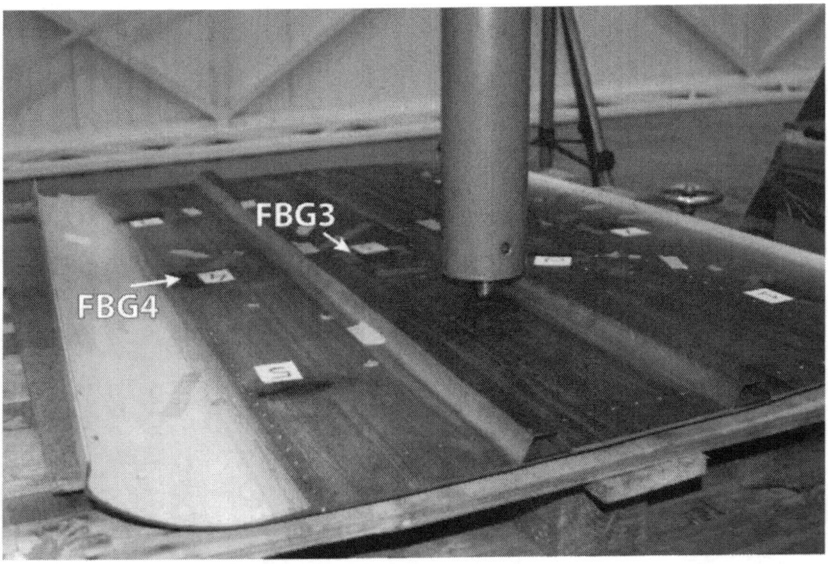

Figure 5. Instrumented specimen for impact detection and certified impact machine used in the tests.

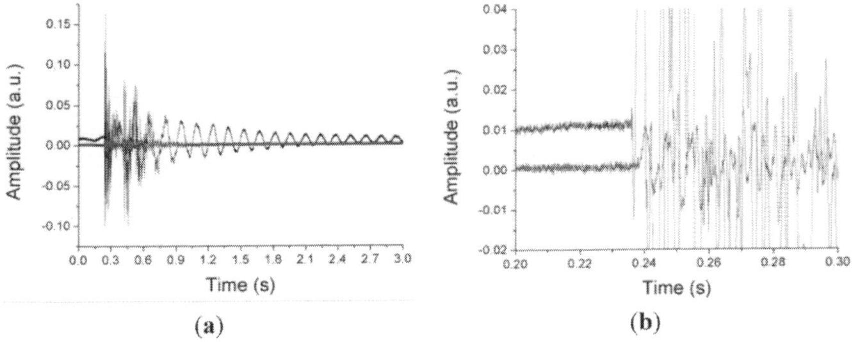

Figure 6. (**a**) Response of two different FBGs (FBG3 and FBG4 shown in Figure 5) to the mechanical waves produced by an impact; (**b**) Detailed view of the arrival of the mechanical waves to the FBGs.

2.2.1. Plastic Optical Fiber Elongation Sensor

In this section we present a plastic optical fiber (POF) elongation sensor as a possible alternative to FBGs for the measurement of strain and vibrations in aircraft structures. The sensor's operating principle relies on the measurement of the phase shift between two sinusoidally-modulated optical signals. These signals travel in two different POFs, one used as reference and the other one bonded to the surface of the aircraft structure under test [51].

The main advantage of this sensor is the use of low-cost components. Conventional POFs are used as the sensitive component of the sensor. Due to the multimode nature of these fibers, the phase and the polarization state are not maintained along the fiber, so the only possible transduction technique must be based on intensity modulation [52]. InFigure 7 a schematic representation of the sensor is depicted. The output of the voltage-controlled oscillator (VCO) is a sinusoidal voltage with an adjustable frequency (f_m). This VCO signal is applied to the transmitter, which is basically a light-emitting diode at 650 nm. The Y coupler divides the optical signal in two identical ones that will be transmitted by the reference fiber and the measuring fiber, respectively. The receiver, essentially a photodiode, transforms into voltages the optical signals coming out from the fibers as:

$$V_{R1} = \sin(2\pi f_m t) \tag{6}$$

$$V_{R2} = \sin(2\pi f_m t + \Delta\phi) \tag{7}$$

When the measuring fiber is elongated or shortened due to the strain applied to the aircraft structure, the phase difference between the signals ($\Delta\phi$) given by Equations (6) and (7) changes. In order to capture the phase difference, the outputs of both receivers are connected to a phase comparator from which the elongation of the fiber (Δl) can be estimated as [53]:

$$\Delta l = \frac{c}{n_{core}} \cdot \frac{1}{360^\circ} \cdot \frac{\Delta\phi}{f_m} \tag{8}$$

where c is the speed of light in vacuum and n_{core} is the refractive index of the fiber core. Once the phase comparator signal is acquired by a data acquisition module (NI-USB3210 from National Instruments), a Labview program processes the signal to obtain the elongation of the fiber.

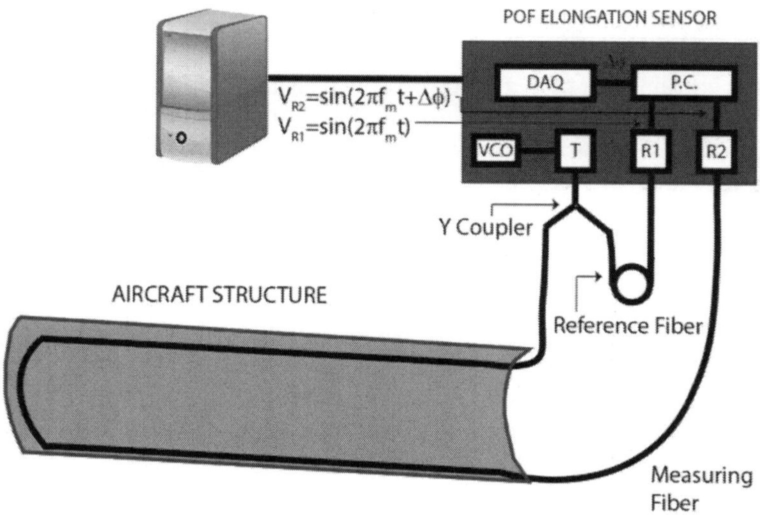

Figure 7. POF elongation sensor components and set-up. VCO: Voltage-controlled oscillator; T: Transmitter; R1: Receiver 1; R2: Receiver 2; P.C.: Phase comparator; DAQ: Data acquisition module.

In contrast to the discrete-point measurement obtained using FBGs, the POF elongation sensor provides a distributed measurement of the strain in the structure, since the overall fiber elongation is taken into account. In previous works [54, 55], the sensor was employed to characterize the response of a wing flap to flexural-type loading conditions. We compared the results of the POF elongation sensor to those provided by FBGs. For these tests, the structure to be assessed was a helicopter (Eurocopter EC 135) tail boom which can be observed in Figure 8. The assembly is composed of a rigid metal structure on one end of the tail boom and a hydraulic actuator on the other one. The load applied by the actuator produces bending of the tail boom and, therefore, stress in the structure. Three different fatigue cycles were applied to the tail boom and each load cycle lasted for 16 s. The forces applied in the fatigue cycles were 0.25, 0.50 and 0.75 kN.

Besides the POF elongation sensor, ten FBGs in two different fibers were instrumented on the surface of the tail boom for the sake of comparison. In the first fiber, four FBGs were installed in the longitudinal direction of the tail boom and the last one in a perpendicular direction. For the second fiber, four FBGs were placed longitudinally and one was used for temperature compensation. The response of the FBGs was measured using the SM130-200 interrogator from MicronOptics.

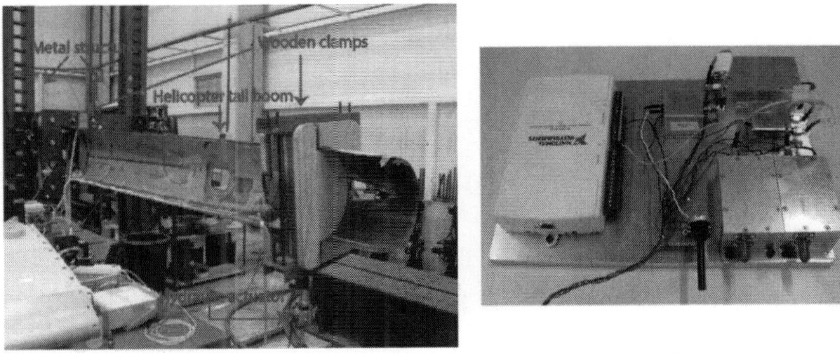

(a) (b)

Figure 8. (**a**) Experimental set-up for the tests performed on the tail boom; (**b**) Prototype of the electronics for the POF elongation sensor.

The results obtained by the POF elongation sensor for several load cycles can be observed in Figure 9. When a force of 0.75 kN or 0.5 kN is applied, the cycles are clearly detected by the sensor. However, for the case of 0.25 kN the noise makes it impossible to distinguish the stress cycles in the post-processed signal. The goal of these tests was to demonstrate the capability of the POF elongation sensor to detect the load cycles on the tail boom. Although this goal was achieved, more tests should be carried out to ensure the reliability of the sensor, and the quality of the output signal must be improved. The offset variations in the displacement signals are due to the drift of the modulation frequency provided by the VCO, which is interpreted by the phase comparator as a change in the phase of the received signals. Hence, the VCO performance must be optimized to stabilize the modulation frequency and avoid changes in the offset of the output signal. Another way to improve the POF elongation sensor would be to bond the reference fiber on the outer surface of the tail boom, so that the phase difference between the two optical signals would increase. Thus, increasing the sensitivity of the sensor. For the sake of comparison, the results obtained for one of the FBGs are plotted on Figure 10. The signals are obviously less noisy than those of the POF elongation sensor. The strain in the structure is represented by the changes of the reflected wavelength in the FBG for the three applied forces. As expected, the strain given by the FBG placed in the perpendicular direction of the tail boom is noticeably smaller than the ones provided by the FBG placed longitudinally.

The FBGs are much more sensitive than the POF elongation sensor and offer an excellent performance for a single-point measurement. On the other hand, the POF elongation sensor provides a distributed measurement of the strain in the whole structure and it is a cost-effective option when the required measurement resolution is not so high.

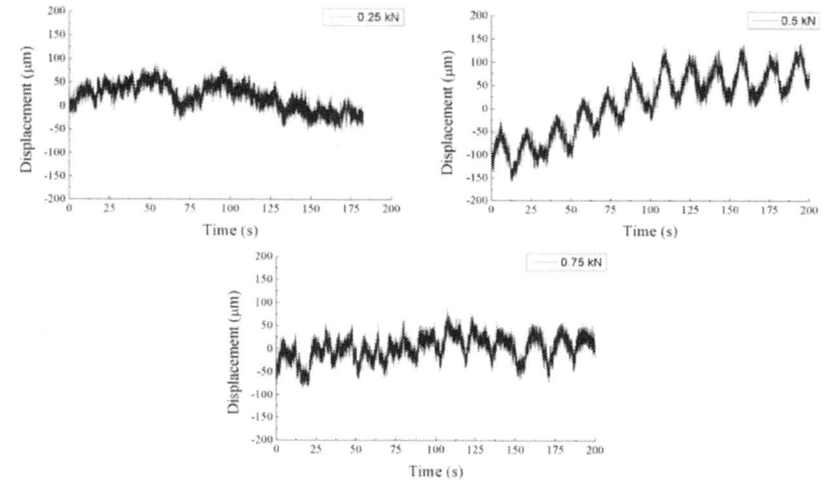

Figure 9. POF elongation corresponding to the three fatigue cycles of 0.25 kN, 0.5 kN and 0.75 kN.

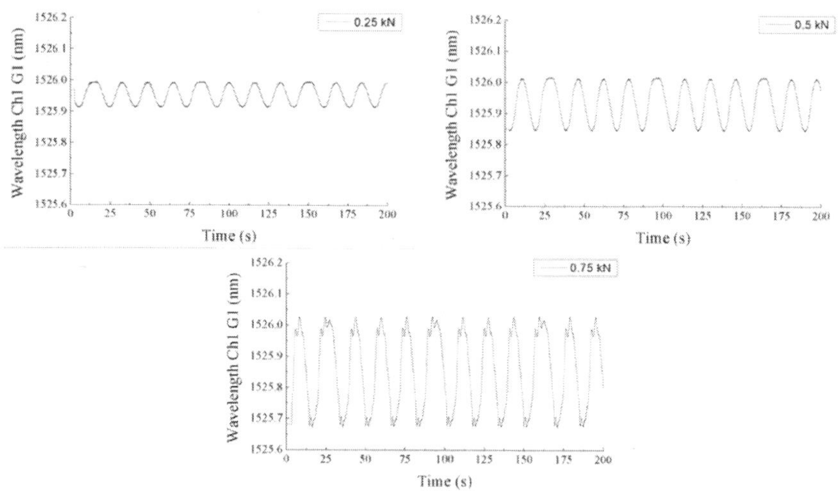

Figure 10. Wavelength variations due to the strain in the tail boom provided by the first FBG during the three fatigue cycles.

2.2.2. Long Period Gratings in Microstructured POF

In this section we present the use of long period gratings in microstructured POFs (mPOFs) as an option to measure strain in aircraft structures instead of FBGs. mPOFs are the counterpart version of the photonic crystal fibers using polymer fibers instead of glass fibers. In the mPOFs a pattern of air channels along the POF guides the light and confers unique properties to the fiber [56]. Among these properties, the single-mode and polarization-maintaining behavior, combined with the intrinsic flexibility and ease of connection of the POF are particularly interesting for sensing purposes. In order to obtain an LPG in an mPOF, a periodic modification of the cladding structure of the mPOF is written along the direction of light propagation [57], so that the core mode is coupled to cladding modes, which are rapidly attenuated, resulting in depressions in the transmission spectrum at wavelengths where the Equation (9) is satisfied [58]:

$$m\lambda = (n_{core}(\lambda) - n_{cl}(\lambda))\Lambda_{LPG} \qquad (9)$$

where m is the order of the interaction, n_{core} and n_{cl} are the effective indices of the core and the cladding, and Λ_{LPG} is the period of the grating. Unlike FBGs, this period is much larger than the wavelength of the light propagating along the mPOF, so LPGs are easier to fabricate. In the case of LPGs, the length of the structure is in the order of 30 mm, whereas the grating period range is typically from 100 μm to 1 mm [59]. The LPG causes the coupling between the core-guided mode and the forward-propagating cladding modes at certain wavelengths, which causes "notches" in its transmission spectrum. The working principle of the LPG is depicted in Figure 11.

Figure 12 shows the mPOF employed to inscribe the LPG that we used in the tests and its normalized transmission spectrum. Since changes in the strain of the mPOF can modify the period of the LPG, the wavelength that matches the phase condition changes. As a result, the depression at 620 nm in the spectrum is sensitive to strain and its wavelength changes accordingly.

In order to detect the changes of the resonant wavelength in the LPG, light from a white source is launched into the fiber and the spectrum of the output light is obtained by using the USB4000 miniature fiber optic spectrometer from Ocean Optics. The experimental set-up is represented in Figure 13 [60].

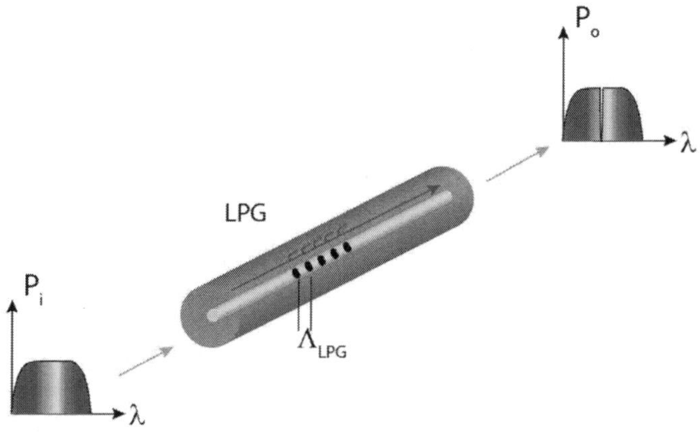

Figure 11. LPG concept and working principle.

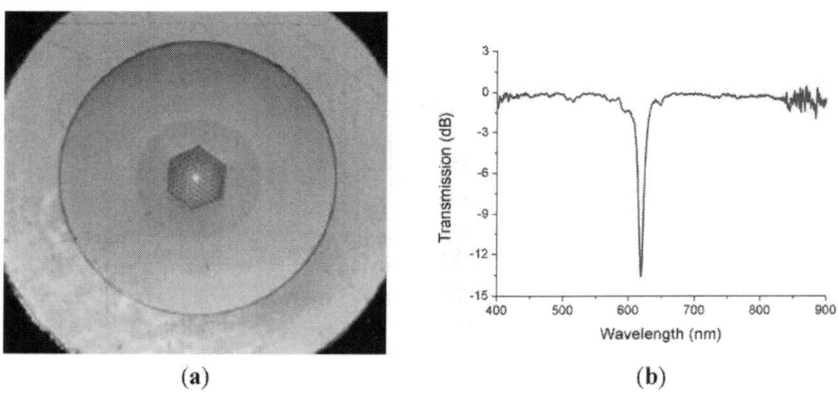

Figure 12. (**a**) Cross section of the mPOF used to inscribe an LPG; (**b**) Normalized transmission spectrum of the LPG used in the tests.

The LPG mPOF with a total grating length of 60 mm was bonded to the surface of a one-meter-length steel plate, whereas on the other surface of the plate an FBG was instrumented as a reference sensor (see Figure 14). The SM130-200 interrogator was employed to capture the response of the FBG. The plate was subjected to a triangular-like tension cycle at a frequency of 0.2 Hz by means of a servo-hydraulic actuator.

As can be observed in Figure 15a, the responses of the sensors are 180° out of phase because of the opposite variation of the wavelength with respect to the strain for FBGs and LPGs. In the case of the LPG, the fluctuations in the signal are due to the resolution of the spectrometer employed in these tests. To smooth these fluctuations a spectrometer with higher resolution can be used, or low-pass filtering can be applied to the signal. The spectrum corresponding to the LPG signal is represented in Figure 15b; the peak corresponding to the fundamental frequency (0.2 Hz) is clearly identified.

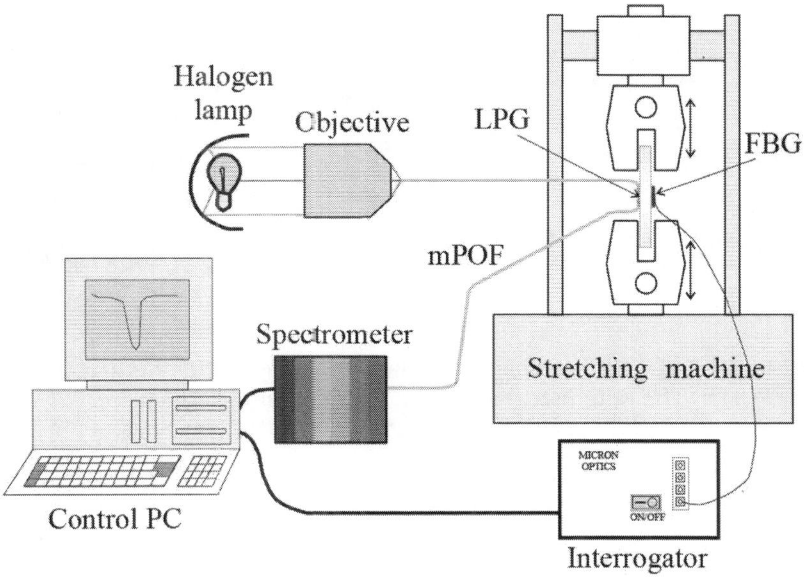

Figure 13. Experimental set-up for the traction-compression tests in a steel plate specimen.

3. SHM IN AIRCRAFT ENGINES

In addition to monitoring the integrity of the aircraft structures it is also essential to ensure the condition of the engines. Optical methods can be employed to monitor different engine parameters such as vibration, temperature, pressure or rotational speed of the engine [61]. They are commonly based on FBGs or interferometric technology but we have developed a reflective intensity-modulated method to assess the performance and SHM of aircraft engines. Several prototypes of this optical sensor are being successfully employed to characterize turbines of aircraft engines in a wind tunnel installation. Other systems based on microwave [62], eddy current [63], or capacitive [64] probes have been developed and in [65] are compared to different optical techniques.

(a) (b)

Figure 14. (**a**) Plate specimen assembled on the stretching machine with the LPG; (**b**) The FBG instrumented on the other side of the plate.

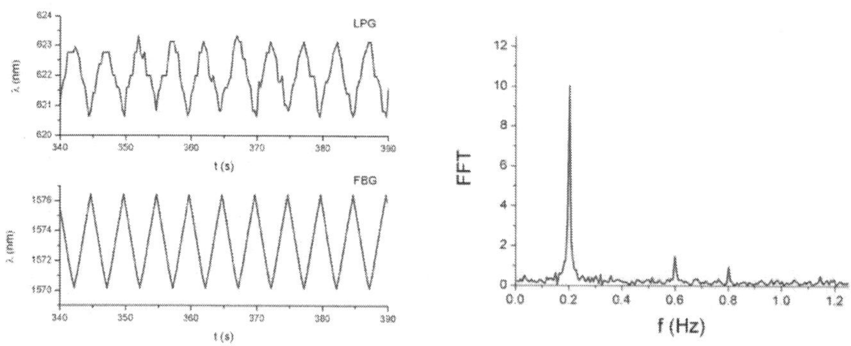

(a) (b)

Figure 15. (**a**) Responses of the LPG and the FBG to a triangular-like tension cycle at a frequency of 0.2 Hz; (**b**) FFT of the LPG response.

There are three different types of blade vibrations that can be analyzed, namely, radial, tangential and axial (see Figure 16). Currently, the sensor is being used to quantify the radial and tangential vibrations.

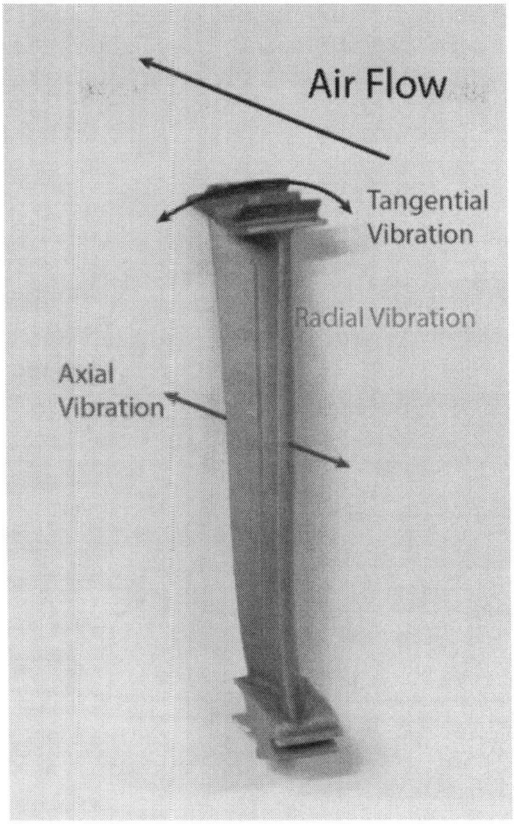

Figure 16. Possible vibrations of the blade during turbine operation: radial (green), axial (red) and tangential (blue). The turbine blade is courtesy of the Aeronautical Technologies Center.

The distance between the tip of the blades of an aircraft engine and its casing is known as tip clearance (TC). The measurement of this distance is equivalent to evaluating the radial vibrations of the blades. This measurement is interesting since it is possible to get a more efficient engine by reducing its TC [66]. Several kinds of sensors, such as capacitive or discharge probe sensors, can be used to carry out these measurements. In Section 3.1 we compile the improvements in the design of an optical sensor for TC measurements of turbine rigs and present another practical application of this optical sensor.

Tip timing is a well-known [67] technique to compute the amplitude of blade vibrations using the differences between the theoretical and the real arrival times of the blades. Since our sensor is placed in a radial position, the tangential vibrations of the turbine blades are also assessed. These vibration amplitudes provide useful information about the turbine operation and the SHM of the blades, since it is possible to detect cracks in the blades or flutter. In Section 3.2 we present a set of tip timing tests for a turbine with 146 blades.

3.1. Tip Clearance Measurements

The main component of the optical sensor is a trifurcated bundle of optical fibers with a common leg on one side and three legs on the other. In Figure 17, the bundle and a cross section view of the common leg are depicted.

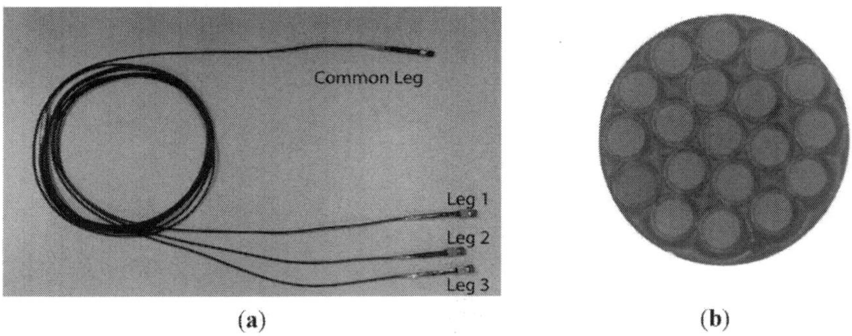

Figure 17. (**a**) Trifurcated bundle of optical fibers; (**b**) Microscope view of the cross section of the bundle common leg.

The central fiber guides the light from a laser module connected at leg 1 to the probe end, so that the blades of the turbine are illuminated. This fiber is surrounded by two rings of receiving fibers which collect the reflected light from the blades. The set of receiving fibers corresponding to the first ring (6 fibers) and the second ring (12 fibers) are gathered in bundle legs 2 and 3, respectively. At the end of these legs two photodetectors are connected to convert the optical signals into voltage. The distance from the sensor tip to the blade is calculated from the quotient of these voltages to minimize the effects of fluctuations of the light source or variations in the reflectivity of the surface [68,69]. Detailed information about the configuration of the sensor and the set-up to perform the measurements is provided in [70]. The calibration curve of the sensor is depicted in Figure 18. Previous works carried out the measurements using the front-slope region (region I), whereas our sensor operates in the back-slope region (region II). In this region, there is a trade-off between the sensitivity of the sensor and its linear range, so that the TC can be measured for larger distances but with lower sensitivity.

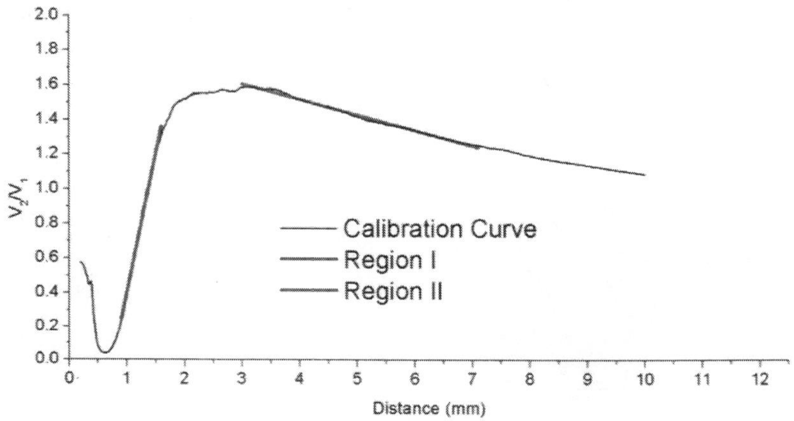

Figure 18. Calibration curve of the sensor showing the two possible regions of measurement: region I (front-slope) in blue and region II (back-slope) in red.

Several configurations have been tested to improve the sensitivity of the sensor. The two main changes have been the reduction of the modal noise at the output of the optical probe and the employment of asymmetric gain for the photodetectors [71]. The evolution of the sensor is summarized in Table 1. Although the third configuration shows the best precision, its maximum operating temperature is 60 °C, which restricts the measurements to low temperatures. In contrast, the bundles of glass fibers can work at temperatures up to 350 °C. Therefore, we decided to use the fourth configuration for future measurements.

The sensor can be used for other purposes beyond the measurement of the TC in the turbines. For instance, three sensors have been used to assess the vibrational behavior of a rotating disk that separates two chambers of different pressures in a real aircraft engine. This disk fluttered under certain circumstances and three sensors were installed to detect these vibrations during some tests carried out in a wind tunnel. The sensors were placed at 120° with respect to each other, as shown in Figure 19. The three ends of the probes were placed 4.8 mm away from the rotating disk.

Figure 21 shows how the amplitude of the vibration changes when the pressure difference varies. The maximum amplitude of the vibrations is achieved at a pressure difference of 2.7 bar. In Table 2 the amplitude, expressed in dB with respect to one micrometer, and the frequency of the disk vibration for all measured pressure differences are listed. The uncertainty in the frequency measurements is 2 Hz, it could be improved employing a higher number of points to perform the FFT. The obtained frequency values match exactly with those given by one strain gauge instrumented on the surface of the disk.

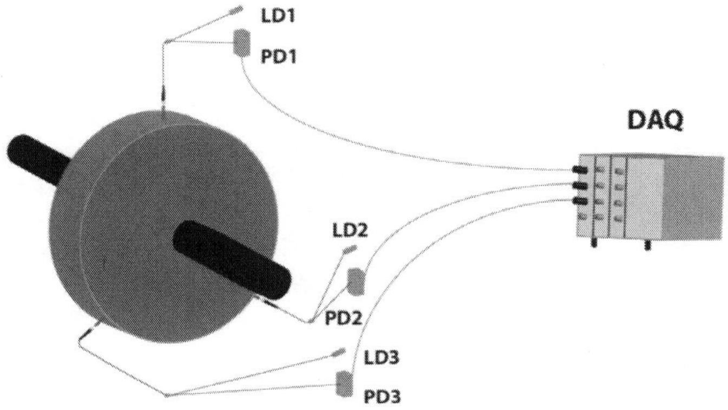

Figure 19. Set-up of the three sensors to evaluate the vibrations in the rotating disk. LDx: Laser diode x; PDx: Photodetector x; DAQ: Data acquisition system.

The disk vibrated even when it was not rotating due to the air flow between the edges of the disk and the casing, so measurements for several rotational speeds from 0 to 3000 revolutions per minute (rpm) and multiple pressure differences at both sides of the disk were carried out. In Figure 20 the signal obtained from one of the sensors and its FFT when the disk is static with a pressure difference of 2.75 bar are shown. The FFT presents a clear peak at 2236 Hz that indicates the frequency of the radial vibration of the disk.

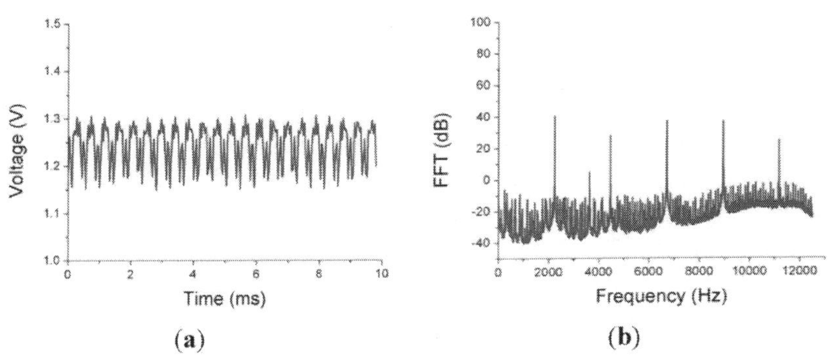

Figure 20. (**a**) Voltage output from one of the sensors when the disk is static and the pressure difference is 2.75 bar; (**b**) FFT of the signal.

Table 1. Characteristics of the different configurations of the TC sensor.

Config.	1st Sensor Configuration	2nd Sensor Configuration	3rd Sensor Configuration	4th Sensor Configuration
Light source (Laser)	7 mW 650 nm	30 mW 650 nm	30 mW 650 nm	20 mW 650 nm
Optical fibers bundle	Multimode $Ø_{core} = 100$ μm NA = 0.22	Scrambler + Multimode $Ø_{core} = 100$ μm NA = 0.22	POF $Ø_{core} = 240$ μm NA = 0.5	Single-mode illuminating fiber $Ø_{core} = 4.3$ μm NA = 0.12 Multimode receiving fibers $Ø_{core} = 100$ μm NA = 0.22
Photodetectors gain	Symmetric gain $G_1 = G_2 = 0.75 \times 10^4$ V/A	Asymmetric gain $G_1 = 0.75 \times 10^5$ V/A $G_2 = 2.38 \times 10^5$ V/A	Asymmetric gain $G_1 = 0.75 \times 10^5$ V/A $G_2 = 2.38 \times 10^5$ V/A	Asymmetric gain $G_1 = 0.75 \times 10^5$ V/A $G_2 = 2.38 \times 10^5$ V/A
Calibration curve	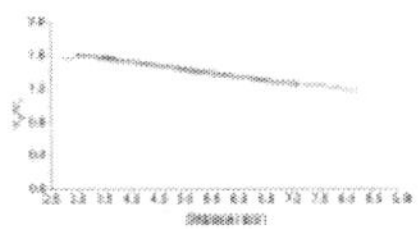 $V_2/V_1 = -0.089d + 1.8783$	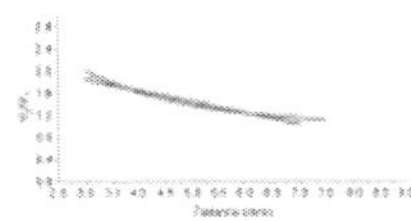 $V_2/V_1 = -0.2002d + 2.4578$	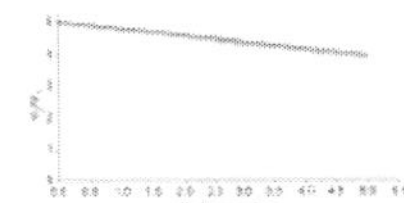 $V_2/V_1 = -0.213d + 5.0064$	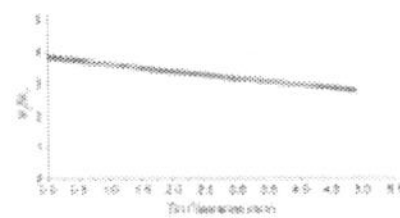 $V_2/V_1 = -0.2167d + 3.8448$
Cross section of the common leg				
Laboratory precision	141 μm	51 μm	33 μm	24 μm
Wind tunnel precision	24 μm	-	25 μm	28 μm

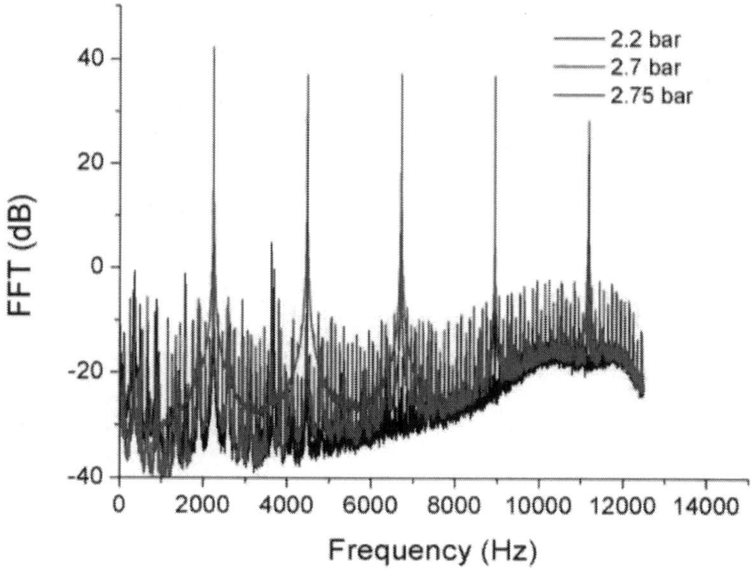

Figure 21. FFTs of the sensor signal obtained at pressure differences of 2.2, 2.7 and 2.75 bar.

3.2. Tip Timing Measurements

As we mentioned before, tip timing is a technique used for the evaluation of the amplitude of tangential vibrations of the blades. Taking into account the difference between the theoretical and the real blade arrival times, the amplitude of the blade deflection is obtained. The entire signal processing to get the deflection of each blade is extensively explained in [70]. The travelling wave spectrum is calculated as the FFT of the deflection values for the blades, and represents the average amplitude of the vibration of the blades at a certain nodal diameter. Figure 22 depicts the deviations of every blade of a 146-blade turbine turning at different rotational speeds. These deviations provide useful information to monitor the turbine blades, since they allow the detection of cracks or flutter when the amplitude of the blade vibration exceeds a predefined alarm value.

Table 2. Amplitude and frequency of the disk vibrations at several pressure differences when the disk is static.

Pressure Difference (bar)	Amplitude (dB)	Frequency (Hz)
2.2	14.3	2238
2.3	19.5	2238
2.4	31.6	2238
2.5	38.5	2238
2.6	32	2236
2.7	42.2	2236
2.75	40.5	2236

Although there are some exceptions, the results show a trend towards increasing the amplitude of the blade vibrations as the rotational speed gets higher. The tests were repeated again four days later. In Figure 23 the results obtained at 3390 and 5570 rpm in this second test campaign are shown. As can be observed, the amplitudes of the vibrations are quite similar for both cases. This fact indicates that the blades of the turbine did not suffer any significant damage between both tests.

4. CONCLUSIONS

Aircraft structures, due to their operating conditions and design principles, require intense inspection and maintenance operations. SHM based on optical sensors could play an important role in these operations provided that two issues can be overcome. The potential economic and life-safety benefits of optical sensors for SHM should be clearly demonstrated to the aviation companies, so that the industry takes decisions towards the extensive use of these devices. The other aspect to be considered is the lack of agreed standards and certification of optical sensors for SHM in aircraft structures, which is an essential condition for the application of optical sensors for SHM on a large scale. FBGs represent the most mature and promising technology for this purpose and, even though they can be employed for other applications, they are mainly used for strain measurements. For this purpose, they present several advantages with respect to strain gauges, like their long-term stability, high multiplexing capability and suitability for being embedded into composite materials. Two other early-stage alternatives for strain measurement in aircraft structures are mentioned in this article. Regarding engines, optical sensors have provided excellent results for the evaluation of aeronautical turbines in ground tests, so the next challenge will be to apply them to aircraft engines in flight.

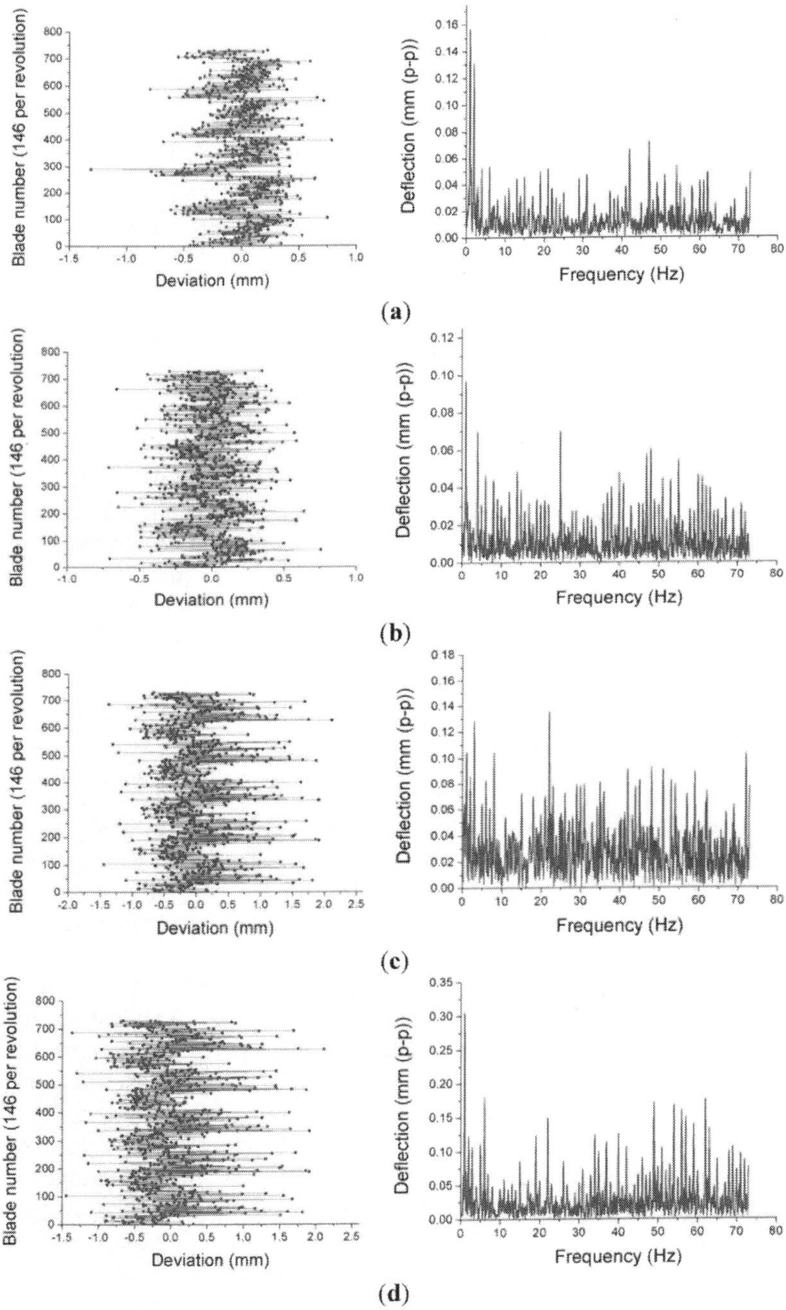

Figure 22. Deviations of each blade form the equilibrium position and travelling wave spectrum of a 146-blade turbine working at 2422 rpm (**a**); 3390 rpm (**b**); 4360 rpm (**c**) and 5570 rpm (**d**).

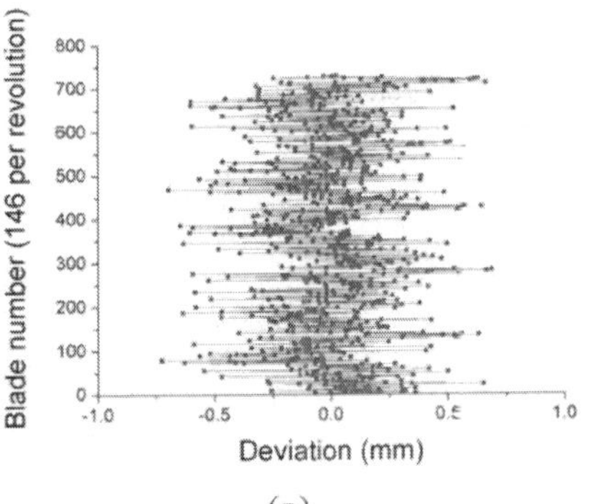

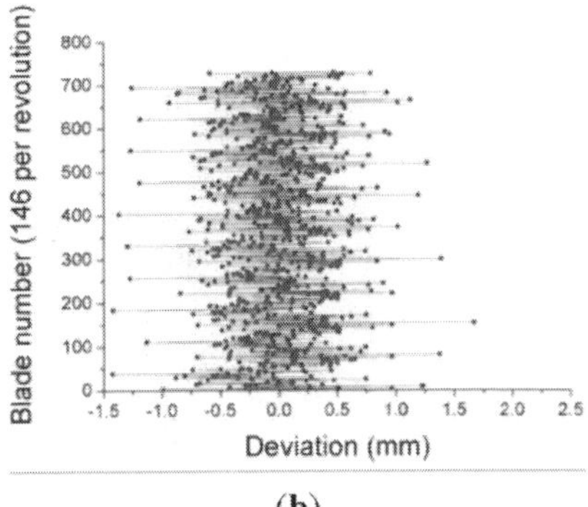

Figure 23. Deviation of each blade form the equilibrium position obtained four days later at 3390 rpm (**a**) and 5570 rpm (**b**).

ACKNOWLEDGMENTS

The authors would like to thank Professor A. Güemes for the permission to use the FBG responses shown in Figure 2. This work has been sponsored by the Ministerio de Economía y Competitividad (Spain) and FEDER funds under project TEC2012-37983-C03-01, the Gobierno Vasco/Eusko Jaurlaritza under projects IT664-13, ETORTEK14/13 and by the University of the Basque Country (UPV/EHU) through programs UFI11/16, US13/09 and EUSKAMPUS.

CONFLICTS OF INTEREST

The authors declare no conflict of interest.

REFERENCES

1. Boller, C.; Meyendorf, N. State-of-the-art in Structural Health monitoring for aeronautics. In Proceedings of the International Symposium on NDT in Aerospace 2008, Fürth, Germany, 3–5 December 2008.

2. Boller, C. Why SHM? A Motivation. Available online: https://www.researchgate.net/profile/Christian_Boller/publication/268151300_Why_SHM_A_Motivation/links/5463efc80cf2837efdb347cf.pdf (accessed on 30 June 2015).

3. Kapoor, H.; Braun, C.; Boller, C. Modelling and optimisation of maintenance intervals to realise Structural Health Monitoring applications on aircraft. In Proceedings of the 5th European Workshop on Structural Health Monitoring, Sorrento, Italy, 28 June–4 July 2010.

4. Farrar, C.R.; Worden, K. An introduction to structural health monitoring. *Philos. Trans. R. Soc. A Math. Phys. Eng. Sci.* **2007**, *365*, 303–315.

5. Roach, D. Real time crack detection using mountable comparative vacuum monitoring sensors. *Smart Struct. Syst.* **2009**, *5*, 317–328.

6. Zhou, G.; Sim, L.M. Damage detection and assessment in fibre-reinforced composite structures with embedded fibre optic sensors—A review. *Smart Mater. Struct.* **2002**, *11*, 925–939.

7. Güemes, A. SHM Technologies and Applications in Aircraft Structures. In Proceedings of the 5th International Symposium on NDT in Aerospace, Singapore, 13–15 November 2013.

8. Grattan, K.T.V.; Sun, T. Fiber optic sensor technology: An overview. *Sens. Actuators A Phys.* **2000**, *82*, 40–61.

9. Claus, R.O.; Gunther, M.F.; Wang, A.; Murphy, K.A. Extrinsic Fabry-Perot sensor for strain and crack opening displacement measurements from −200 to 900 degrees C. *Smart Mater. Struct.* **1992**, *1*.

10. Güemes, A.; Fernandez-Lopez, A.; Fernandez, P. Damage Detection in Composite Structures from Fibre Optic Distributed Strain Measurements. In Proceedings of the 7th European Workshop on Structural Health Monitoring, Nantes, France, 8–11 July 2014.

11. Mrad, N.; Xiao, G.Z. Multiplexed fiber Bragg gratings for potential aerospace applications. In Proceedings of the International Conference on MEMS, NANO and Smart Systems, Banff, AL, Canada, 24–27 July 2005.

12. Takeda, N. Fiber optic sensor-based SHM technologies for aerospace applications in Japan. In Proceedings of the 15th International Symposium on: Smart Structures and Materials & Nondestructive Evaluation and Health Monitoring, San Diego, CA, USA, 14–18 March 2008.

13. Nishino, Z.T.; Chen, K.; Gupta, N. Power modulation-based optical sensor for high-sensitivity vibration measurements. *IEEE Sens. J.* **2014,** *14,* 2153–2158.

14. Read, I.; Foote, P.; Murray, S. Optical fibre acoustic emission sensor for damage detection in carbon fibre composite structures. *Meas. Sci. Technol.* **2002,** *13.*

15. Yari, T.; Nagai, K.; Hotate, K. Monitoring aircraft structural health using optical fiber sensors. *Mitsubishi Heavy Ind. Tech. Rev.* **2008,** *45,* 5–8.

16. Majumder, M.; Gangopadhyay, T.K.; Chakraborty, A.K.; Dasgupta, K.; Bhattacharya, D.K. Fibre Bragg gratings in structural health monitoring—Present status and applications. *Sens. Actuators A Phys.* **2008,** *147,* 150–164.

17. Guo, H.; Xiao, G.; Mrad, N.; Yao, J. Fiber Optic Sensors for Structural Health Monitoring of Air Platforms. *Sensors***2011,** *11,* 3687–3705.

18. Mrad, N.; Guo, H.; Xiao, G.; Rocha, B.; Sun, Z. On the use of a compact optical fiber sensor system in aircraft structural health monitoring. In Proceedings of the Photonic Applications for Aerospace, Transportation, and Harsh Environment III, Baltimore, MD, USA, 1 May 2012.

19. Gholamzadeh, B.; Nabovati, H. Fiber optic sensors. *World Acad. Sci. Eng. Technol.* **2008,** *42,* 335–340.

20. Kahandawa, G.C.; Epaarachchi, J.; Wang, H.; Lau, K.T. Use of FBG sensors for SHM in aerospace structures. *Photon. Sens.* **2012,** *2,* 203–214.

21. Hill, K.O.; Meltz, G. Fiber Bragg grating technology fundamentals and overview. *J. Lightw. Technol.* **1997,** *15,* 1263–1276.

22. Frövel, M.; Pintado, J.M.; Güemes, A.; del Olmo, E.; Obst, A. Multiplexable fiber Bragg grating temperature sensors embedded in CFRP structures for cryogenic applications. In Proceedings of the 3rd European Workshop on Structural Health Monitoring, Granada, Spain, 5–7 July 2006.

23. Davis, C.; Tejedor, S.; Grabovac, I.; Kopczyk, J.; Nuyens, T. High-strain fiber bragg gratings for structural fatigue testing of military aircraft. *Photon. Sens.* **2012,** *2,* 215–224.

24. Ecke, W.; Latka, I.; Willsch, R.; Reutlinger, A.; Graue, R. Fibre optic sensor network for spacecraft health monitoring. *Meas. Sci. Technol.* **2001,** *12.*

25. Wood, K.; Brown, T.; Rogowski, R.; Jensen, B. Fiber optic sensors for health monitoring of morphing airframes: II. Chemical sensing using optical fibers with Bragg gratings. *Smart Mater. Struct.* **2000**, *9*, 170–174.

26. López-Higuera, J.M. *Handbook of Optical Fibre Sensing Technology*; Wiley and Sons: Chichester, West Sussex, UK, 2002.

27. Lemaire, P.J.; Atkins, R.M.; Mizrahi, V.; Reed, W.A. High pressure H_2 loading as a technique for achieving ultrahigh UV photosensitivity and thermal sensitivity in GeO_2 doped optical fibres. *Electron. Lett.* **1993**, *29*, 1191–1193.

28. Martinez, A.; Khrushchev, I.Y.; Bennion, I. Direct inscription of Bragg gratings in coated fibers by an infrared femtosecond laser. *Opt. Lett.* **2006**, *31*, 1603–1605.

29. Meltz, G.; Morey, W.; Glenn, W.H. Formation of Bragg gratings in optical fibers by a transverse holographic method.*Opt. Lett.* **1989**, *14*, 823–825.

30. Hill, K.O.; Malo, B.; Bilodeau, F.; Johnson, D.C.; Albert, J. Bragg gratings fabricated in monomode photosensitive optical fiber by UV exposure through a phase mask. *Appl. Phys. Lett.* **1993**, *62*, 1035–1037.

31. Malo, B.; Hill, K.O.; Bilodeau, F.; Johnson, D.C.; Albert, J. Point-by-point fabrication of micro-Bragg gratings in photosensitive fibre using single excimer pulse refractive index modification techniques. *Electron. Lett.* **1993**, *29*, 1668–1669.

32. Xu, M.G.; Archmbault, J.L.; Reekie, L.; Dakin, J.P. Thermally-compensated bending gauge using surface-mounted fibre gratings. *Int. J. Optoelectron.* **1994**, *9*, 281–283.

33. Haran, F.M.; Rew, J.K.; Foote, P.D. A strain isolated fibre Bragg grating sensor for temperature compensation of fibre Bragg grating strain sensors. *Meas. Sci. Technol.* **1998**, *9*, 1163–1166.

34. Song, M.; Lee, S.B.; Choi, S.S.; Lee, B. Simultaneous measurement of temperature and strain using two fibre Bragg gratings embedded in a glass tube. *Opt. Fibre Technol.* **1997**, *3*, 194–196.

35. Patrick, H.J.; Williams, G.M.; Kersey, A.D.; Pedrazini, J.R. Hybrid fibre Bragg grating/long period fibre grating sensor for strain/temperature discrimination. *Photon. Technol. Lett.* **1996**, *8*, 1223–1225.

36. Xu, M.G.; Archmbault, J.L.; Reekie, L.; Dakin, J.P. Discrimination between strain and temperature effects using dual-wavelength fibre grating sensors. *Electron. Lett.* **1994**, *30*, 1085–1087.

37. James, S.W.; Dockney, M.L.; Tatam, R.P. Simultaneous independent temperature and strain measurement using in-fibre Bragg grating sensors. *Electron. Lett.* **1996**, *32*, 1133–1134.

38. Song, M.; Lee, B.; Lee, S.B.; Choi, S.S. Interferometric temperature-insensitive strain measurement with different diameter fibre Bragg gratings. *Opt. Lett.* **1997**, *22*, 790–792.

39. Jung, J.; Nam, H.; Lee, J.H. Park, N.; Lee, B. Simultaneous measurement of strain and temperature by use of a single Bragg grating and an erbium-doped fibre amplifier. *Appl. Opt.* **1999**, *38*, 2749–2751.

40. Jung, J.; Park, N.; Lee, B. Simultaneous measurement of strain and temperature by use of a single fiber Bragg grating written in an erbium:ytterbium-doped fiber. *Appl. Opt.* **2000**, *39*, 1118–1120.

41. Dunphy, J.R.; Meltz, G.; Varasi, M.; Vannucci, A.; Signorazzi, M.; Ferraro, P.; Voto, C. Embedded Optical Sensor Capable of Strain and Temperature Measurement using A Single Diffraction Grating. U.S. Patent No. 5, 399,854, 21 March 1995.

42. Ferreira, L.A.; Farahi, F.; Arau, F.M. Simultaneous measurement of strain and temperature using interferometrically interrogated fiber Bragg grating sensors. *Opt. Eng.* **2000**, *39*, 2226–2234.

43. Lo, Y.L.; Kuo, C.P. Packaging a fiber Bragg grating with metal coating for an athermal design. *J. Lightw. Technol.* **2003**,*21*, 1377–1383.

44. Tian, K.; Liu, Y.; Wang, Q. Temperature-independent fiber Bragg grating strain sensor using bimetal cantilever. *Opt. Fiber Technol.* **2005**, *11*, 370–377.

45. Viegas, D.; Pinto, N.M.P.; Romero, R.; Frazão, O.; Costa, A. Birefringence effect in uniform fibre bragg gratings. Available online: https://www.researchgate.net/profile/Diana_Viegas/publication/228903206_Birefringence_effect_in_uniform_fibre_Bragg_gratings/links/0fcfd50a529adefaea000000.pdf (accessed on 30 June 2015).

46. Güemes, A. Fiber Optics Strain Sensors. Available online: http://ftp.rta.nato.int/public/PubFullText/RTO/EN/STO-EN-AVT-220/EN-AVT-220-03.pdf (accessed on 30 June 2015).

47. Montero, A.; Saez de Ocariz, I.; Lopez, I.; Venegas, P.; Gomez, J.; Zubia, J. Fiber Bragg Gratings, IT Techniques and Strain Gauge Validation for Strain Calculation on Aged Metal Specimens. *Sensors* **2011**, *11*, 1088–1104.

48. Frieden, J.; Cugnoni, J.; Botsis, J.; Gmür, T. Low energy impact damage monitoring of composites using dynamic strain signals from FBG sensors–Part I: Impact detection and localization. *Compos. Struct.* **2012**, *94*, 438–445.

49. Gomez, J.; Jorge, I.; Durana, G.; Arrue, J.; Zubia, J.; Aranguren, G.; Montero, A.; López, I. Proof of Concept of Impact Detection in Composites Using Fiber Bragg Grating Arrays. *Sensors* **2013**, *13*, 11998–12011.

50. Vella, T.; Chadderdon, S.; Selfridge, R.; Schultz, S.; Webb, S.; Park, C.; Peters, K.; Zikry, M. Full-spectrum interrogation of fiber Bragg gratings at 100 kHz for detection of impact loading. *Meas. Sci. Technol.* **2010**, *21*.

51. Luber, M.; Poisel, H.; Loquai, S.; Neuner, C.; Bachmann, A.; Ziemann, O.; Hartl, E. POF strain sensor using phase measurement techniques. In Proceedings of the 16th International Plastic Optical Fibres Conference, Torino, Italy, 10–12 September 2007.

52. Kalymnios, D.; Scully, P.; Zubia, J.; Poisel, H. POF sensors overview. In Proceedings of the 13th International Plastic Optical Fibres Conference, Nürnberg, Germany, 27–30 September 2004.

53. Neuner, C. Faseroptischer Längensensor Basierend auf Laufzeitmessung. Master's Thesis, Georg Simon Ohm University of Applied Science Nuremberg, Nuremberg, Germany, 2007.

54. Durana, G.; Kirchhof, M.; Luber, M.; de Ocáriz, I.S.; Poisel, H.; Zubia, J.; Vázquez, C. Use of a novel fiber optical strain sensor for monitoring the vertical deflection of an aircraft flap. *Sens. J.* **2009**, *9*, 1219–1225.

55. Gomez, J.; Zubia, J.; Aranguren, G.; Arrue, J.; Poisel, H.; Saez, I. Comparing polymer optical fiber, fiber Bragg grating, and traditional strain gauge for aircraft structural health monitoring. *Appl. Opt.* **2009**, *48*, 1436–1443.

56. Van Eijkelenborg, M.; Large, M.C.J.; Argyros, A.; Zagari, J.; Manos, S.; Issa, N.A.; Bassett, I.; Fleming, S.; McPhedran, R.C.; Sterke, M.; *et al.* Microstructured polymer optical fibre. *Opt. Express* **2001**, *9*, 319–327.

57. Van Eijkelenborg, M.; Padden, W.; Besley, J.A. Mechanically induced long-period gratings in microstructured polymer fibre. *Opt. Commun.* **2004**, *236*, 75–78.

58. Hiscocks, M.P.; Van Eijkelenborg, M.; Argyros, A.; Large, M.C.J. Stable imprinting of long-period gratings in microstructured polymer optical fibre. *Opt. Express* **2006**, *14*, 4644–4649.

59. James, S.W.; Tatam, R.P. Optical fibre long-period grating sensors: Characteristics and application. *Meas. Sci. Technol.* **2003**, *14*, 49–61.

60. Durana, G.; Gomez, J.; Aldabaldetreku, G.; Zubia, J.; Montero, A.; de Ocdriz, I.S. Assessment of an LPG mPOF for strain sensing. *IEEE Sens. J.* **2012**, *12*, 2668–2673.

61. Wild, G. Optical fiber Bragg grating sensors applied to gas turbine engine instrumentation and monitoring. In Proceedings of the Sensors Applications Symposium, Galveston, TX, USA, 19–21 February 2013.

62. Violetti, M.; Skrivervik, A.K.; Xu, Q.; Hafner, M. New microwave sensing system for blade tip clearance measurement in gas turbines. In Proceedings of the IEEE Sensors, Taipei, Taiwan, 28–31 October 2012; pp. 1–4.

63. Dowell, M.; Sylvester, G. Turbomachinery prognostics and health management via eddy current sensing: Current developments. In Proceedings of the Aerospace Conference, Aspen, CO, USA, 6–13 March 1999.

64. Sheard, A.G. Blade by blade tip clearance measurement. *Int. J. Rotat. Mach.* **2011**, *2011*.

65. Russhard, P. The use of blade tip timing technologies to assess and monitor rotor blade health from design to production. In Proceedings of the Test Cell and Controls Instrumentation and EHM Technologies for Military Air, Land and Sea Turbine Engines, Rzeszow, Poland, 20–23 April 2015.

66. Wiseman, M.W.; Guo, T.H. An Investigation of Life Extending Control Techniques for Gas Turbine Engines. In Proceedings of the American Control Conference, Arlington, VA, USA, 25–27 July 2001.

67. Heath, S.; Imregun, M. A survey of blade tip-timing measurement techniques for turbomachinery vibration. *J. Eng. Gas Turbines Power* **1998**, *120*, 784–791.

68. Cao, S.Z.; Duan, F.J.; Zhang, Y.G. Measurement of rotating blade tip clearance with fibre-optic probe. *J. Phys. Conf. Ser.* **2006**, *48*, 873–877.

69. Ma, Y.; Li, G.; Zhang, Y.; Liu, H. Tip Clearance Optical Measurement for Rotating Blades. In Proceedings of the International Conference on Management Science and Industrial Engineering (MSIE), Harbin, China, 6–11 January 2011.

70. García, I.; Beloki, J.; Zubia, J.; Aldabaldetreku, G.; Illarramendi, M.A.; Jiménez, F. An Optical Fiber Bundle Sensor for Tip Clearance and Tip Timing Measurements in a Turbine Rig. *Sensors* **2013**, *13*, 7385–7398.

71. García, I.; Zubia, J.; Berganza, A.; Beloki, J.; Arrue, J.; Illarramendi, M.A.; Mateo, J.; Vazquez, C. Different configurations of a reflective intensity-modulated optical sensor to avoid modal noise in tip-clearance measurements. *J. Lightw. Technol.* 2015, *12*, 2663–2669.

13

NEW METHODS OF ENHANCING THE THERMAL DURABILITY OF SILICA OPTICAL FIBERS

Karol Wysokiński [1,*], Tomasz Stańczyk [1], Katarzyna Gibała [2], Tadeusz Tenderenda [1,3], Anna Ziołowicz [4], Mateusz Słowikowski [4], Małgorzata Broczkowska [4] and Tomasz Nasiłowski [1,3]

[1] InPhoTech, ul. Slomińskiego 17/31, 00-195 Warszawa, Poland

[2] P.H. ELMAT, Rogoźnica 312, 36-060 Głogów Małopolski, Poland

[3] Institute of Applied Physics, Military University of Technology, ul. Kaliskiego 2, 00-980 Warszawa, Poland

[4] Polish Centre for Photonics and Fibre Optics, Rogoźnica 312, 36-060 Głogów Małopolski, Poland

ABSTRACT

Microstructured optical fibers can be precisely tailored for many different applications, out of which sensing has been found to be particularly interesting. However, placing silica optical fiber sensors in harsh environments results in their quick destruction as a result of the hydrolysis process. In this paper, the degradation mechanism of bare and metal-coated optical fibers at high temperatures under longitudinal strain has been determined by detailed analysis of the thermal behavior of silica and metals, like copper and nickel. We furthermore propose a novel method of enhancing the lifetime of optical fibers by the deposition of electroless nickel-phosphorous alloy in a low-temperature chemical process. The best results were obtained for a coating comprising an inner layer of copper and outer layer of low phosphorous nickel. Lifetime values obtained during the annealing experiments were extrapolated to other temperatures by a dedicated model elaborated by the authors. The estimated copper-coated optical fiber lifetime under cycled longitudinal strain reached 31 h at 450 °C.

KEYWORDS

optical fiber sensors; harsh environments; microstructured optical fibers; metal-coated optical fibers; thermal degradation of silica; thermal reliability of optical fibers

1. INTRODUCTION

Optical fibers have revolutionized the telecommunication industry, due to their remarkable properties, such as, for example, large data transfer rate, possibility of multiplexing many signals in one fiber and immunity to external conditions [1]. Furthermore, in the last three decades, optical fibers were widely investigated for applications in the sensing industry. This is mainly because of their compact size, flexibility, possibility of both point and distribution sensing and often low cost [2]. Furthermore, no electric current is needed for their operation, which together with their temperature performance (compared to existing electronic solutions), makes them perfect for sensing in explosion endangered zones.

Microstructured optical fibers have all of the advantages of standard optical fiber, and additionally, by the possibility of modifying the air-hole microstructure, they offer an additional degree of freedom in fiber design, making it possible to fabricate fibers dedicated for specific sensing applications (*i.e.*, pressure, curvature, temperature, strain, torsion, vibration, gas and liquid, as well as electric and magnetic fields sensing [3]). What is more, similarly to traditional fibers, there are also many different methods of sensing with microstructured fibers, which include, e.g., polarization changes [4,5], change of refractive index [6], interference [7] or fiber Bragg gratings [8]. Apart from the physical parameters, the concentration of chemical and biochemical entities can also be detected using microstructured optical fibers. This includes sensors of pH [9], humidity [10], gases, such as methane [11] or ammonia, and even biochemical sensors, which can detect molecules, like proteins [12] or DNA [13]. The principle of operation of chemical sensors can be based on absorption within a hollow core [11], evanescent field absorption [14], fluorescence [15] or Raman spectroscopy (including surface-enhanced Raman spectroscopy [16]).

Optical fibers intended for use at high temperatures need to be coated with a thick layer of metal, which will separate the fibers from the external environment. This is crucial due to the degradation of silica in such conditions, which results in a severe deterioration of the mechanical parameters. However, in extreme environments (*i.e.*, temperatures reaching 500 °C), the most frequently used metal coatings oxidize and leave the surface of the fiber unprotected. This may not be a major issue in steady-state high temperature tests; however, in the majority of industrial applications, fibers are frequently stretched or bent, which is caused by manifold factors, like, for example, thermal expansion of the sensor case. Therefore, it is important to perform long-term experiments under strain, because they will better correspond to the real-life operation of optical fiber devices.

In this work, we present our approach to deposit high-temperature-resistant metal on optical fibers by a low temperature chemical method. We furthermore examine the impact of a high-temperature conditioning on mechanical parameters of bare silica fibers and their metal coatings. Sample fibers were annealed for a long time at high temperatures, and after each hour, they were stretched to see the influence of both strain and temperature. Additionally, since temperature reliability tests are time consuming, we have elaborated a model enabling one to extrapolate the obtained results to other temperatures.

2. DEGRADATION OF OPTICAL FIBERS AT HIGH TEMPERATURE

2.1. Degradation of Silica

Fiber optic sensors used at temperatures exceeding 350 °C cannot be embedded in a polymer coating, due to the thermal degradation of plastics in such conditions [17]. Silica is well known for its inertness at low temperatures and is a valuable material for laboratory equipment. Nevertheless, SiO_2 undergoes the reaction of hydrolysis, resulting in a formation of Si–O–H silanol groups [18], according to Equations (1) and (2):

$$H_2PO_2^- + H_2O \rightarrow H_2PO_3^- - 2H^+ + 2e^- \quad (1)$$

$$\equiv Si\text{-}O\text{-}Si\equiv + OH^- \rightarrow \equiv Si\text{-}O^- + HO\text{-}Si\equiv \quad (2)$$

The latter equation proves that alkaline solutions are particularly aggressive due to the strong corrosive properties of OH^- ions [19]. The hydrolysis reaction worsens the mechanical parameters of fibers, because of the presence of cracks and flaws on the fiber surface. It was shown that high air humidity has a strong negative impact on silica fiber bending strength [20]. In regard to aging and fatigue, it was also proven that the increase of temperature with constant relative humidity causes the deterioration of mechanical properties [21]. This can be attributed to a higher reaction rate at higher temperatures, since the reaction rate constant usually is given by the Arrhenius expression. Likewise, the most popular optical fiber fatigue models employ typical chemical reaction dependencies [20,21]. It was shown that the silica hydrolysis reaction order was two with respect to the air humidity [21]. All of the fatigue models assume that water reacts with silica, which produces fractures on the fiber surface, which finally worsen its mechanical parameters. This is an intuitive postulate, since it is well known that even a small crack on a rigid object becomes a weak point, at which breakage is likely to occur. This property is widely used in the cleaving of optical fibers, glass capillaries, ceramic tiles, *etc.*

2.2. Metal Coatings and Their Oxidation at High Temperatures

Considering the issues mentioned above, using bare optical fibers is not a recommended choice for high temperature applications. This is why metal-coated fibers are used for such conditions instead [22]. The deposition of metals onto the fiber is usually performed by an adaptation of the Ohno continuous casting process [22,23]. This process is based on drawing the fiber through a liquid metal near its melting point, which enables molten metal to solidify on the fiber surface. This means that the choice of possible metal coatings is limited to those with a relatively low melting point. Therefore, the most popular and commercially available coating materials are copper, aluminum and gold. While gold coated fibers are the least corrosion susceptible, their high price makes them also the least interesting for commercial applications. Furthermore, aluminum has a low melting point (660 °C for pure metal), which excludes it from extreme temperature applications. On the other hand, copper, with its melting point of above 1000 °C and relatively low price, seems to be the best candidate for high temperature applications. However, the high oxidation rate of copper is a factor that should be taken into account.

After heating copper-coated fibers to temperatures exceeding 500 °C in air, the metal coating changes its color to black, and the fiber becomes extremely brittle. Three distinct temperature ranges of copper oxidation kinetics can be distinguished [24]: 350–550 °C, 550–850 °C and 850–1050 °C. They all differ in the mechanism and oxidation rate. In general, the mechanism of the oxidation process is based on the outward diffusion of copper atoms from the metal substrate to the surface [24]. Long exposure to an oxidizing atmosphere produces cracks within the oxide layer [24], which finally becomes porous [25]. Both the porosity of the oxide layer and the diffusion of copper atoms to the surface contribute to the destruction of copper-coated fibers. Oxygen reacts with the metal layer to produce copper oxides. As the thickness of the oxide layer grows, cracks and pores can be formed on its surface. Eventually, the porosity of this layer enables aggressive agents, like water, to reach the fiber surface and develop cracks, which finally results in the deterioration of the mechanical properties of a fiber.

2.3. Enhancing the High Temperature Performance of the Optical Fibers

Surface modification can prevent the inevitable destruction of copper-coated optical fibers. The most straightforward approach could be an electrochemical deposition of a thin noble metal layer. However, copper exhibits a high interdiffusion rate with precious metals, like gold, platinum or palladium [26,27], which makes it impossible to use such materials without a suitable middle layer acting as a diffusion barrier [26]. Another way is to deposit thin (approximately 1 μm) films of pure nickel, chromium or electroless Ni–P alloy, which were reported to provide good results for blocking the diffusion of copper, even at elevated temperatures up to 550 °C [27]. What, however, needs to be taken into account is that nickel and copper are mutually soluble at higher temperatures of the order of magnitude of 750 °C [26]; therefore, the nickel layer needs to be relatively thick to work in such conditions. It is even possible to plate bare silica optical fiber with nickel (with an appropriate underlayer), which omits the problems related to susceptibility to corrosion of copper [28], but it is technologically challenging to use this method to deposit metals on longer (e.g., 1 m) fragments of optical fibers or on more complex sensing transducers. There are also other methods of enhancing the corrosion resistance of copper, like chromate conversion coating or coloring, already known for decades. However, all of the compounds that usually build these layers are not thermally resistant. For instance, the black layer of copper sulfide oxidizes in air at 500 °C, and a synthetic patina can be unstable even at room temperature.

Since the deposition of nickel seems to be the best means to improve the copper-coating resistance to high temperatures, we verify this approach in our work. There are two main methods of nickel plating—electrolytic and electroless. In the electrolytic method, nickel is reduced from a nickel salt solution by electrolysis; whereas the electroless method employs a chemical reagent to reduce nickel ions from the solution to the metallic form. The major advantages of the electrolytic method are: the possibility of reaching high deposition rates, the fact that the product is a chemically pure metal and the opportunity of tuning the mechanical parameters by solution composition and current density. The biggest disadvantage of this method is the uneven distribution of the coating thickness, dependent on the shape of the coated object. This property is not a major issue for optical fibers, since it is possible to arrange the anodes in an appropriate way around the fiber. However, when a more complex object (e.g., sensor head consisting of a fiber and a transducer) is to be coated, achieving a uniform distribution of electric field becomes extremely difficult. In the electroless method, this problem is not encountered, because the reduction reaction takes place on the entire activated surface; hence, even very sophisticated surfaces may be uniformly coated. Along with the additional possibility of the deposition of nickel on non-conductive surfaces, the electroless plating process becomes a very attractive method of enhancing the high temperature resistance of both bare silica and copper-coated optical fibers, as well as more complex fiber components.

2.4. Electroless Nickel Plating and the Preparation of Samples

For a proper operation, the electroless method requires three basic components: a metal salt, a reducing agent and a catalyst. Nickel plating solutions, obviously, must contain Ni^{2+} ions; however, presence of other metals can yield the formation of alloys. The reducing agent reduces nickel ions to the metal form and simultaneously is being oxidized, according to Equation (3) (the reaction takes place only at the surface of a catalyst):

$$Ni^{2+} + red \rightarrow Ni + ox \tag{3}$$

The electroless plating method is different from other chemical reduction methods, because the deposited metal is a catalyst of the reaction. This means that once a proper catalyst is deposited at the surface of an object, the metal layer will be continuously growing in an electroless plating bath. Eventually, thick layers of metal up to 1000 μm can be obtained [29].

It is believed that 99% of the world's electroless nickel was being plated using sodium hypophosphite as the reducing agent [30]. Reaction Equation (4) describes such a process:

$$Ni^{2+} + 4H_2PO_2^- + H_2O \rightarrow Ni + P + 2HPO_3^{2-} + H_2PO_3^- + 3H^+ + 3/2H_2 \tag{4}$$

According to Brennel and Riddell, this can be divided into several steps [30]. The first one is an anodic reaction of water and hypophosphite:

$$H_2PO_2^- + H_2O \rightarrow H_2PO_3^- + 2H^+ + 2e^- \quad (5)$$

At the same time, three distinct cathodic reactions utilizing electrons released in the latter process can occur:

$$Ni^{2+} + 2e^- \rightarrow Ni \quad (6)$$

$$2H^+ + 2e^- \rightarrow H_2 \quad (7)$$

$$H_2PO_2^- + 2H^+ + e^- \rightarrow P + 2H_2O \quad (8)$$

The last equation shows that apart from the Ni metal, also P is formed during the plating process. Phosphorous is incorporated into the nickel layer forming Ni–P alloy. These equations also explain the influence of pH on the coating composition. The content of phosphorous usually resides within 2%–13% [30]; however, lower and higher concentrations can also be achieved. It is noteworthy that there are also other equivalent mechanisms describing the process of electroless nickel plating [31].

As stated above, electroless plating requires in its initial stage the presence of an appropriate catalyst. It is known that the most suitable ones are the noble metals, with palladium providing the highest plating rates [32]. The deposition of Pd on a copper substrate can be performed simply by dipping a fiber in a water solution containing $PdCl_2$ and HCl for a few minutes.

Electroless nickel can be deposited not only on metal substrates like Cu, but also on insulators. This approach can possibly lead to omitting the problem of the copper oxidation and deposition directly on bare silica fiber. Nevertheless, more sophisticated methods of the substrate pre-treatment are required.

The properties of electroless Ni alloy are strongly dependent on its phosphorous content. Corrosion resistance changes along with P percentage [30], e.g., low phosphorous alloys are resistant to high temperatures and alkali solutions, but they are prone to acids, whereas high phosphorous alloys are acid-resistant, but corrode in alkali environments and oxidize quicker at high temperatures, due to the oxidation of P. From the mechanical point of view, the most important properties of the metal layer are: hardness, ductility and internal stress. As the strong mechanical wear of the fibers in fiber optics is not commonly encountered, the best alloy for such applications should have low hardness and high ductility. This is evident in high-phosphorous samples [30]; however, high temperature conditioning causes a rapid increase of the hardness of such a layer. Internal stress should be as low as possible, and this is encountered in medium-phosphorous (*ca.* 10% P) alloy. However, the same alloy has a very low ductility, which can lead to the evolution of cracks in the coating during bending. On the other hand, low-phosphorous deposits have high hardness, moderate ductility and high internal stress [30]. What is more, after annealing, the properties of electroless nickel can change, as it becomes harder and more brittle [33]. All of these factors show that it is necessary to optimize the P content in the alloy to obtain coatings with the best performance at high temperatures. In our work, we experimentally evaluated three different nickel-phosphorous alloys in order to determine which gives the best results during long-term exposure to high temperatures.

3. EXPERIMENTS AND RESULTS

3.1. Sample Preparation

Electroless nickel was deposited on approximately 1-m long copper-coated optical fibers (sole or attached to a metal membrane) according to a procedure consisting of the following steps: degreasing in NaOH, etching in HNO_3, activation in $PdCl_2$/HCl solution and electroless plating. Each step was followed by rinsing. Solution composition and plating principles can be found in [31].

Firstly, we have deposited a thin, approximately 2 μm, layer of nickel alloys with low (3.5%), medium (10%) and high (13%) phosphorous content on commercially available copper-coated fibers. The amount of phosphorous in the deposit was controlled by adjustment the pH and the concentrations of nickel and hypophosphite ions. Usually, the amount of Ni^{2+} and $H_2PO_2^-$ is relatively high, which is convenient for the long-term use of the plating bath, because above a certain level of concentrations, the deposit composition is constant (it does not change as the reduction reaction takes place). This means that the crucial variable for P percentage is the pH value for a given type of bath. Alkaline solutions give low P content, while weakly acidic baths give a medium amount of P, whereas more acidic solutions give a high phosphorous alloy.

As presented in the following paragraphs, annealing at 600 °C indicated that medium-phosphorous samples were much more fragile than the two other alloys. Subsequently, approximately 5-μm thick layers of electroless nickel (high and low P) were deposited on two copper-coated fiber samples, which were then examined for longitudinal strain resistance at high temperatures of 600–700 °C. In the next step, a 5-μm layer of low-phosphorous nickel was deposited on a copper membrane (which can be used as sensor transducer [34]) with copper-coated fibers electrochemically bonded to them. For the minimum bending radius tests, a 10-μm layer of low-phosphorous electroless nickel was examined.

3.2. Tensile Strength of Copper Coated Fibers at Room Temperature

For the first set of tests, we used common single mode fibers coated with a 20-μm layer of copper, which in terms of high temperature degradation, can be considered an equivalent of any type of silica optical fiber (e.g., microstructured, polarization maintaining, *etc.*). Along with the standard copper-coated fibers, we have also examined more complex fiber optic transducers (*i.e.*, copper-coated fibers electrolytically bonded to a metal membrane [34]). The tensile tests were performed on a linear translation stage, to which the fibers and copper membranes were attached.

The result of the tensile strength measurement is a maximum relative elongation (*i.e.*, longitudinal strain) at which the sample breaks. What needs to be noted is that during the tensile strength tests, a breakage always occurred on the fiber itself and not at the membrane-to-fiber joint, which proves the high strength of such a connection and the fact that the tensile strength of the fiber is more critical. Our first goal was to check the safe value of strain that can be applied to a copper coated fiber without breaking it. Taking into account the strength of 8 measured samples (Figure 1a), we have set the safe longitudinal strain for the subsequent tests below 10 mε. The low tensile strength of copper-coated fibers, together with the relatively wide spread of results (Figure 1a), in comparison to standard polymer-coated fibers (with reported elongations of 6.5% at rupture [35]) can be attributed to the strain introduced to the fiber during high temperature processing (*i.e.*, deposition of copper). Therefore, not only the longitudinal strain, but also the annealing time should be considered during the analysis of the further experiments.

Subsequently, we have measured the transmission during mechanical stretching. The results shown in Figure 1b indicate that there is no significant difference in the intensity of the transmitted light until the sample breaks, which means that such optical fibers can be used in sensing applications, in which the fiber is subjected to continuous longitudinal strain without any characteristic decay in the fiber transmission.

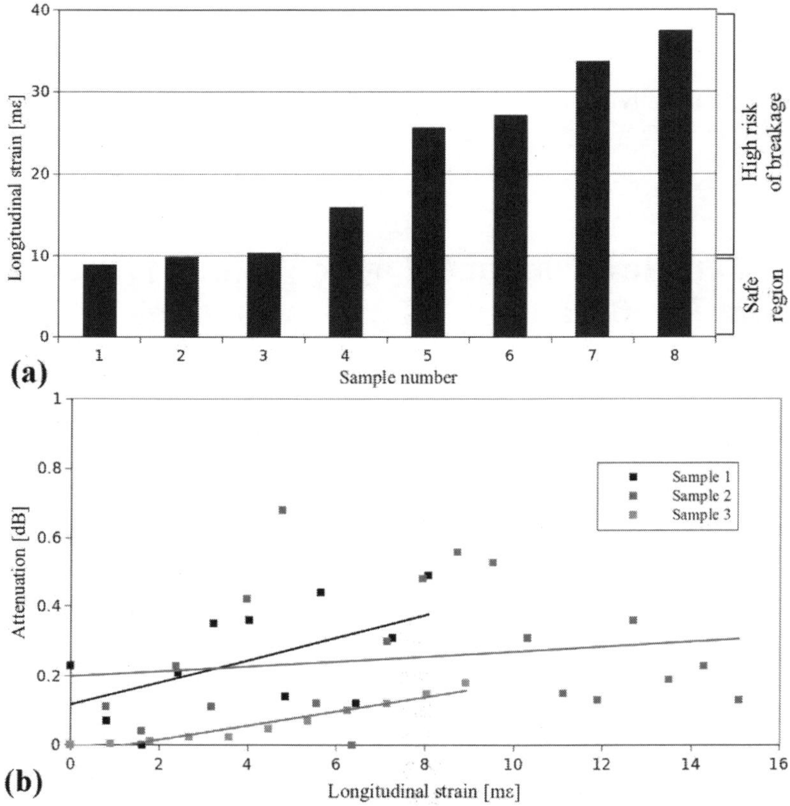

Figure 1. (**a**) Tensile strength of 8 copper-coated optical fibers with longitudinal strain values at breakage given on the *y*-axis. (**b**) Attenuation as a function of longitudinal strain.

3.3. High Temperature Annealing of Metal-Coated Fibers

3.3.1. Annealing of Copper Coated Optical Fibers

As temperature is the most critical factor decreasing the reliability of copper coated fibers, we have performed high temperature annealing tests in a furnace. The applied temperature was modified and increased during the test in order to speed up the process of degradation (with the temperature profile depicted in Figure 2 as a blue line). Furthermore, the fiber samples were coupled with a VCSEL (Vertical-Cavity Surface-Emitting Laser) source and a power meter in order to monitor changes in transmission during the heating process (with the measurement density increased at the temperature transitions). As one can see in Figure 2, no significant change in the fiber transmission can be noticed until the end of the process, after over 70 h of annealing at temperatures of 500 °C and higher. What was, however, observed was the extreme decay of the mechanical resistance of the samples. The copper surface changed its color to black, and the fibers became extremely brittle (which was related to the copper surface oxidation process). No manual operation was then possible, making the samples useless for further operation, since even a small strain applied to a fiber resulted in a breakage.

Taking into account real-life sensor operation conditions, at which the fiber may be simultaneously subjected to strain and high temperature, we propose an additional test of strain after annealing, in which the samples are taken out of the furnace oven every hour and exposed to strain at room temperature. The longitudinal strain was applied to the samples gradually until reaching the maximum value of 7.5 mε (according to the results presented in Section 3.1). Such a test allowed one to check *in situ* the condition of the fiber and simulate real life sensor operation under maximum strains.

3.3.2. High Temperature Durability of Repeatedly Stretched Copper-Coated Optical Fibers

Evaluation of the electroless nickel influence on the optical fiber high temperature performance should be preceded by checking the properties of the optical fibers only coated with copper. We have annealed 20 cm-long samples of copper-coated fibers at different temperatures, and after every hour (±1 min) of annealing, they have been stretched to a value of 7.5 mε. The results of the measurements are presented in Table 1.

These results show that the time of life of the copper-coated fibers decreases non-linearly while the temperature increases. Such a test gives plenty of information about the destruction kinetics. Further work, including a detailed analysis of these results and an extrapolation model, can be found in Section 4.

3.3.3. Tensile Strength Measurements of Annealed Metal-Coated Fibers

Initial annealing experiments conducted with a thin, approximately 2 μm, layer of electroless nickel indicated that low and high phosphorous coatings endure longer at high temperatures than medium phosphorous deposits. Therefore, we have tested two samples of copper-coated fibers with an overlayer of low and high-phosphorous nickel (with an additional two samples coated only with copper for reference). The thickness of the coating was approximately 5 μm. During every experiment, samples were stretched after each hour of annealing. The tensile strain was applied to the fiber gradually with a step of approximately 0.05 mε up to approximately 7.5 mε.

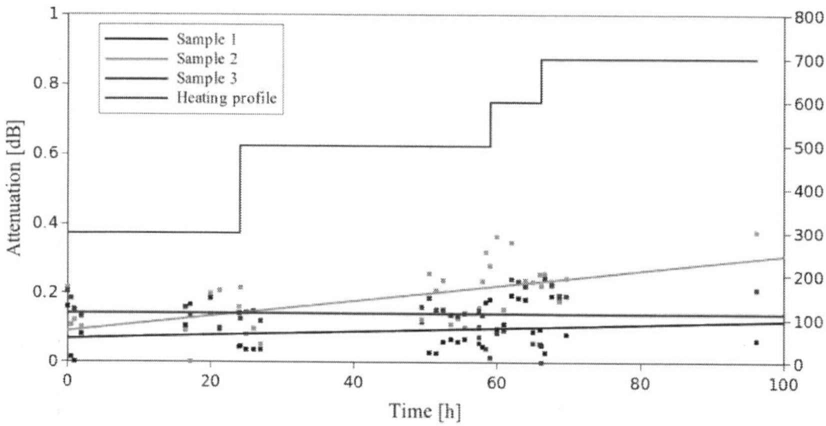

Figure 2. Heating profile and attenuation change over the time of measurement of copper-coated fibers.

Table 1. The results of the tensile stress experiments performed every hour during annealing at varied temperatures. All of the samples were Cu-coated optical fibers, which were repeatedly stretched every hour. Each value represents the time of annealing, after which the sample broke.

Sample Number	Hours at 500 °C	Hours at 550 °C	Hours at 600 °C	Hours at 650 °C
1	21	–	–	–
2	–	15	–	–
3	–	15	–	–
4	–	–	9	–
5	–	–	–	5
6	–	–	–	5

The results are presented in Table 2. It can be seen that nickel coated fibers (5 μm of Ni) are more resistant to high temperature, since they endure one hour longer at 700 °C. Furthermore, the low-phosphorous alloys had a higher value of tensile strength at breakage. Low P deposits exhibit a lower oxidation rate at high temperatures in air, but they are much less ductile and have higher internal stress than high P alloys. However, as proven in the experiment, the latter is not an important issue in fibers coated with copper prior to nickel; hence, low P alloys seem to be more reliable for usage in high temperature sensing applications.

Table 2. Tensile strength of metal-coated fibers after two-step annealing, firstly at 600 °C and then after increasing the temperature to 700 °C.

Sample	Hours at 600 °C	Hours at 700 °C	Strain at breakage (mε)
Cu-coated	3	3	6.6
Cu-coated	3	3	7.6
Cu-coated + low P Ni	3	4	7.1
Cu-coated + high P Ni	3	4	3.7

When the temperature was raised to 700 °C, we noticed that the oxidation process increased drastically, even for nickel-coated fibers. We have performed a subsequent set of tests with copper-coated fibers electrochemically attached to the metal membrane. Furthermore, as high P nickel coatings showed lower temperature resistance, we have performed the tests only for two low P samples (and two pure copper-coated fibers for reference). Samples coated solely with copper withstood only 2 h in the third stage of the experiment (at 600 °C), while fibers with 5 μm-thick nickel overlayer endured 2 h more, which again proves the positive effect of electroless nickel on the fiber high-temperature resistance. Moreover, the fiber lifetime depends on the heating profile, as the time to breakage varies between the tests presented in Table 2 and Table 3.

Table 3. Tensile strength of metal-coated fibers after three-step annealing, firstly at 400 °C, then at 500 °C and subsequently at 600 °C.

Sample	Heating Steps			Strain at Breakage (mε)
	Hours at 400 °C	Hours at 500 °C	Hours at 600 °C	
Cu-coated	5	2	2	4.6
Cu-coated	5	2	2	7.9
Cu-coated + low P nickel	5	2	4	5.2
Cu-coated + low P nickel	5	2	4	10.5

It is important to highlight the fact that the samples in this experiment were more complex than in the previous one, since it also contained the metal membrane. Fabrication of such samples requires a series of pre-treatment processes, including etching and electroplating, which may affect the high temperature durability of the fiber, because etching decreases the coating thickness and electroplating increases it. Thus, there is a very high chance of obtaining spots with too low or too high a thickness, which will change the fiber lifetime at high temperatures. Nevertheless, this experiment provides information about the high temperature durability of optical fiber devices manufactured by electrochemical methods.

3.3.4. Minimum Bending Radius Tests

Another way to evaluate the mechanical properties of an optical fiber in real-life sensing applications in harsh environments is to verify its transverse strain resistance, by performing a bending test. The principle of such a test is a simplified measurement technique firstly reported by Kurkjian [36,37] with the test scheme shown in Figure 3. It is noteworthy that the test refers solely to the fiber mechanical properties and should not be confused with the telecommunication fiber minimum bending radius test, at which transmission is monitored.

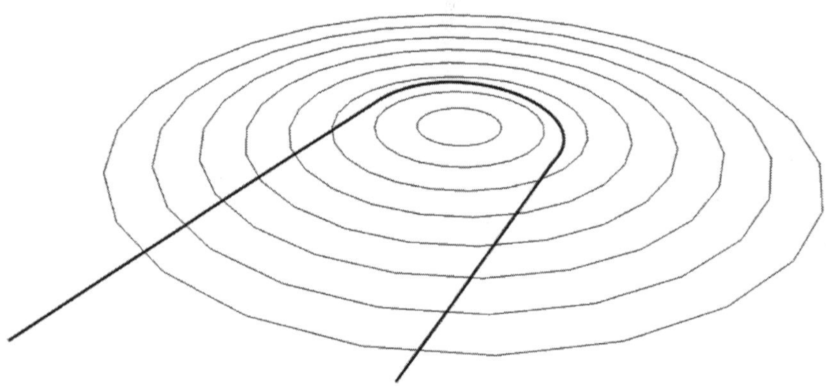

Figure 3. A principle of the minimum bending radius measurement.

Results of the measurements are given in Table 4, which contains the values of the minimum bending radius before and after annealing for (1) 3 h at 500 °C and (2) 2 h at 600 °C. Each result was obtained by calculating the average from three single measurements. Beside copper- and nickel-coated fibers (with a low P nickel layer of approximately 10 μm), the results for bare fiber and acrylate-coated fiber at room temperature are given for comparison. Before annealing, a very low value of 1 mm was noted for acrylate-coated fiber, copper-coated fiber and copper-coated fiber with a nickel layer. Due to the inevitable destruction of bare and acrylate-coated fibers during annealing at high temperatures, they have not been tested for the change of minimum bending radius. After annealing at 500 °C, fibers with a nickel overlayer gave better results than fibers coated with copper only. After 2 h at 600 °C, the difference between these two fiber types got even bigger, which seems to be intuitive, since nickel serves as an additional barrier between fiber and air.

Table 4. Results of the minimum bending radius (MBR) tests (before annealing, 500 °C for 3 h, 600 °C for 2 h). All numbers are given in millimeters.

Sample Type	MBR Before Annealing	MBR After 3 h in 500 °C	MBR After 2 h in 600 °C
Bare	4	–	–
Acrylate-coated	1	–	–
Copper-coated	1	1.5	2
Copper-coated + 10-μm NiP	1	1	1

The positive effect of nickel coating affects not only the mechanical parameters of a fiber, the morphology after heating changes, as well, and the change is clearly visible. Samples coated with copper and copper with nickel were heated to 500 °C for 3 h, after which they looked entirely different. Electroless nickel coating after annealing is covered with a colorful oxide layer with good adhesion to the substrate, while copper-coated fibers have a black layer of copper oxides, which easily comes off the surface of the fiber.

Minimum bending radius tests give additional information about the mechanical properties of the fibers before and after the annealing. This technique has lower resolution than the tensile stress test; however, the information it provides can be particularly useful for manual operation with the optical fibers. One can use this parameter to judge if fibers can be bent before or after the heat processing.

4. EXTRAPOLATION OF RESULTS TO OTHER TEMPERATURES

4.1. Basic Properties of the Extrapolation Model

The evaluation of the usefulness of metal-coated optical fibers for long-term operation at high temperatures needs verification of the possibility of extrapolating the results obtained during annealing at higher temperatures. The purpose of that is to conduct accelerated wear experiments, in which detailed knowledge of the mechanisms and kinetics of the oxidation of metals is essential.

We postulate that lifetime t of nickel plated copper-coated optical fibers at high temperatures can be described as:

$$t = t_1 + t_2 + t_3 \tag{9}$$

where t_1 is the time needed for the degradation of the nickel layer, t_2 is the time required for the degradation of the copper coating and t_3 is the time needed for the degradation of the silica fiber. This model is simplified, because it does not take into account the influence of diffusion and the influence of screening the copper coating by oxidized nickel. In the first case, the discrepancy caused by omitting the diffusion is relatively small, due to the small values of interdiffusion coefficients, thanks to which nickel can be used as a barrier coating for Cu, even at high temperatures [27]. The screening effect is also insignificant due to the high porosity of the oxidized nickel layer and the relatively high rate of NiP alloy oxidation [38].

The results of only the stretching tests will be taken into account during the extrapolation. The major reason why we have chosen this technique is that it provides better resolution than the minimum bending radius test. What is more, the bending test procedure required breaking every fiber, which is inconvenient during the experiments; therefore, it must have been changed for such measurements.

Our initial experiments have shown that a bare fiber breaks at a strain within the range of 1.4–3.2 mε after annealing at 600 °C for an hour. What is more, the fiber stretching was performed with a single hour interval; thus, the lifetime of a bare silica fiber should not be taken into account in our model, since it is of the order of magnitude of the experimental uncertainty. Eventually, t_3 in Equation (9) can be recognized as negligible, hence:

$$t \approx t_1 + t_2 \tag{10}$$

4.2. Oxidation of Metals According to the Parabolic Law

The kinetics of the metal oxidation can be described by many dependences, out of which is parabolic (Equation (11)), which is the most commonly recognized one, due to its simplicity, self-consistency and good correspondence to the experimental results [39]:

$$w^2 = kt \tag{11}$$

where w stands for weight gain per surface unit, k is a parabolic rate constant of oxidation and t is time. The oxidation rate of a sample decreases over time, due to the elongation of the route that reagents need to go through to get to the reaction site. The oxidation rate tends to continuously decrease as the oxide layer thickness increases. Furthermore, k changes with temperature, with the dependence given by the Arrhenius equation [39]:

$$k = Ae^{\frac{-E_a}{RT}} \tag{12}$$

where A is a constant ($g^2 \cdot cm^{-4} \cdot s^{-1}$) and E_a is activation energy (kJ/mol).

In order to extrapolate the experimental results obtained for a certain temperature to other temperatures, experimental lifetimes must be multiplied by the factor k_{T1}/k_{T2}, which is a consequence of dividing Equation (11) for temperature T_1 by the same equation for temperature T_2 (in both cases, the coating thickness is the same; thus, $w_1 = w_2$). After taking into account Equation (12), the quotient of two values of k at different temperatures is equal to:

$$k_{T_2}/k_{T_1} = e^{\frac{E_a}{RT_1} - \frac{E_a}{RT_2}} \tag{13}$$

The results reported in [38,39,40] clearly indicate that there are different oxidation mechanisms for different temperature ranges. In each temperature range, the activation energy is different, because E_a increases with temperature. Additionally,E_a strongly depends on the metal composition. Literature data are usually given for pure metals, which are much more corrosion resistant than common alloys. Our observations indicate that copper used for fiber coating is not chemically pure, as it gets covered with a dark layer after immersion in 5% NaOH solution for a few minutes, which does not occur with pure Cu. Furthermore, metal-coated fiber samples were repeatedly stretched during our experiment, which does not correspond to the methodology reported for weight gain measurements, where no strain is applied [38,39,40]. This factor in an evident way influences the results, since the oxide layer residing on the top of the metal coating is brittle and prone to breakage during stretching. Directly underneath such fractures, oxidation takes place more vigorously, as the access to the metal surface is not hindered by the oxide layer.

Therefore, it is necessary to calculate the E_a specific for our experiments and to evaluate the whole model by checking the lifetime values for measurements from Table 2. It is important to emphasize the fact that due to the conditions of the experiments, this will not be an activation energy according to the strict definition, but an effective value of E_a. It will have an approximate character, as different oxidation mechanisms may occur at different temperatures.

4.3. Least Square Calculation of the Kinetic Parameters

We have performed a series of annealing experiments with concurrent stretching conducted at different temperatures, which was presented in Table 1. Thus, we can calculate the basic parameters of the parabolic equation of oxidation and apply it to our data. Taking into account Equation (11) with respect to Equation (12), one can obtain:

$$\ln t = \ln \frac{w^2}{A} + \frac{E_a}{RT} \quad (14)$$

The above equation clearly indicates that the dependence between $\ln t$ and $1/T$ should be linear. In Figure 4, we have plotted these two factors.

Data plotted in Figure 4 between 550 and 650 °C are evidently linear, which is justified by the high value of R^2 equal to 0.995. The resulting effective activation energy is here equal to 69.2 kJ/mol. This value is 27% lower than for 4 N copper reported for the temperature range of 550–800 °C (95 kJ) [39], which again indicates that the metal protecting the optical fiber is an alloy. Apart from that, stretching can also influence E_a, and therefore, this parameter is valid only when the stretching threshold for different experiments is the same. In our case, all of the samples were stretched up to a value of 7.5 mε. The value of A also can be affected by stretching; however, we have not determined the exact value of this parameter, but rather, a value of w^2/A. As was already stated, w^2 depends on the coating thickness, which should be the same in every measurement. In this case, w^2/A was equal to 6.2×10^{-4} $g^2 \cdot cm^{-4} \cdot s^{-1}$.

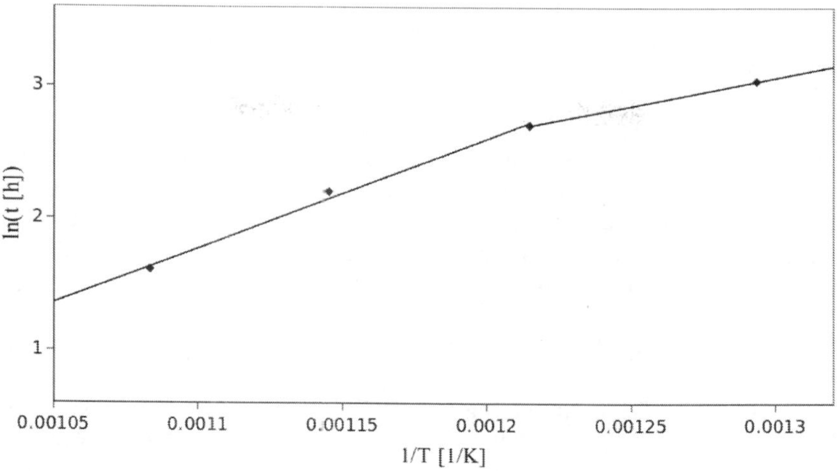

Figure 4. Dependence between the time of breakage and the annealing temperature of copper-coated fibers.

Similarly, we have calculated the value of effective activation energy for the data below 550 °C, which yielded 35.6 kJ/mol, while w^2/A was equal to 8.3 × 10^{-2} $g^2 \cdot cm^{-4} \cdot s^{-1}$. The E_a value is 31% lower than the one reported for 4 N copper (52 kJ/mol) for the 350–550 °C range [39]. The low amount of measurements for this region exact caution during further analysis. However, the activation energy is lower than the one reported in the literature, similarly to the higher temperature range, which indicates, that this approximation should be quite accurate. What is more, other authors also report that copper exhibits different oxidation mechanisms under and above approximately 550 °C, which results in two different values of activation energy [24,39].

Data obtained during experiments with electroless nickel appear to be insufficient to determine the E_a and w^2/A, since only two lifetime values for distant temperatures (difference of 200 °C) could be taken into account.

4.4. Model Evaluation

The model described above can be evaluated by comparing the experiment from Table 2 with the time that the fiber should withstand at high temperature, according to the extrapolation model. The relative degradation of the fiber coating can be calculated with the following equation:

$$\frac{w_1}{w_{\max}} = \sqrt{e^{\frac{-E_a}{RT}} \cdot t} \qquad (15)$$

where w_1 stands for weight gain for a certain experiment, w_{max} is a maximum possible weight gain and t is the time of annealing since the beginning. The cylindrical shape of the coating can be neglected, because calculated values of w_1/w_{max}only refer to the thickness of the oxidized layer, since the shape of the coating has not been taken into account.

At 600 °C, w_1/w_{max} = 0.591 after 3 h of annealing, *i.e.*, 59.1% of the coating thickness will be oxidized. The same value of relative degradation is obtained for 1.125 h of annealing at 700 °C. After heating the fiber for two more hours at 700 °C, the relative destruction is equal to 0.985. According to our model, the whole metal layer should be oxidized to enable air to reach the silica fiber surface; therefore, one hour more at 700 °C is needed to destroy the fiber. Thus, we obtained that the fiber should break after 3 h at 600 °C and 3 h at 700 °C, which was the actual result of the experiment. Therefore, the extrapolation model can be considered reliable.

Many manufacturers provide information that copper-coated fibers can operate at 450 °C. The elaborated model can be also used for determining the lifetime of the optical fibers at this temperature. After taking into account Equation (15), one obtains 31 h of operation with repeatedly applied stress at 450 °C.

The experiment with fibers binding to the metal membrane should not be evaluated this way, due to the reasons that were already explained in Subsection 3.3.3.

5. CONCLUSIONS

We have presented an analysis of the long-term thermal durability of copper-coated silica fibers and a method of enhancing such durability by the deposition of an additional layer of electroless nickel-phosphorous alloy. We have experimentally verified that low phosphorus alloys are characterized by the best temperature resistance, with a significant fiber lifetime (under extreme temperature and cycled longitudinal strain) increase in comparison to sole Cu-coated fibers. We also developed a model of fiber lifetime prediction, which allows one to extrapolate the obtained results to other temperatures. The calculated effective (specific for the performed test) activation energies for Cu prove that temperature and cyclic longitudinal strain play fundamental roles in the destruction mechanism.

ACKNOWLEDGMENTS

The work described in this paper was partially supported by the Innovative Economy Programme as projects POIG.01.04.00-06-117/12 and POIG.01.03.01-06-085/12, by the National Centre for Research and Development within the research project PBS1/B3/12/2012 and by the European Commission under the 7th Framework Programme through the "Space" action of the "Cooperation" Programme, BEACON, Grant Number 607401.

AUTHOR CONTRIBUTIONS

Tomasz Nasiłowski conceived of the project. Tadeusz Tenderenda coordinated the project and, together with Karol Wysokiński and Tomasz Stańczyk, designed the experiments. Tomasz Stańczyk, Karol Wysokiński, Anna Ziołowicz and Mateusz Słowikowski performed the experiments. Karol Wysokiński deposited metals on the optical fibers. Katarzyna Gibała developed the technology of bonding optical fibers to metal membranes. Małgorzata Broczkowska helped with analyzing the data and gave professional advice. The manuscript was written by Karol Wysokiński and revised by Tadeusz Tenderenda and Tomasz Nasiłowski.

CONFLICTS OF INTEREST

The authors declare no conflict of interest.

REFERENCES

1. Agrawal, G.P. *Fiber-Optic Communication Systems*, 4th ed.; Chang, K., Ed.; John Wiley&Sons, Inc.: Hoboken, NJ, USA, 2010.

2. Lopez-Higuera, J.M. *Handbook of Optical Fibre. Sensing Technology*, 1st ed.; John Wiley&Sons, Inc.: Hoboken, NJ, USA, 2002.

3. Frazão, O.; Santos, J.L.; Araújo, F.M.; Ferreira, L.A. Optical sensing with photonic crystal fibers. *Laser Photonics Rev.* **2008**, *26*, 449–459.

4. Martynkien, T.; Szpulak, M.; Statkiewicz, G.; Golojuch, G.; Olszewski, J.; Urbanczyk, W.; Wojcik, J.; Mergo, P.; Makara, M.; Nasilowski, T.; *et al.* Measurements of sensitivity to hydrostatic pressure and temperature in highly birefringent photonic crystal fibers. *Opt. Quantum Electron.* **2007**, *39*, 481–489.

5. Tenderenda, T.; Skorupski, K.; Makara, M.; Statkiewicz-Barabach, G.; Mergo, P.; Marc, P.; Jaroszewicz, L.R.; Nasilowski, T. Highly birefringent dual-mode microstructured fiber with enhanced polarimetric strain sensitivity of the second order mode. *Opt. Express* **2012**, *20*, 26996–27002.

6. Wang, R.; Yao, J.; Miao, Y.; Lu, Y.; Xu, D.; Luan, N.; Musideke, M.; Duan, L.; Hao, C. A reflective photonic crystal fiber temperature sensor probe based on infiltration with liquid mixtures. *Sensors* **2013**, *13*, 7916–7925.

7. Monzon-Hernandez, D.; Minkovich, V.P.; Villatoro, J. High-temperature sensing with tapers made of microstructured optical fiber. *IEEE Photonics Technol. Lett.* **2006**, *18*, 511–513.

8. Geernaert, T.; Kalli, K.; Koutsides, C.; Komodromos, M.; Nasilowski, T.; Urbanczyk, W.; Wojcik, J.; Berghmans, F.; Thienpont, H. Point-by-point fiber Bragg grating inscription in free-standing step-index and photonic crystal fibers using near-IR femtosecond laser. *Opt. Lett.* **2010**, *35*, 1647–1649.

9. Yang, X.H.; Wang, L.L. Fluorescence pH probe based on microstructured polymer optical fiber. *Opt. Express* **2007**, *15*, 16478–16483.

10. Konstantaki, M.; Pissadakis, S.; Pispas, S.; Madamopoulos, N.; Vainos, N.A. Optical fiber long-period grating humidity sensor with poly (ethylene oxide)/cobalt chloride coating. *Appl. Opt.* **2006**, *45*, 4567–4571.

11. Pinto, A.M.R.; Lopez-Amo, M. Photonic crystal fibers for sensing applications. *J. Sens.* **2012**, *2012*.

12. Ruan, Y.; Schartner, E.P.; Ebendorff-Heidepriem, H.; Hoffmann, P.; Monro, T.M. Detection of quantum-dot labeled proteins using soft glass microstructured optical fibers. *Opt. Express* **2007**, *15*, 17819–17826.

13. Candiani, A.; Bertucci, A. Giannetti, A.; Konstantaki, M.; Manicardi, A.; Pissadakis, A.; Cucinotta, A.; Corradini, R.; Selleri, S. Label-free DNA biosensor based on a peptide nucleic acid-functionalized microstructured optical fiber-Bragg grating. *J. Biomed. Opt.* **2013**, *18*.

14. Segawa, H.; Ohnishi, E.; Arai, Y.; Yoshida, K. Sensitivity of fiber-optic carbon dioxide sensors utilizing indicator dye.*Sens. Actuators B* **2003**, *94*, 276–281.

15. Chu, C.-S.; Lo, Y.-L.; Sung, T.-W. Review on recent developments of fluorescent oxygen and carbon dioxide optical fiber sensors. *Photonic Sens.* **2011**, *1*, 234–250.

16. Viets, C.; Hill, W. Fibre-optic SERS sensors with conically etched tips. *J. Mol. Struct.* **2001**, *563–564*, 163–166.

17. Stolov, A.A.; Simonoff, D.A.; Li, S. Thermal stability of specialty optical fibers. *J. Lightwave Technol.* **2008**, *26*, 3443–3451.

18. Rayss, J.; Widomski, J.; Poskościelny, W.M. The influence of water and temperature on the strength of optical fibers.*Optoelectron. Rev.* **1995**, *3–4*, 107–109.

19. Matthewson, M.J.; Kurkjian, C.R. Environmental effects on the static fatigue of silica optical fiber. *J. Am. Ceram. Soc.***1988**, *71*, 177–183.

20. Armstrong, J.L.; Matthewson, M.J.; Kurkjian, C.R. Humidity dependence of the fatigue of high-strength fused silica optical fibers. *J. Am. Ceram. Soc.* **2000**, *83*, 3100–3109.

21. Mrotek, J.L.; Matthewson, M.J. The combined effect of temperature and humidity on the fatigue parameters and reliability of optical fiber. *Proc. SPIE* **2004**, *5465*, 268–273.

22. Bogatyrev, V.A.; Semjonov, S. Metal-coated fibers. In *Specialty Optical Fibers Handbook*; Mendez, A., Morse, T.F., Eds.; Academic Press: Waltham, MA, USA, 2007; pp. 491–512.

23. Ohno, A. Continuous Metal Casting. U.S. Patent 4,515,204, 15 December 1982.

24. Zhu, Y.; Mimura, K.; Isshiki, M. Oxidation mechanism of copper at 623–1073 K. *Mater. Transact.* **2002**, *43*, 2173–2176.

25. Wan, Y.; Wang, X.; Sun, H.; Li, Y.; Zhang, K.; Wu, Y. Corrosion behavior of copper at elevated temperature. *Int. J. Electrochem. Sci.* **2012**, *7*, 7902–7914.

26. Pinnel, M.R. Diffusion-related behaviour of gold in thin film systems. *Gold Bull.* **1979**, *12*, 62–71. [Google Scholar] [CrossRef]

27. L.B.H. Diffusion barriers for gold plated copper. *Gold Bull.* **1975**, *8*, 52.

28. Maier, R.R.J.; Havermann, D.; MacPherson, W.N.; Hand, D.P. Embedding metallic jacketed fused silica fibres into stainless steel using additive layer manufacturing technology. *Proc. SPIE* **2013**, *8794*.

29. *Electroless. Nickel Plating: A Guide*; Artistic Plating Company, Inc.: Milwaukee, WI, USA.

30. Parkinson, R. Properties and Applications of Electroless Nickel. Available online: http://www.nickelinstitute.org/en/ Technical Literature/Technical%20Series/PropertiesandApplicationsofElectroles sNickel_10081_.aspx (accessed on 29 April 2014).

31. Mallory, G.O. The fundamental aspects of electroless nickel plating. In *Electroless Plating: Fundamentals and Applications*; Mallory, G.O., Hajdu, J.B., Andrew, W., Eds.; Noyes Publications/William Andrew Publishing, LLC: Norwich, NY, USA, 1990; pp. 1–56.

32. Krulik, G.A. Tin-palladium catalysts for electroless plating. *Platin. Metals Rev.* **1982**, *26*, 58–64.

33. Keong, K.G.; Sha, W.; Malinov, S. Hardness evolution of electroless nickel–phosphorus deposits with thermal processing. *Surf. Coat. Technol.* **2003**, *168*, 263–274.

34. Krisch, H.; Lau, M.; Tournillon, S. Optical fibre sensor for measurement the large range of strain and frequency vibrations of flat diaphragm. *Proc. SPIE* **2010**, *7653*.

35. Holleis, S.; Hoinkes, T.; Wuttke, C.; Schneeweiss, P.; Rauschenbeutel, A. Experimental stress–strain analysis of tapered silica optical fibers with nanofiber waist. *Appl. Phys. Lett.* **2014**, *104*.

36. Matthewson, M.J.; Kurkjian, C.R.; Gulati, S.T. Strength measurement of optical fibers by bending. *J. Am. Ceram. Soc.* **1986**, *69*, 815–821.

37. Matthewson, M.J.; Kurkjian, C.R. Static fatigue of optical fibers in bending. *J. Am. Ceram. Soc.* **1987**, *70*, 662–668.

38. Tomlinson, W.J.; Wilson, G.R. The oxidation of electroless Ni–B and Ni–P coatings in air at 800 to 1000 °C. *J. Mater. Sci.* **1986**, *21*, 97–102.

39. Zhu, Y.; Mimura, K.; Lim, J.-W.; Isshiki, M.; Jiang, Q. Brief review of oxidation kinetics of copper at 350 °C to 1050 °C.*Metall. Mater. Trans. A* **2006**, *37A*, 1231–1237.

40. Peraldi, R.; Monceau, D ; Pieraggi, B. Correlations between growth kinetics and microstructure for scales formed by high-temperature oxidation of pure nickel. II. Growth kinetics. *Oxid. Metals* 2002, *58*, 275–295.

INDEX